Robert L. Ferguson

PRACTICAL TECHNIQUES FOR THE RECORDING ENGINEER

PRACTICAL TECHNIQUES FOR THE RECORDING ENGINEER

A STREAMLINED TECHNIQUE FOR SPEED, ACCURACY AND DOCUMENTATION

SECOND EDITION

SHERMAN B. KEENE

Edited By Greig Stout

S.K.E. Publishing
P. O. Box 2519
Sedona, AZ 86336
(602) 282-1258

Library of Congress Catalog Card Number: 81-148444

International Standard Book Number: 0-942080-00-9

==
Library of Congress Cataloging In Publication Data

Keene, Sherman (1947-)

PRACTICAL TECHNIQUES FOR THE RECORDING ENGINEER
A Streamlined Technique for Speed, Accuracy and Documentation

1. Sound Recording and Reproducing 2. Recording Techniques
3. Recording Equipment Set-Up/Alignment/Operation I. Title

1981 621.389'3 81-148444

ISBN 0-942080-00-9

==

Second Edition

PREFACE

This book is a complete guide for recording engineers. Recording engineering is an art, requiring as much creativity on the part of the engineer, as on the part of the artists. Sound engineering is also a science, requiring careful preparation, a methodical approach, record-keeping, and technical know-how including the fields of acoustics, mechanics, electronics, music, and psychology. In this one book, the reader can find in-depth knowledge covering all levels of these subjects.

In contrast with other "conventional" books about recording, this book includes only information from the various sciences which will actually be useful to the reader. There is no technology for its own sake to be found herein.

The wealth of information presented in these 28 chapters may appear overwhelming. For the readers covering the information for the first time, a chapter by chapter approach will enable the reader to apply the knowledge as it is learned. The more advanced sound engineers using this book as a reference, will enjoy the comprehensive table of contents and index sections.

This is the first recording engineering book to actually give practical techniques used by top engineers. "Trade secrets" and "inside moves" usually passed on by a chief engineer to a favorite assistant are available here for the first time.

The efforts of authors of other books on the subject of recording engineering cover only a fraction of the numerous steps required for performing successful engineering. These books do not tell readers the realities of actual session practices such as:

* What documentation to keep and where to keep it

* How to encourage top engineers to share their knowledge with you

* Why the "wrong" mike may be the "right" mike

* How to trace and clear common session hums and noises

* Grasping the "ins and outs" of complex professional equipment through the use of simple block diagrams

* The (seemingly) long lost art of making exceptional safety copies, and high quality tape copies

* How to improve your sound by using acoustics

* How to perform session work with resolved and interlocked tape machines

* Special effects which can be created with or without special effect equipment

* The physics of the musical scale -- how to correlate tape speed to elapsed time and musical notes

* How to determine the proper session set up

* How to "patch around" session audio problems

* How to minimize production hold-ups

* How to set up recorders with and without Dolby noise reduction

* How to eliminate microphone malfunctions even before the session begins

* How to handle unexpected session changes when the session is already under way

As of this, the third printing, over fifty recording schools, colleges, and universities have already adopted the use of this book. The Teacher's Manual for this textbook supports a three level, 24 module, two textbook course, complete with lesson plans, homework, verbal review questions, final exams, suggested experimental session content, and a student workbook. Having taught recording engineering courses around the world, I know the questions aspiring engineers ask. This book answers them.

TABLE OF CONTENTS

Basic 5: "Basic Audio Theory"

Basic 6: "Basic Types of Console Operation"

Basic 7: "Operating Procedures #2"

Basic 8: "Tape Machines #1"

INTERMEDIATE ENGINEERING

Intermediate 1: "Acoustics #1"

Intermediate 2: "Console Symbols for Diagrams"

Intermediate 3: "Operating Procedures #3"

Intermediate 4: "Musical Terminology"

Intermediate 5: "Intermediate Audio Theory"

Intermediate 6: "Tape Machines #2"

Intermediate 7: "Operating Procedures #4"

Intermediate 8: "Acoustics #2"

ADVANCED ENGINEERING

Advanced 1: "Acoustics #3"

Advanced 2: "Tape Machines #3"

Advanced 3: "Advanced Audio Theory"

Advanced 4: "Special Effects #1"

Advanced 5: "Special Effects #2"

Advanced 6: "Advanced Audio Systems"

Advanced 7: "Distortion"

Advanced 8: "Audio Psychology"

The Neve 8108

The NECAM System

The Lexicon Model 224

The AMPEX ATR124

LIST OF ILLUSTRATIONS

BASIC CHAPTERS

INTERMEDIATE CHAPTERS

ADVANCED CHAPTERS

ABOUT THE AUTHOR

Sherman Keene has been associated with many top recording studios and artists in his career in audio engineering. Since 1969, he has engineered projects at, Wally Heider Recording; the Record Plant, L.A. and New York; Sunwest Recording, L.A. (where he was chief engineer for two years); Paramount Recording, L.A.; Electric Lady Land, N.Y.; and Bolic Sound, L.A. At one time or another, he has recorded Linda Ronstadt, Jimmy Spheres, Jim Webb, the Buffalow Springfield, Tierra, Alice Cooper, Frank Zappa, Ike and Tina Turner, The Turtles, and other top recording artists and groups.

Sherman, an accomplished musician, plays piano, guitar, banjo, bass, and cello. For two years he was the bassist/vocalist for the rock group "Spirit", performing and recording with them. The world famous group, "Spirit", with Randy California, Ed Cassidy and "Barry" (Sherman's middle name) Keene recorded two albums for Mercury Records -- "The Spirit of 76" and "Son of Spirit." He also performed on stage with the Ike and Tina Turner Review.

Sherman worked for two years as Frank Zappa's personal mixer. While working with Frank, Sherman performed large-band, sound reinforcement in concerts from coast to coast, as well as in Europe. Many times, these 52 (or more) input shows would include a separate simultaneous mix for four channel recording as well as musically "scored" special effects performed from his console. As Frank's engineer, he recorded these albums -- "Just Another Band From L.A.", "Overnight Sensation", "The Filmore East Album", "200 Motels", and "Apostrophe."

In addition to engineering recordings, he has designed special-effect recording equipment, live performance equipment, and installed multitrack studios. As an accomplished technician, he has supervised maintenance on out-board equipment, multitrack tape machines, consoles, and microphones. As a technical writer, he writes articles for audio magazines, training manuals, school curriculums, and books on the audio field.

In 1979 at the invitation of Tamco, a Japanese company which supplies audio engineering equipment and audio engineers to recording studios in Tokyo, Sherman travelled to Japan for two months giving lectures and demonstrations of "West Coast" recording techniques. (Most Japanese studios do not have their own recording engineers, but hire them for each session as needed -- from companies like Tamco). Near the conclusion of his stay, Sherman lectured to the assembled chief recording engineers of all Tokyo studios.

At the end of 1982, Sherman traveled to Caracas, Venezuela to deliver his course, in person, to the engineering staff of Telearte Recording Studios. Telearte is a world-class, state of the art, multichannel recording studio.

The Telearte seminar included bi-lingual lectures, question and answer periods, daily written homework, review sessions and practical, hands-on recording session work each day for five weeks.

Sherman has created and published a very specialized program for training recording engineers. It is the first practical program to offer a published Teacher's Manual which incorporates two textbooks, and a student's workbook. He has been an active educator since 1978, when he organized the first in-house engineer training program at Wally Heider's Studios. Since then, he continues to give lectures, classes and seminars, at colleges and universities around the United States and abroad. He has created a correspondence program, with which he is currently training students around the world. Many of his former students are now engineering and, in some cases, producing, large, big-money sessions in various locations around the world.

FOREWORD

I have sometimes been credited as being a pioneer in sound recording techniques. When I was learning sound engineering, there were no teachers, no schools, and no books dedicated to this subject. Then, the process of becoming an engineer was to learn by doing, occasionally trading information with other engineers. That was the only way into the field. Trial and error learning was, and still is, a very time consuming way to learn the techniques which are needed even more today, than they were needed then.

The few books that finally came along on the subject of sound engineering, were quite dry and technical -- not aimed at the more personal, artist/technician world of actual recording work.

In the early 80's, I was president (again) of the studio in Hollywood which bears my name. At that time, we were fortunate to have Sherman Keene on our staff. Sherman had created the first in-house, in-depth, formal training program for aspiring recording engineers. He was applying his program to the training of our apprentice engineering staff. The training they received put them far in advance of what they would have obtained from leaning over someone else's shoulder and/or sneaking in for a look at on-going sessions. Sherman's comprehensive sound engineering program revolved around his recently finished text book and teacher's manual -- the first ever, aimed at teaching a streamlined technique of recording.

Sherman brought many of our studio's fledgling engineers into bright, successful careers as sound engineers. His book, "Practical Techniques for the Recording Engineer", has filled the educational gap from which the recording industry had been suffering for decades.

By Wally Heider

President, Wally Heider
Recording Studios
Hollywood, California

DEDICATION

To my sweet wife

Rhea Gwendolyn

whose encouragement saw me through
the making of this book.

My thanks to Charles Sullivan, Frank Zappa, and Ike Turner for giving me the room to be creative and for what they taught me.

My special thanks to Greig Stout for his editing work and his other valuable contributions.

ENGINEER'S TOOL KIT

1. A suitable attache style case
2. A calculator with an inverse key, a log (base 10) key, an antilog key, an x squared key, a square root of x key
3. Can of Magna-See Solution
4. A set of numbered clothespins 1-32
5. Heavy and medium point colored marking pens
6. Q-Tips
7. A small, leak proof container of isopropyl alcohol (water-based)
8. Editing scissors and razor blades
9. At least two each of:
   In line phase reversals
   5, 10, 15 dB adjustable in line pads
   A line to mike level pad
   Ground lifts (XLR)
   3 to 2 A.C. adaptors
10. In line mike level signal oscillator
11. A battery operated cable test box (with XLR and RTS connectors - LED readout)
12. An inexpensive sound pressure level meter (Radio Shack makes a good one)
13. A 1/4 inch, 1/2 inch and 2 inch editing block (for well to do mixers who are making it big and going out on the road to obscure studios)
14. A head degausser (Handimag is excellent)
15. Carefully degaussed razor blades
16. Small set of tools (plastic boxed screwdrivers, nut drivers, and allen wrenches)
17. A tape strobe (Dubbings makes good ones)
18. A small penlight flashlight (studios get very dark sometimes)
19. An inexpensive volt/ohm/amp meter
20. A small digital stopwatch (wrist type is best)
21. A dual male phone to stereo female phone adaptor cable (for plugging into two track machines and listening with headphones)
22. A set of china markers of different colors (for marking the console faders in different parts of the mix in progress)

Consider the artist coming into the studio and seeing it for the first time. The studio should give the artist a feeling of at-homeness and security in which to work but also relax. Clean up the ash trays, trash bins, mike stands, etc. On the weekend paint the music stands and brillo the mike booms. If there is a horrible stain on the floor from a previous session, set up over it using a throw rug. Tidy up all the desk accessories -- pads, grease pencils, pens, phones, etc. -- in the control room. Think about what the artists will see when they are playing in the studio and aim the band at the most visually attractive area. All these cosmetic considerations only apply to those studios which are not new or newly redecorated; new or like new studios have already been cosmetically treated to account for artists looking at them favorably.

SETTING UP THE STUDIO

To set up thoughtfully and thoroughly give yourself plenty of time, realizing that you must also be able to set up with a minimum amount of time as well. This is because usual 1 hour "pads" (blank book time between sessions) are sometimes used up by inexperienced engineers who misjudge the time that will be required.

The basic steps are:

1. Cleaning up the studio and control room
2. Deciding on, and laying the floor plan
3. Wiring in the mikes, phones, AC, etc.

In laying the floor plan consider the artist's desire to see each other and in some cases hear each other while recording. Usually the best floor plans feature the drummer as the central figure with the other instrument areas radiating away from him. Part of the reason for this is that electric instrument players like to face their amps while they play and if the above setup is used, then they are facing both their amps and the drummer. Set up the singer's live mike for "reference vocals" so that the wrong end of the mike is facing the loudest instrument areas, with the deadest wall area in front of the mike (behind the singer). During the basic tracking session, a rough vocal track is recorded along with the rhythm tracks so that the musicians, producer and engineer will know where they are in the song. Hence the name "reference vocals." Generally even though the producer would like the band to face him in the control room during recording, the best plan is the one which frees the artist from looking at anyone outside the group while recording. There's always time for looking at and listening to the producer later. If it sounds like making a really good setup is difficult -- that's because it is!

When wiring in the mikes remember to start at the floor and work your way up. If you are going to use rugs, then these must be laid first. Set up the things that are likely to be moved around by the artist last (phones, baffles, music stands, etc.) so that these items and their cables will be on top. When the wires and amps get pulled to make way for alterations, mikes and stands and the like won't come tumbling down. Always protect the mikes. If there's any way to damage the mikes due to your setup then you can rely on the artist to find the way.

Perform the setup in these steps:

Decide upon the floor plan and throw down the rugs where you want them. Lay in the AC power boxes to the areas where you will have the amps, the music stand lights, and electric instruments. It's unlikey that AC power cables will fail during a session, that's why they go in at the bottom.

Lay in cue cables from the point where you will want the phones, to the nearest wall, then along the wall to where the cue outputs are located. Cover your tracks with a walkway rug so that it will be next to impossible for any person to trip over your cue or other cables, and thereby tow down a mike stand with an expensive mike on it. Sometimes when a mike stand is knocked over or towed down by its cable, it falls onto a very expensive musical instrument -- do not let that happen. Plan the cue cable run pathway thinking that you will want to run the mike cables by the same route as well, one next to the other in a flat array, not bunched together in a large bundle. In the case of the cue cables leave some coiled at the artist's end in case he wants to move around or away from his loaction; leave the majority of the cable coiled at the wall with the feeding end of the coil (male XLR connector) towards the top.

Now move the baffles (large, portable panels used to prevent sound from going places where it isn't wanted) into place and partition the electric and acoustic areas off as per your floor plan. Place the baffles over the cue and AC cables. For instrument areas completely enclosed by baffles, be sure to leave a doorway so that the player can enter his booth. In the case of the drum booth, leave a baffle door wide enough so that the drums can be moved in without danger of knocking something over. It is best to leave the mike stands right up against the baffle wall so that they won't be knocked over.

Having the instrument areas defined, now move the chairs and music stands into place. Attach the music stand lights and plug them into the AC. With the furniture in place, you will see the setup clearly for the first time. If there are oversights in the setup, now is the time to do something about it. Each guitar, including bass guitar, needs a chair and a music stand, preferably with a music stand light. It is always nice if the main house lights can be dimmed for a ballad so that the singers and musicians can "get into it" without the sheet music becoming difficult to read.

Place the mike stands where they will be used and readjust the baffles so that there is the greatest possibility that the artist won't knock the stand over while recording. In the case of an acoustic guitar baffle area, for example, make the exit to the left and provide a space to the player's right so that upon finishing a track the player can lay his guitar down to the right and leave the baffle cage to the left. In this case, the mike stand should be to the artist's right, in front of the "safe area" where his guitar is to be placed. Don't forget to put an extra mike stand at about 10 feet in the air in the geographical center of the setup for your use as a listening mike to hear what the musicians are saying to you. Lay in the mike cables leaving a minimum amount at the base of the mike stand. The majority of the extra cable should remain coiled up at the mike input wall. Hold the cable in one hand and throw several barber pole loops over the mike boom so that the female end hangs near the end of the boom with several inches to spare for plugging in the mike (the boom is the cross piece that is held by the vertical mike stand). At the wall end of the mike cable, plug the male end into the mike input panel, according to the prearranged mike input list.

Now bring out the mikes. Look each mike over carefully and check the pattern switch, the capsule pad switch, and the battery meter (if it's a battery mike). Fix the mike to the stand. Before proceeding to adjust them for heigth and placement, loosen all the clamps and fasteners so that the boom is completely loose. Slide the boom through the boom clamp until the boom balances perfectly, considering the weight of the mike and the weight of the mike boom. With the boom balanced, clamp down the fasteners and procede to aim and adjust the mike. In the case of the artist's room talk back mike, adjust it for circular pattern if possible and no pad. Use a straight stand.

Retreat to the control room and close all the studio and control room doors. Bring each mike up cautiously to full volume through the console. Listen carefully for any buzzes, clicking, hisses, hums, radio interference, etc. This is the only time that you can safely bring the mikes up to full volume, since when the artists arrive there will likely not be any quiet time in the studio again until they leave. Guard against anyone making a loud noise during full volume tests since a really loud noise will probably launch your studio's woofer voice coils right out of their gaps. A loud noise includes quickly closing an air tight studio door. You must protect the monitor system since you always need it to work with in the studio. After checking the quietness of the mikes, have the second engineer go around and tap on the mikes one by one, if there's even the slightest reason to suspect that there may be any mistakes in the mike input list.

Using white paper tape, put a strip on the console just over the faders and mark the fader strip with the names of the instruments using the instrument code.

If your phone conference with the client was thorough and you already know what the track allocation will be, then you can put a white paper tape strip over the monitor system individual track level controls with the name of the instrument to be on that monitor position (track allocation is how many tracks are to be recorded on the multitrack machine and what goes on each track).

Have at least one extra high quality mike standing by somewhere in the studio with a stand and enough cable to reach any of the anticipated musician areas. Have at least one extra cue distribution box and a couple sets of phones standing by. (A cue distribution box is a headphone box which is plugged into the end of the cue distribution cable. It is what the headphones are plugged into.)

If the setup is only for a vocal, then put some extra energy into making the vocal area special. Enclose the vocalist in tall, good-looking baffles. Lay down a nice throw rug. Have your best stereo phones out for use. Off to one side, have a small table for wine, cigarettes, or whatever. Supply a comfortable high stool, which is the correct heigth, so that the artist's head is at the same altitude, with the mike, whether he's sitting or standing.

THE INSTRUMENT CODE

This is a shorthand code for use in naming instruments. The code is useful for taking notes when talking to the producer about the instrumentation for the upcoming sessions. The code is used to write the mike input list. The mike input list is a four column list laid out in this fashion:

STUDIO XL #	INPUT NAME	MIKE MODEL #	FADER USED #

The mike input numbers are taken from the number of mike inputs available on the console, the second column the names of the instruments (in shorthand code), the third column is the make and/or model number of the mike you or the producer specify, and the fourth column for the eventual console channel that the mixer may crosspatch the mikes to. The instrument code is also used for writing the names of the inputs across the console input section (where the faders are). If 1 inch paper leader is used instead of paper tape, then the strip can be coiled up in a 1 inch splicing tape box with those of the mixdown fader strips (song by song) and used again. Many times valuable notes are written on this input strip especially during mixing, which might otherwise be lost. The code is used to mark the monitor system pots (pots=potentiometer -- or simple volume control, usually rotary). This code can also be used to mark paper taped clothespins, which can be clipped onto the male mike cable ends where they plug into the mike input wall, so that the cabling can be re-plugged without having to chase down each wire to find out what mike it is; usually though, the clothspins should be marked on directly with the numbers (1-32 etc.), which lets you just clip them on without having to write the instrument names. The code is also written on the track sheets, which keep track of what is recorded on each track of the multitrack master.

THE INSTRUMENT CODE

AG	Acoustic guitar
EG	Electric guitar
DG	Direct guitar
DEG	Direct electric guitar
MEG	Miked electric guitar
ELG	Electric lead guitar
ARG	Acoustic rhythm guitar
AKB	Acoustic keyboard
AKB lo	Acoustic piano lo strings microphone
EKB	Electric keyboard
MB	Miked bass (guitar or upright)
DB	Direct bass
K	Kick drum
SN	Snare drum
HH	Hi hat
T1	Tom-tom number one
T2,3	Additional tom-toms
OH	Overhead drum (mike)
OHR	Right overhead mike

CO	Cello
VA	Viola
VN	Violin
BV	Bass viol
TP	Trumpet
TB	Trombone
TS	Tenor sax
AS	Alto sax
BS	Bass sax
etc.	

When the producer calls, make a shorthand list of the upcoming instrumentation using this code. Ask if he has any mike preferences. Ask how many toms and cymbals there will be to the drum set and how many of the guitarists may want to use acoustics instead of their electrics. This is far superior to forgetting to ask for preferences and finding them out when he arrives -- after you've already set up the mikes you've chosen and checked. Ask the producer what he is planning for track allocation and write this down too.

Now make up the mike input list. While imagining the console input section and how you would like the inputs to appear, draw the left column numbered 1 through 24 (or 36 or how ever big your console input section is). Next, position the instruments by microphone usage down the next column. The third column is your or the producer's mike choices by model number. If you use one mike for each drum and there are 7 tom-toms, then you need:

The Mike Input List

Input XLR#	Instrument Mike Name
1	K
2	SNT (top snare mike)
3	SNB (snare bottom mike)
4	HH
5	T1
6	T2
7	T3
8	T4
9	T5
10	T6
11	T7
12	CYM 1
13	CYM 2
14	AKB lo
15	AKB hi
16	DEG#1
17	DEG#2
etc.	

Continue this list including all the instruments. Now try to fit the mike supply you have with the list of desired mike inputs. You have to learn the mike inventory of your studio of course. If you have enough mikes you're done. If not, you'll have to use less mikes on the instruments until you can mike the session. When you finalize the mikes you'll use, fill in the third column, which in part will look like this:

Input	Instrument	Mike Model#
=====	==========	===========
1	K	Sennheiser 441
2	SNT (top snare mike)	Sony c37a
3	SNB (snare bottom mike)	Shure sm57
4	HH	Sony ecm50
5	T1	Sen 421
6	T2	Sen 421
7	T3	Sen 421
8	T4	Sen 421
9	T5	Sen 421
10	T6	Sen 421
11	T7	Sen 421
12	CYM 1	AKG 421
13	CYM 2	AKG 421
14	AKB lo	Neumann U47 VF14
15	AKB hi	PZM plate
16	DEG#1	Jensen dir
17	DEG#2	Jensen dir
etc.		

Add a column for the fader the mike was crossed patched to and other columns for the EQ used and you'll have a very complete and useful document. Update the list (use pencil) if the mikes get changed during the session. Be sure to sketch a rough drawing of the studio floor plan on the reverse side of the mike input page and include this page with the billing information with the client's work order.

Put a copy of this list in the front office (traffic office). Even if you get a flat tire on the way to the studio, you can have someone else at the studio begin your set up and know that at least they'll use the right mikes, have them plugged into the right inputs, and use the correct floor plan for the session setup in your absence.

For use on the track sheets there are additional entries that help save time and mistakes. One is the use of the word "erase." This word is written in addition to the instrument code of what is already recorded on the track and means that the instrument indicated is present on the track but needs to be erased. Be sure to only write erase when everyone agrees that the instrument is to be erased. Also the use of dates along with overdubbed instruments' entries on the track sheet helps to keep straight which instrument was overdubbed first. This is especially useful when it comes to lead vocals of which there are sometimes 10 or more on a single 24 track tape.

THE RECORDIST'S CODE

This code is used by the tape machine operator/legend keeper during tracking sessions to keep up with a fast engineer. The code is written down on the paper legend, which is then taped to the multitrack tape box. The legend goes with the tape. Every box with recorded tape in it must have a legend. This code is used to preserve information about what's on the tape so that users of the tape will not have to guess what's on it.

The Recordist's Code
====================

HS = head slate -- the engineer recorded the take number at the beginning of the roll. (The slate is the term for a voice announcement that is recorded on all the tracks of the machine along with a low frequency tone. The tone allows for rapid location of each take by counting "beeps" as the tape is fast forwarded. The beeps are the slate tones when heard at fast tape speeds in rewind and fast forward.)

TS = tail slate -- the engineer recorded a slate but after the tune

NS = no slate -- the engineer forgot to record one so don't go try to find this take by its slate

FS = false start -- tune just got going and then the artist flubbed

LFS = long false start -- artist got about halfway through the song and then flubbed

INC = incomplete -- the artist got almost all the way to the ending before he flubbed

C = complete take

PB = played back this take to the artist or producer

M = chosen as the master take of all takes

EOT = end of tape. If they ask you if there's enough tape for one more on this reel -- there isn't

OTM = out to master (this take is missing and is on the new master reel

O/D (date) = overdubbed track

ERASE = this track is to be erased at a convenient time

The recordist's job is to operate the machine, keep accurate take sheets, time each take, and write the time and other information (C, INC, etc.) on the take sheet. When the engineer says, "Record please," then the recordist looks on the take sheet and responds with, "Recording (presses record) take ..." so that the engineer will slate the take correctly. The recordist should note pertinent data on the box -- if there are tones on the reel, what kind of tape, what level the tape is at, which reel this is of the project's total number of reels, etc. The usual reason for lack of proper slating is that the engineer forgets the take number. With your prompting, will slate the take correctly as a matter of course.

GETTING LEVELS AND EQ

At this stage of the book this procedure is very simplistic. One thing though, this same procedure outlined here, when used in the future, after you have a great deal of experience, will allow you to do tracking sessions with the most rapidity of any technique.

To begin getting levels and EQ, you must have already done the studio setup and miking of the instruments as outlined in the preceding section.

First NORMALIZE the console for tracking:

Pull all patch cords out of the patch bay.

Turn all buss reverb return pots down and switch all return switches to off.

Cancel all buss assign switches.

Turn down all input module reverb send pots.

Turn down all monitor system reverb send pots.

Cancel all input module EQ.

Select mike input on the mike/line switches of all input modules having mike inputs for this session; select line input for all the rest of the modules.

Refer to your track allocation notes and push the buss assign switches on each input module so that the desired track allocations occur.

Assure that the program masters and submasters are at the proper setting (ask someone for now).

Punch the speaker assign buttons or adjust the pans in the monitor system so that the instruments on the busses come out of the speakers where you want them to.

Turn up the individual buss monitor and buss cue channel pots for all the console busses to which you have assigned audio (the active busses)

Turn up the monitor and cue master to 12 o'clock.

Put the monitor system into buss or program (pgm) mode.

Push the monitor solo on the first fader on the left hand side of the console that has a mike coming in.

Push up the fader cautiously while watching the proper VU meter (on the meter where you bussed it) for the reading. Adjust the mike pad so that the fader is at either of three settings while the VU meter is reading properly:

1. At the fader setting that is different in color to the rest of the fader markings or

2. The fader setting of -17 dB or

3. The fader setting at the zero area where the minus numbers go through zero on the way to positive numbers at the top of the fader scale.

In a case where the pad steps either give a fader setting that is too high or too low during good VU indications, use the pad setting that puts the fader too low, for example, set to -22 rather than -17. Using higher than optimum fader settings increases the risk of console mike amplifier noise.

Now with the pad set, adjust the EQ.

If the EQ changes the level on the VU meter, then reset the pad and fader so that the fader is again at the recommended setting loss or slightly below (meaning a higher fade amount indicated -- closer to you).

Having finished getting the level and EQ of the first mike, solo the next input position and un-solo the present one.

Perform the same operation on the various inputs all down the console.

In the case of mixing two or more mikes to one buss, solo each mike individually and get the levels and EQ on each, and then solo all the contributors and push up the faders of the contributing input modules until you arrive at a blend. You may use the solo for mixing because the solo system mixes sound in exactly the same way as the console busses do.

After finishing the last fader you may immediately begin to make multitrack recordings even though there is not yet a finished monitor mix. The part of the session required to make the recording is finished, only the listening system is not yet properly adjusted. You may adjust the monitor system at any time, whether recording or not.

Keep in mind that when the artist is told that the machine is recording, the levels that you carefully got during levels and EQ will probably go up at least 4 to 6 dB due to the excitement of the moment.

When mixing several sounds live, work on the blend with the lowest voices first, saving some VU room for the rest of the section.

Always concentrate on the monitor mix while making basic tracks. The monitor mix should always sound just like you want it to. From the monitor mix, you should be ready at any time to make a "rough mix" recording for the producer on a two track or other stereo recorder.

After tracking begins (recording), do not use solos. This is for two reasons: 1) it ruins your monitor mix, which is being recorded on tracks 22 and 23 of the 24 track machine and 2) it is a rude thing to do to the producer. Don't solo during a recording pass.

TAPE RUN IN - THE SLATE - TUNING NOTES - THE COUNT OFF

Described as a procedure, these activities make more sense:

1. The engineer says, "Record please," to the recordist.

2. The recordist looks at the take sheet and sees take 5 is next and says, "Take five," while pressing record.

3. The engineer presses the slate button and says, "We are rolling on take five."

4. The engineer presses the talkback button and says, "Rolling."

5. The piano player, having been coached by the engineer, plays a single note in the key of the song. This note will later be very useful for tuning other instruments to.

6. The drummer hits his sticks together while giving a verbal count off.

The slate, which comes out of all the recording busses at about -15 to -20 on the VUs when the slate button is pushed, is a voice recording at low levels mixed with a 30 or 40 hertz tone. The mike for the voice part is from the talkback system mike on the console and provides the song's title and the take number. The low frequency tone enables the recordist to easily find the slates while winding the tape at high speed. This is because the lo frequency tones on the tape passing by the heads at high speed create a magnetic field which the play head on the machine can pick up with ease and creates the "beep" sound which signals the location of each slate. The tone and voice should only be at about -15 dB on the VU meter. The slate must be transferred to all rough mixes until the master takes are selected. This lets the artist or producer listening at home to to his cassette know which take he's listening to. When the master reel is edited together or when head leaders are installed before a master reel is made, the slate is edited out by the insertion of the leader. If time code is to be laid on for interlocking machines later, do not leader the tune any closer than 30 seconds ahead of the down beat.

The count off is useful for recording a musical introduction if the original tune was recorded without one.

Always record a 60 hertz sync tone on track 24 in case there's something wrong with the tape machine speed.

If you have to use the room listening mike to record the count off, assign the mike to the drum overhead track (buss) and be sure to close the fader and mute it before the music starts.

BREAKING DOWN

This operation follows all but continuing sessions. First, rush to the aid of the mikes before someone knocks over one of the mike stands and damages the mike. Everyone likes to own fine recording mikes, especially home recording enthusiasts disguised as musicians. Clear away partly consumed beverage cans and bottles before they overturn and get on your mike cables. If one does overturn clean the cable off with a damp sponge. There's nothing like coiling up mike cables late at night, tired and grumpy, and discovering that the mike cables have a gummy cola residue on them.

It's best to leave a vocal overdub setup in the studio since it's easy to break down; it comes in handy for a quick overdub especially if you like to use a tube mike, which should warm up thoroughly anyway, and a setup looks good to studio visitors.

To save the console settings, dictate all of the the console settings into a small cassette recorder. Dictate slowly so that you can listen to the recording and re-set the controls without having to stop the tape. You can never tell when the artist will call saying he wants another mix just like the one you just did for him except for some minor detail.

Think about these procedures and you will prepare yourself for operating with them. As you gain experience more and better ideas will come to you, which you can try out and add to your technique. This is what I did.

HOW TO AVOID ALIENATING FUTURE SOURCES OF HELP AND GUIDANCE

When you are just beginning to learn about engineering, there is an overwhelming urge to have strong opinions about engineering subjects while in conversation with senior engineers. Resist these urges completely. Listen! don't speak. Ask lots of questions. Be flattering if you have to, or do some of the engineers dirty work, but get on his good side.

Never argue with anybody. The person you argue with might be an engineer who you will have to work with soon, and you don't want to push any wrong buttons at the start.

Ask, ask, ask. Reserve strong opinions until you really know what you're talking about. Many engineers are very secretive about their knowledge and may not want to part with any of it. This is especially true if you are coming on as a know it all.

Make it very clear, though, that the dirty work you may volunteer to do, and which may be accepted by the "heavy engineer," is strictly for consideration of learning periods, which he is going to provide at an opportune time. Your responsibility will be to insist on these learning periods. If no learning occurs, end the free help.

CONDENSERS, DYNAMICS, RIBBONS, PZM'S AND CONTACTS

There are two vastly different schools of thought as regards choosing microphones. One school says, "There is only one perfect mike to record any given instrument." The other school says, "Any mike is a good mike." Surely when faced with not having your favorite mike, there comes a time when any mike is a good mike. For the beginning study of audio engineering, any mike will do for now.

Condensers are by far the most useful and accurate of all mikes. The reason that the mike has such good response is because of the extreme lightness of the element -- the capsule. Air pressure changes in the air have an easy time of moving the element. Condensers also have the feature of hardness. (Hardness may be defined as the ability of a mike to produce an electrical output which exactly resemble sound waveforms in the air.) When a snare drum is hit there are some really amazing hi frequency noises that come out of the instrument. The condenser mike will produce all these amazing highs in electrical form.

Dynamic mikes are the workhorse of the industry. The newer models approach condensers in their ability to sound live and real.

Dynamic mikes are in reality little speakers hooked up backwards so that you provide the sound and the mike provides the electricity. Just like a speaker, there's a voice coil, a cone, and a gap for the voice coil to go into. These mikes are extremely durable and will stand up to tremendous abuse. The sound quality of dynamics, however, is degraded by abuse just like any other mike. Since the element is heavier (more massive) than the element of the condenser mike, the high frequency response is not as good as that of the condenser. This seeming fault can be a blessing in disguise because sounds with undesirable amounts of highs can be recorded without fancy EQ, making the recorded sound more solid.

Ribbons are the oldest professional recording mikes. The ribbons were around long before the condensers. These mikes get their name from the tiny metal corrugated ribbons that lie in a very strong magnetic field found in the area between the horns of a powerful u shaped magnet. Whereas the dynamic gets its output from moving many wires through a weak magnetic field, the ribbon gets its output from moving only a small piece of wire (crushed into a ribbon) through a very strong magnetic field. This strong magnetic field is dangerous to recorded tapes. Don't set a ribbon mike on your 2 or 24 track master while carrying it to the studio. You may erase a small radius of the tape.

Ribbons sound very good when used on low frequency instruments, such as bass, horns and voices. They tend to sound very warm. Because they accentuate the roundness of male speaking voices, this mike is well used for narration.

To a small extent, the ribbon seems to compress volumes after a certain loudness is reached. At a certain volume, the output of the ribbon does not increase as fast as the output of the instrument. This compression tendency can be used to advantage on some very loud sounds, like lead guitar, to keep the instrument on the VU meter.

Older ribbon mikes are bi-directional. The sound and pickup from the back is the same as from the front of the mike. This means that when using any bi-directional mike, including the ribbon, you will have guaranteed leakage pickup, unless you have the back of the mike pointed where there's no loud sound source. Some ribbons have an adjustable door to prevent this pickup of sound from the rear of the mike. It slides over the ribbon body and blocks off the sound coming into the rear of the mike from reaching the back of the ribbon.

Since only the signals coming into the front of a bi-directional mike produce properly phased electrical signals, any sounds coming in from the back will produce electrical signals out of phase with those coming in from the front. If you speak into the side of a ribbon, your voice pressure waves will push the ribbon both directions at once and the two air pressure forces on the ribbon will produce almost no output. This is a way of listening to "room sound" without hearing yourself since the sounds of your voice's reflections will still be heard coming back from the acoustic environment.

PZM microphones are a remarkable "new" type of mike. The letters stand for Pressure Zone Microphone. This mike excels in making recordings free from the damaging influences of reflected sounds. Its operation hinges on the physical phenomenon whereby when sound strikes a boundary (a wall or other solid surface), the action of the air changes from a to/fro type movement to a movementless pressure buildup. Since the air pressure energy simply cannot move the wall, it causes a pressure zone instead which sends the sound energy rebounding off in a new direction similarly to a billiard ball hitting the rail of the pool table. At the instant that the acoustic energy strikes the boundary, the air molecules come to a dead stop. If a microphone element (in the case of a PZM, a small electret condenser element) happens to be in this pressure zone, then the mike will not suffer from frequency cancellations caused by the arrival at the mike of reflections from the boundary involved -- the mike IS the boundary involved. Without the notches in the audio, which would be caused by such early reflections, the sound derived is more solid and cleaner. The distance from mike (which faces INTO the plate) to the plate must be very tiny -- on the order of 1/16 inch.

The physical design of a PZM is a plate of metal with a small condenser mike element facing down onto the plate at about a 1/16 inch distance. The sound picked up by this mike (which for good low frequency results is best used taped to a large surface like the control room window or wall or underside of an acoustic piano lid) is unlike any other mike. Vocalists, who complain of not being able to tell whose voice is whose in the headphones, can immediately tell themselves apart when the PZM is used instead of the "good" mike that was being used. For recording room ambiance, I have heard no better mike. Ambiance is the quality of a recording that gives it the sound of a place -- an actual room where the instruments are playing. For leakage free piano recording with drums and electric instruments, there is no easier way than to tape two PZM's to the underside of the keyboard lid, throw on some blankets, and baffle the piano's underneath areas -- and there you have it. One of the beauties of this type of piano setup is that the piano is completely moveable (with only some slight caution with the mike cables), but there are no mike stands to worry about, and no expensive mikes to jar or bump. A percussionist can wear one as a "medallion" on his chest (using a neck strap) and thus will always be "on mike" while running from one percussion instrument to another.

Contact microphones are meant to be attached directly to the bodies of instruments. These devices are transducers or pickups, but not really microphones. In the case of the Barcus-Berri Pickup, the pickup acts as a capacitor. When there is music in the instrument, the stresses and strains of the instrument's body are sensed and converted into electrical signals. Because these devices do not rely on air pressure phenomenon, there is very little possibility of stray acoustic pickup. This feature is the most important one. For miking acoustic guitars under loud music conditions, the contact pickup is unbeatable. Sometimes an artist will start out a multitrack recording with just his acoustic guitar and live vocal. Under these conditions, he may wish to save the guitar track and throw away the vocal track. Not having any vocal leakage on the guitar track is therefore nearly as important to him as the sound of the guitar. When this situation comes up think of the contact pickup.

Other good reasons to abandon "good mikes" and use a pickup are when...

1. You've already used up all your "good" mikes, like when the client begins to drop some "by the ways" on you: "Oh, by the way, all four electric guitarists brought their acoustics. I'm sorry I forgot to tell you."
2. The acoustic guitar player wants to move around a lot while recording
3. The acoustic player wants to sit near the drums while recording
4. There is so much wind (outside) that your favorite mike won't work
5. You are already using a good mike but you want a backup for a crisper, leakage-free sound
6. You want a third input for a stereo miking application, where you'll have a left mike, a center (pickup), and a right mike

Even though the preceding examples center around an acoustic guitar, other instruments also apply. The Barcus-Berri Guitar Pickup also works well on: snare and kick drum, hi hat, acoustic piano, cello, bass, electric bass, and other percussion instruments.

When using the mastic stick-on material supplied with the pickup, don't use too much as this will reduce the brightness of the sound.

MICROPHONE PICKUP PATTERNS

Much is written about pickup patterns in many of the technical books available. In truth the only way to realize what's going on with mikes is to use them in an experimental recording session environment.

The pickup pattern of most mikes resembles a cone shaped field of view similar in nature to looking through a camera lens. As with a camera, pay attention to where the mike is aimed and what is in the background. The more cardioid the pattern, the narrower the cone of "view." Anything making sounds outside the pickup area tends to be "ignored" by the mike. A sound outside the pickup cone is not totally ignored, only partly. The real control of what the mike can and can't hear comes from correct mike placement, attention to proximity effect, pattern selection and mike aiming.

You don't always want to use a very tight mike pattern. A shotgun mike would not be a good choice for close miking a vocal choir since at close distances the mike's narrow "field of view" would "hear" only a few singers. Under these conditions, there would be a fidelity recording made of only those singers in the field of view; there would not be a fidelity recording made of the singers outside the field of view. To pickup the entire group, using a shotgun mike, you would have to move the mike back to great distances from the choir.

Some mikes such as the Neuman U-87 provide variable pickup patterns by use of their pattern switches. The U-87 can be a cardioid mike, a circular mike, and a figure eight mike. In the figure eight position, the U-87 behaves just like some ribbon mikes. However, there is not a door on the U-87 for closing off the rear capsule.

When using mikes try to take advantage of the patterns available. For miking two hi tom-toms without getting snare drum leakage, you can "tune out" the snare by using the U-87 in the figure eight mode and aim the dead zone at the snare and the live ends at the proper toms.

There are several fine stereo mikes available, usually at great expense. These mikes can make really amazing stereo recordings. To place these mikes, wander around the studio until you find a spot where the stereo sound is best; put the mike right where your ears are. You can also "place" the mike by positioning the mike and moving the players. If a horn section sounds good enough to record in stereo, but you don't have a good mix in the studio, have the players move around until you hear the mix you want. This can only be done with a group's own horn section since union studio horn players are usually not into standing or moving around very much while they play. It is not really necessary to have a stereo mike to make fine stereo recordings. Use any two identical good quality cardioid condensers and place them ear distance apart, with the listening axis pointed outwards from the central viewing angle by about 45 degrees.

For the album "Just Another Band From L.A.", which I recorded for Frank Zappa, I used only two U-87 mikes for the entire band. The results were excellent, according to my own judgment, Frank's, and the band's. For this stereo recording, three hours of mike placement were required in order to find the ideal location and to coach the band as to how to not change their stage volumes beyond certain limits.

Especially in the beginning, whenever you run across any new mike, experiment and find what the mike sounds good on. Also find out what it sounds like when it distorts. Learn the sounds of the various distortions that happen in the studio.

CHOOSING MIKES FOR DIFFERENT INSTRUMENTS

There are as many theories on mike choice as there are recording engineers. Here are a few guidelines:

1. Use the highest quality mikes on the instruments having the greatest musical range (range of frequencies)
2. Use high sensitivity mikes on soft sources to prevent console noise

3. Use overload immune mikes on loud sources to prevent mike and console input stage distortion

A high sensitivity mike is one which delivers a strong electrical output even in the presence of a very soft acoustic sound. These are also known as hot mikes. Since the output of these mikes is high, they will have enormous output when used on loud sound sources. With loud sources the high output mikes will surely distort and their signal strength may very well distort the console's input stages as well.

By overload immune mikes I mean mikes that can withstand a very loud volume from the instrument (like a kick drum or three Marshall guitar amps) without distorting. These mikes typically have no interior amplifier so they do not have very much output at low volumes. When overload immune mikes are used on quiet sources, the input section of the console must be turned up to very high gain to produce a good VU reading. Doing this will most likely produce console noise.

When hearing mike overload use a more overload immune mike. When hearing console noise, use a more sensitive mike.

Use condensers for:

Acoustic guitar, acoustic piano, large groups of voices, single voices not closer than 4 inches, strings, horns (at a reasonable distance), percussion instruments (remember: for good playback hardness, keep VU levels reasonable)

Use dynamics for:

Loud sources, non-full range sources (lo fi instruments like an electric piano speaker output), sources with undesirable hi frequency energy, electric guitar, electric piano, cymbals, toms, bass amps, kick drum, hand held applications, low strings, low horns.

Use ribbons for:

Low frequency roundness, narration, low horns, screechy vocals, and acoustic sources with excessive highs.

In all cases experiment to find the best mike. For an odd sounding vocalist, line up several types of mikes and have the vocalist sing while moving from one mike to another after each song section. This kind of experimentation will teach you more about the mysteries of mikes than any book.

TONES, TRANSIENTS, AND NOISE

*** Please refer to Illustration B3-1 (Basic lesson 3 Illustration #1) -- Sherman's Table of Signals and Noises ***

Tones are the first of the three arbitrary divisions that I divide all musical sounds into. A tone is defined as that part of a musical sound that is a continuous string of waveforms. (Musical waveforms are what you'd have if you made a graph of air pressure during a short time period, like a second with the air pressure displayed vertically and the time horizontally). A tone is regularly repeated over and over. In a second, an organ note (Concert A) of 440 cycles per second would graph as 440 identical waveforms.

Simple tone waveforms are produced by such instruments as the electric organ and the synthesizer if sine waves are selected. The sine wave is the simplest tone with no overtones whatsoever. Other instruments such as the violin or the trumpet have horribly complicated waveforms. These waveforms are basically tones since they repeat regularly. A complex sound with many overtones may be simplified by using filtering and EQ so that only the simple tone waveform remains, but then the sound is indistinguishable and it is not possible to tell what kind of instrument produced the sound in the first place. It is very difficult, however, to take a simple tone and add overtones and raggedness to it to produce a realistic classical or other instrument sound. This is, after all, a good thing since recording the actual instruments is then necessary, requiring recording studios and engineers.

The repeat rate of the waveform per second gives rise to the frequency of the wave. (Frequency is the number of complete waveforms in each second and is called hertz, meaning waveforms per second). Musicians call the frequency of the note its pitch. The pitch of middle C on the piano is 261.63 Hertz, meaning that the sine part of the waveform repeats slightly more than 261 times each second. The A from the oboe that the orchestra tunes to is 440 hertz or cycles per second.

*** Please refer to Illus. B3-2 -- Sam's Piano Keyboard Frequency Graph ***

Transients are the hitting or striking sounds that emanate from such instruments as drums, blocks, and percussion instruments of all kinds. Transients are characterized by a very short duration time base. The energy of the transients is usually many times greater than the energy of the tone part of the sound (if there is any tone part). These characteristics of transients make for problems during recording. Since transients are so short, the VU meter is unable to even move before the hi level sound is over. This causes the VU meter to always read too low when reading transients. In some cases the VU meter may be reading as much as 18 dB too low.

Because the acoustic piano is played by hitting the strings with little hammers, it is classified as a percussion instrument and sure enough, transients come forth when the strings are miked near the hammers.

Musical noises, which are not tones or transients, are desirable sounds that come from musical instruments. They are sometimes quite harsh and grating on the nerves if they are recorded out of proper proportion to the rest of the musical sounds due to poor recording technique.

Musical noises can be broken down into two types: audible noises and subliminal noises.

ACOUSTIC ENERGY

Can be divided into ...

MUSICAL SOUNDS	&	NON-MUSICAL SOUNDS
Acoustic energy produced by musicians		Acoustic energy produced by non-musicians or EQuipment Eg: hiss in the air from noisy guitar amps left on, air conditioner rumble, subway train rumble, passing sirens fan wind hitting the mike

Musical sound can be further divided into:

ACOUSTIC SIGNALS	&	ACOUSTIC LEAKAGE
Musical sounds the engineer wishes to record		Musical sounds the engineer does not wish to record (at least with the mike in question) Eg: vocal sounds crossing from a loud vocalist into the mike of a soft spoken vocalist

Acoustic signals are divided into:

MUSICAL NOISE	MUSICAL TRANSIENTS	MUSICAL TONES
Non tone, non transient scrapings, grindings, and screech-ings which are a neces-sary part of some instrument's sounds Eg: the violin	A short timebase, hi frequency, hi energy pulse usually associated with the very begin-ning of a percussive note or sound	A regularly repeating waveform making up the major part of the instrument's characteristics of pitch, vibrato, and quality or tone (timbre)

Musical noises are further divided into:

AUDIBLE NOISES	&	SUBLIMINAL NOISES
Clearly apparent noises which are expected and recorded by conventional means		Subtle noises which the player is likely to hear but not really identify with the sound of the instrument until he notices that there is "something missing" when he hears a playback of the recorded music Eg: good clean pick noises coming from a bass guitar's strings

>>>>>>>>>>>>>>>> Acoustic To Electrical >>>>>>>>>>>>>>>>>>
>>>>>>>>>>>>>>>> Transducer Conversion >>>>>>>>>>>>>>>>>>>>

>>>>>>>>>>>>>>>>>>> (MICROPHONES) >>>>>>>>>>>>>>>>>>>>>>>>>

Acoustic/electric energy transducers (mikes) convert pressure/rarefaction phenomenon in air into electrical energy in wire

Mike output energys can be divided into:

ELECTRICAL SIGNALS	&	ELECTRICAL NOISE
Desired electrical waveforms		Unwanted electrical waveforms Eg: buzz, electronic hiss, cb radio interference

Electrical noises can be divided into:

EQUIPMENT NOISE	&	OPERATOR NOISE
Noise caused by malfunctioning recording EQuipment Eg: mikes, console systems, guitar amp noise, equipment which hums or is oscillating		Noise caused by misuse of console or other systems Eg: moving a dirty fader during a take, bussing reverb to the wrong buss causing the snare reverb to be recorded on the bass track, changing EQ during a take and causing a huge "pop"

Illus.
B3-2

Courtesy of
Howard w.
Sams & Co.
Inc

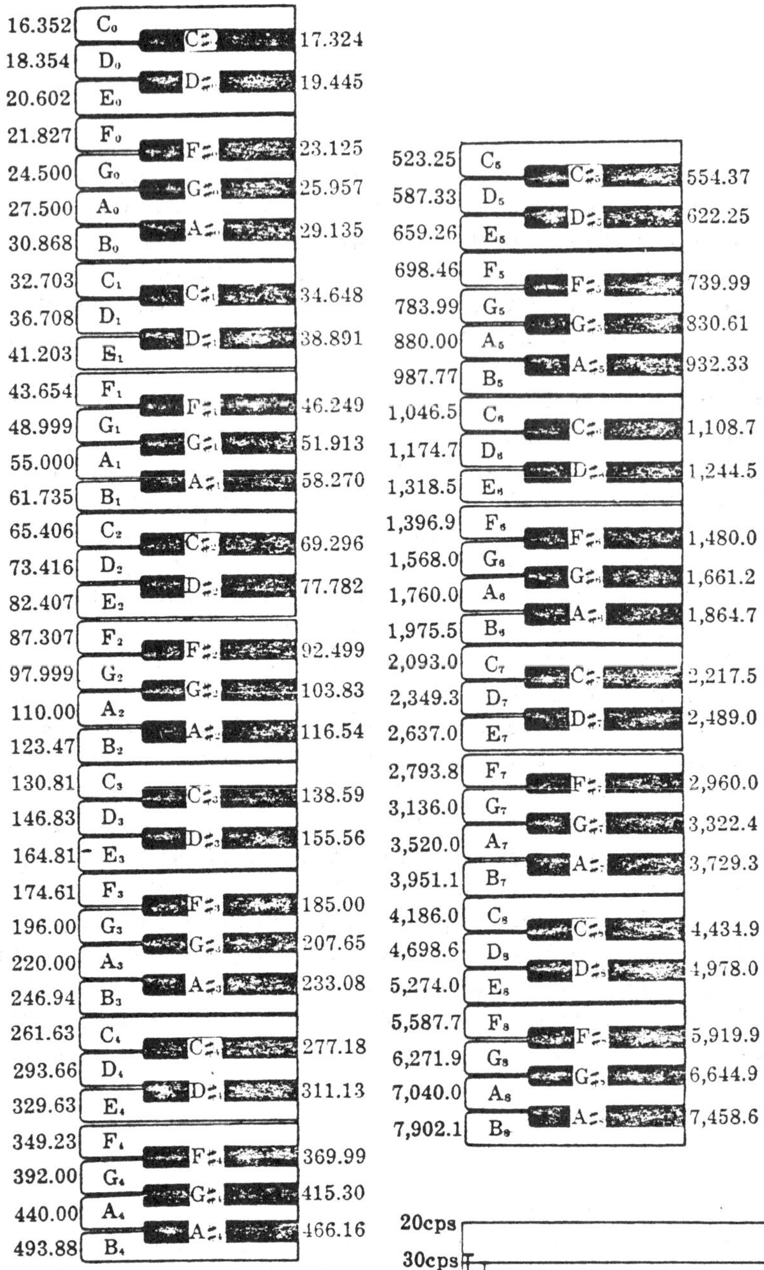

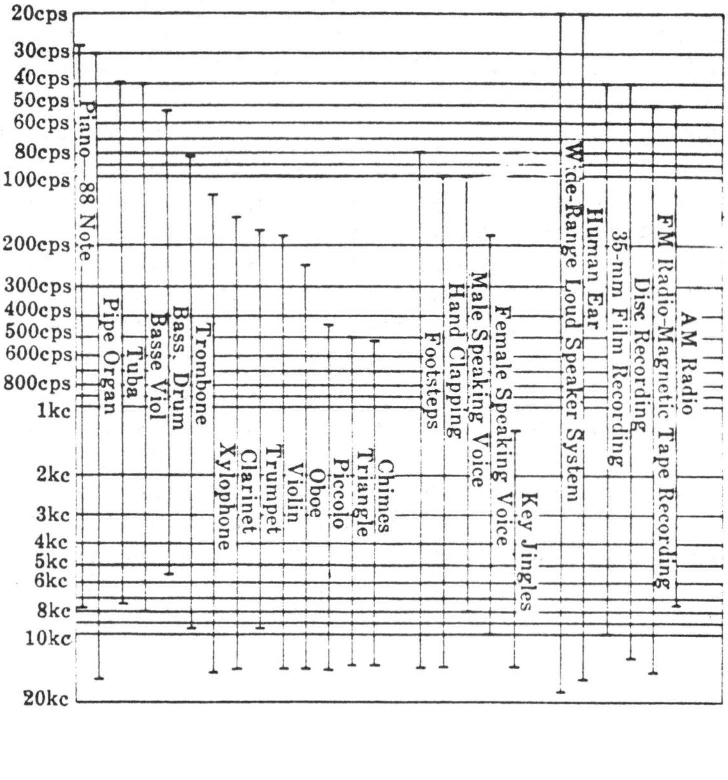

The Audio Frequency of Piano Notes and musical Instruments

subliminal noises.

Audible noises belong to the complete sound of the instrument, such as the collection of grindings, scrapings and scratchings produced by a violin bow when it is pulled across the string. To increase volume, a violinist applies rosin to the bow to increase its friction with the strings (really he is just trying to get a better grip on the strings for more volume or easier playing). If the violin is recorded without these noises, the playback will be lifeless and unrealistic. If the mike is placed too close to the bridge, and thereby a recording is made with too much of the noise, then the violin will sound screechy and unpleasant musically.

Subliminal noises are noises that can easily be overlooked in the recording process. While they may not actually be missed, their absence can cause the sound to be less distinct and impressive than if they were present. A good example of this kind of situation would be the snapping noises coming from a saxophone or the clicking noises coming from electric bass or guitar strings when played with a pick. The bass player hears these pick noises while he plays (if he has his right earphone off) by direct sound radiation into his ear from the pick, but when he comes in to hear a playback, the noises are completely missing. It is a subliminal noise below the attention of the players and the engineer. If, however, a barcus-berri pickup is applied to the metal of the bass bridge, these subliminal noises can easily be included (only just enough to be barely perceptible) in the sound derived from the bass mike. When the bass player hears the sound with the subliminal noises restored, he will probably find the presence and clarity of what he played and where he actually picked the notes a perceivable improvement to his sound. If he records with another engineer who doesn't use a subliminal pickup, the bassist will probably now notice the lack of clarity and the engineer will not be able to duplicate your sound with any amount of EQ or effects. It's the case of not realizing you like something until it's gone.

ACOUSTIC ENERGY

Acoustic energy is the pressure and rarefaction of the atmosphere which produces what humans perceive as sound. Any sound that is audible to a human is recordable by the use of some kind of mike. Sounds that are not audible to humans are also recordable, but the original sound would not really be considered musical since no human could hear them.

Music produced in the studio by musical instruments or other types of sound producing equipment gives rise to two types of acoustic energy: acoustical signal and acoustical noise.

Acoustical signals are desired by the engineer. These acoustic energies are usually direct sound from the instruments. (Direct sound is that sound coming directly from the source without having any reverb from the room or reflections from nearby surfaces).

Acoustical noises are unwanted acoustical energies. One source of acoustical noise would be a humming guitar amp, which was radiating hum into the studio while vocals were being recorded. One track's worth of this acoustic noise recorded onto tape would probably be OK, but several track's worth, when played back at the same time during a mixdown would show up as a noticeable hum. The wind draft noises produced by a rotating Leslie speaker if picked up due to too close a placement of the mike or lack of a wind screen, would be an acoustical noise. Air conditioning systems produce hissing and rumbling sounds that would go onto a recording with the desired instrument if not controlled. When the snare drum appears on what's supposed to be a tom-tom track due to the tom mike "overhearing" the snare, the presence of the snare sound is acoustical noise. In a case such as this, any EQ or reverb, etc., that you use on the tom recording affects the snare sound as well. When a sound reaches a mike that is not supposed to "hear" it, the phenomenon is known as leakage.

PROPERTIES OF INSTRUMENT'S SOUNDS

Some instruments produce simple tones. Such instruments are the electric organ and the synthesizer when sine waves are being played. (Sine waves are the simplest waveform with no overtones or harmonics). Simple tone instruments are defined as those that produce nearly overtone free sounds with no attack noises. Use of EQ damages these sounds the least.

Some instruments produce sounds with striking attack noises. Examples are the electric bass played with a pick, piano, harpsichord, chimes and bells. The attack sound is needed and necessary. Without these noises, the instrument sounds distant and/or distorted. The attack noise is usually much greater in volume than the tone part, but the ear fails to notice this. A good microphone does notice however, and produces attack voltage waveforms to match the nature of the sound. Attack noises are not, however, usually necessary to the appreciation of the sound and, additionally, can cause distortion, which will definitely disturb the quality of the sound. This quality of attack noises gives rise to the science of level setting for making magnetic tape recordings without saturating the tape.

Some instruments produce only percussive sounds with no tone content: blocks, claves, etc.

The more ragged or irregular the waveform of an instrument is, the more hi frequency energy is present. Instruments with ragged waveforms require more "VU room", that is, they need lower VU readings to record with the certainty that no tape distortion will occur. Instruments with ragged waveforms are: strings, horns, reeds, cymbals and voices. It is with these instruments that the most changes in the original quality and characteristics of the sound occur when EQ is used.

*** Please refer to Illus. B3-3 -- "Envelopes and the ADSR" ***

ADSR

Attack, decay, sustain and release are the four terms used by synthesizer makers to describe the envelope of the particular instrument's sound. (Envelope is a picture of an instrument's volume over time; the volume history is the vertical part of the graph, the time is the horizontal part.) Imagine the envelope by drawing a line from left to right across a page while moving your pen up for loud and down for soft instrument volumes. An organ has no sound until you press a key, whereupon full volume comes forth immediately. The note remains at full volume until the key is released, whereupon the note disappears immediately. An electric bass, on the other hand, has an attack sound that rises sharply, drops off immediately to a lesser volume, and then gradually fades away over a long time. A wood

Illus. B3-3

Envelopes and the A D S R

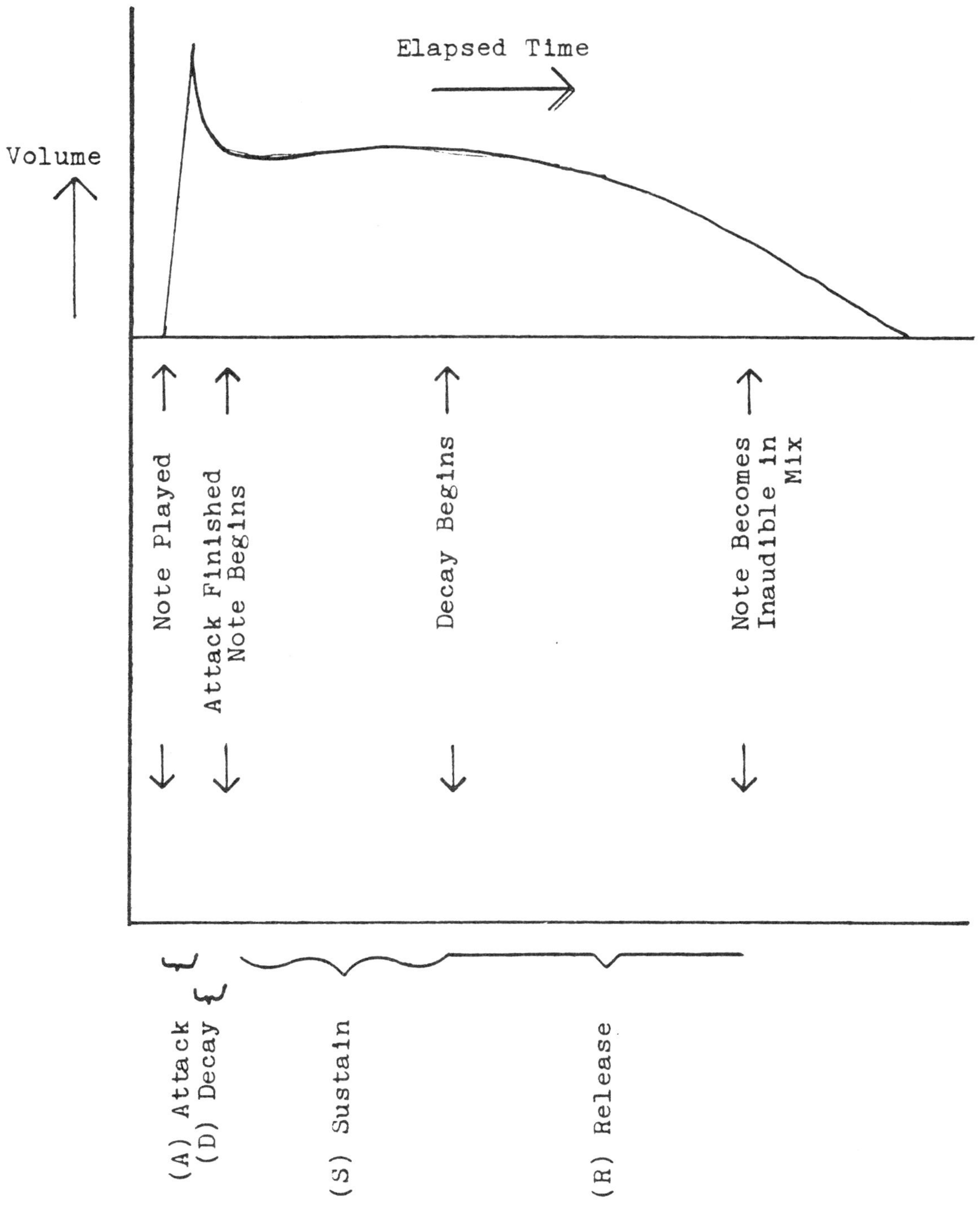

block produces only a quick vertical line in that it is a transient sound and only lasts a very short time, and then is completely quiet again.

The attack is the time it takes for the initial rising of the note to reach approximately 60% of its full volume. Decay is the time it takes for the note to drop from its attack volume to its sustain volume (the sustain volume is the main body of the note and occurs just past the attack noises). Sustain is the amount of time the note lasts before the release. The release is the time it takes for the sustain volume to drop away to dead quiet. The synthesizer device that provides these times is called the A.D.S.R. -- for Attack, Decay, Sustain, Release. Go to a music store and have a salesman demonstrate one if you never have seen one work.

*** Please see Illus. B3-4 -- "Frequency Distribution" ***

Frequency distribution is a picture of volume vertically across frequency horizontally, i.e., a graph with the horizontal axis labeled 20 Hertz at the left and 20,000 Hertz at the right. The vertical axis is the volume and is very soft at the bottom and very loud at the top. A bass instrument's frequency distribution has much bass frequency content but little high frequency content. A cymbal's energies would be located more to the hi frequency side of the envelope graph. A single note from an organ would be a straight vertical line over the position on the graph, representing the frequency of the note being played.

WHERE DOES THE MUSIC COME FROM? WHERE TO PUT THE MIKE?

All engineers have their own methods for mike placement. While you are the second engineer, it is best to follow the directions of your first engineer so that you can steal his ideas and improve on them later.

Violins and violas are miked in front about two feet away and about two feet over the level of the instrument. For a more cutting sound, get closer. Cellos and bass violas are miked about two feet from the floor and one foot from the "F" hole.

Horns are miked in front of the bell at about one foot with a good overload immune mike. Because they are loud, it should be possible to mike them from farther away with no leakage from other instruments except other nearby horns. Always seat the louder horns in front so they won't "leak" over the shoulders of players sitting in front of them and into their mikes.

Reeds are miked from the center of the instrument at about 8 inches.

Single vocals are miked slightly above and 4-6 inches away from the mouth. For warmer vocal sounds, go to one foot and point the axis of the mike capsule at the chest instead of the mouth.

Choral voices should be miked from a distance of about 15 feet , or wherever the group sounds good to you when you walk around listening for mike placement areas. If you put the mike near the conductor, and if the conductor really has any control over the choir, you will get a great mix, since he will be "mixing" the vocal strengths of the sections by conducting

Illus. B3-4

FREQUENCY DISTRIBUTION

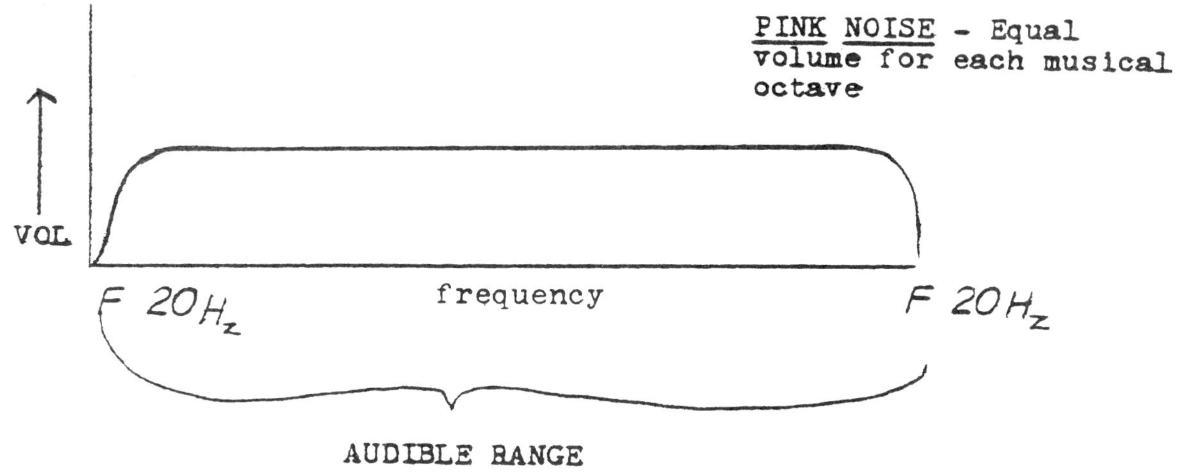

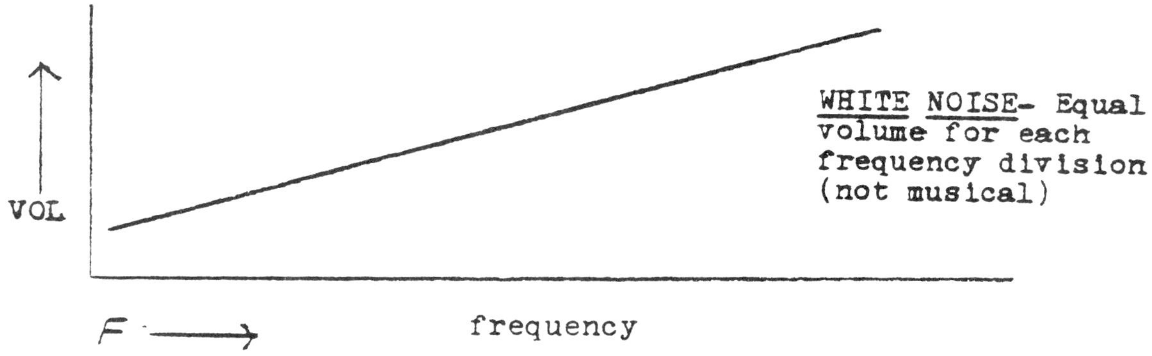

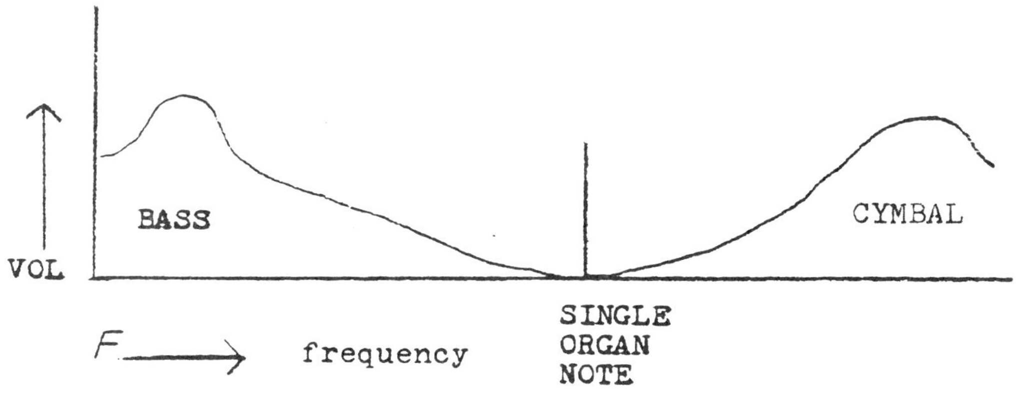

their dynamics. You can use many mikes (eg. as in one mike for each vocal section) and then mix the multiple mikes together, but no console can mix instruments or voices together like air can. Air is the best mixer. Sometime set up many mikes and one good stereo mike in a good acoustical environment and compare them to see what I mean.

MUSIC COLLECTION DISTANCE

Music collection distance is the minimum distance from the source where, if you put a mike, you get the "full effect" or sound of the instrument. With a violin, if you mike too close, you get a harsh, squeaky sound. Moving farther back gives the mike the ability to hear the warmer sounds coming from the body of the instrument. Miking a Marshall amp stack with one mike at a few inches distance is, to me, ridiculous because the mike is not even "aware" of the largeness of the sound caused by the many speakers, which give the emotional impact the guitarist hears when standing 10 or 15 feet from the amp. The excitement produced in the listener by listening to many speakers (instead of just one) is called the choral effect and is lost by close miking.

If you put a mike where the sound sounds good to you, or where the player is standing while "appreciating" or setting up his amplifier's sound, your recordings will pack a larger impact than if you mike inside of the music collection distance.

ZEROING THE CONSOLE AT SESSION'S END

*** Proper understanding of this section requires use of an audio console or a very good picture of one. Please refer to chapter 6 of "Modern Recording Techniques" by Robert Runstein ***

This operation will assure you that no signals come out of the console to tape, reverb chambers, special effects, etc., without your having activated the system yourself. Remember: you can blow up reverb chamber speakers without knowing it since the chamber may be very far away from the control room and by not listening to the reverb return signal, you might not be aware of the catastrophic volumes going on inside the chamber. These are the things you need to do to normalize:

Turn down the control room monitor master
Turn down all individual monitor volume controls
Turn down the studio monitor master
Turn down the cue masters
Pull down all faders
Pull all patches (if there are any patches in the mike out/mike pre in patch bay -- pull the mike out patch first)
Turn console oscillator off and its pot down
Un-punch all buss assign switches
Pull down all reverb send pots
Un-punch all reverb assign switches
Set all mike/line switches to line
Set all EQ controls to "flat"
Set all EQ in/out switches to out
Set all buss sub masters to the proper setting
Set the buss masters to the proper setting
Turn all buss reverb return switches off
Turn down all buss reverb return volume controls down
De-select all speaker assigns
Turn down all individual monitor volume controls

USE OF THE PADS AND TRIMS

The mike pad (sometimes with an associated trim) is the console control for setting how much gain there is in the mike input circuit.

In older consoles, the amount of the gain of the mike pre was fixed at where the mike pre performed best electronically, and the actual gain of the input channel was thus pre-set. One would therefore carefully select the mike to be used for proper output level and then add additional amounts of pad before the mike pre, if the selected mike was still a little too hot. On newer console designs, there is still a pad, which is usually positioned before the mike transformer in the circuit, but the gain of the modern mike pre amp is now adjustable over a wide range. The hottest to the most insensitive mikes can be used with ease. This means that un-informed recording engineers can choose the entirely wrong mike and record with it anyway.

A "pad" is a set of resistors that preserve the audio format of the signal (like impedance for example) while turning down the level without changing the EQ. A trim, on the other hand, is usually an amplifier gain control, which is connected to the actual mike preamp. In some consoles the pad and trim are combined in one switch.

At this stage, it is best to consider the use of the mike input module as a procedure without much concern for the details of what's happening electronically.

After assigning the channels to the desired busses, and confirming that the channels are properly labeled with the correct mike names, solo the farthest lefthand fader and activate the channel's monitor solo. Slowly push up the fader, while watching the VU meter that it is assigned to.

Observe the fader setting when the VU meter is reading the indication that you want (near to, but not into the red).

If the fader is set too low (closer to you -- a higher amount of attenuation on the fader scale), reduce the gain of the channel (or increase the pad) so that the fader can be pushed up to the correct setting (see Basic 1).

If you use all the pad (or the lowest gain setting) and you still can't raise the fader up to its proper setting loss position, then you definitely have chosen too hot a mike.

If you use all the gain (or remove all the pad), and the recording VU meter indicated level is too low, then you have used a mike that was too overload immune -- its output level is too low.

Now solo the next input module to the right and un-solo the one you just finished.

Set up the gain on the new module.

Continue until you finally un-solo the last module after finishing with it, and the levels are all set.

Now, even without having a completed cue or monitor mix setup, you could make a non EQ recording. EQ will be discussed later.

Headroom is a term used to describe the number of dB between the nominal level and the maximum output the device is capable of delivering. (Nominal level is the level of a sine wave present when an audio device is reading zero on its VU meter. When a console is reading 0 VU, and it can actually put out +24 dB, then its headroom is 24 DB. Only sine waveforms are used since VU meters only read true levels with steady state sine waves.

When you adjust the console mike input system you are in reality adjusting the amount of headroom that you have so that you do not introduce distortion. In almost all consoles, when the sub master and master faders are set correctly, and you get a good level on the VU meter with the fader set to its proper position, you have a good 24 dB of headroom. The more sine wave-like a recording waveform becomes, the less headroom you actually need when VU meters are used to judge levels; peak meters read correct levels with any kind of waveform. Recording a steady state sinewave requires no headroom -- the sine wave can be just below distortion and will never distort.

If you get your levels with the sub master, master, or fader set above their proper setting, then there will be excess amounts of headroom. Excessive amounts of headroom will lead to tape recordings with console electronic noise on them. Un-needed headroom should be avoided, along with operating with too little headroom, which risks distortion.

BALANCED INPUTS

Balanced low-impedance inputs allow the engineer to use long mike input cables without picking up radio interferece and hum. This property is especially useful when doing public address (sound reinforcement) or remote recording work in areas where there is a radio transmitter nearby. Also remote recording situations, having far less than controlled grounding conditions, benefit from balanced input circuits.

Having balanced mike inputs requires two things:

1. The console have mike input transformers (or active balanced inputs)

2. The using of only professional, three wire, (balanced) low impedance mikes

These mikes have two signal wires, neither of which has anything to do with the shield or ground. The wire that has the electrically positive signal when the mike is in a pressure wave is called the hi or (+) wire. The other wire has negative going electricity when the mike is in a pressure wave and is called the lo or (-) wire. The third wire is braided and surrounds the two signal wires and is called the shield. (Shielding is the process of draining off unwanted signals radiating into the wire by connecting the shield to mother earth).

Shielding doesn't get rid of all the interference that is commonly encountered, so another electronic technique must be used, which is called common mode rejection.

Common mode rejection refers to the use of balanced inputs, where an input transformer requires any signal to exist in a push-pull fashion between the hi and low signal wires before any signal is passed to the other side of the transformer. Any disturbance coming in on both signal wires at the same time (that is a signal common to both wires, like radiated hum) cannot pass through the transformer and therefore will not pass any signal to the console.

Disturbances radiated into the mike cables from a radio station, a humming guitar amp, or CB transmitters driving by outside the studio making illegally powerful transmissions, are first drained away by the shield to a large degree. Any disturbances making it through to the signal wires inside the cable cause common mode electricity to flow to the mike input transformers. With both signal wires pushing or pulling together, the transformer will block the disturbance. The mike signals, however, are always of the push-pull variety and always derive output signals from the mike transformers. The use of shielding and common mode rejection with console input stages allows operation with very long mike cables.

INBOARD EQUALIZATION

All EQ found in the console input modules or elsewhere in the console is inboard EQ. All other EQ is outboard EQ and must be patched.

Console EQ should be used to adjust for response errors of the mike, such as having the mike too close, too far away, or making up for lack of proper sound quality from electric instruments. The superior way to get the sound you want is to change mike placement and use the right mike. Overuse of EQ will cause your audio projects to have a less clear and less solid sound.

Voices that already blend well in the air will not blend properly if recorded with the wrong mikes and EQ'd to get a "good sound."

Using overly close mike placement and then using EQ to return to a "natural" sound will always produce a less natural sound than changing the mike placement.

Trying to repair a recording made with too much EQ by adding equal and opposite "reparative" EQ will further ruin the quality of the sound.

In most cases it is not possible for an engineer to undo EQ damage done at another studio with a console having parametric EQs. (Parametric EQs are advanced types where both the frequency, amount, and width of the boost or cut frequency area are continuously adjustable).

Some producers, notably jazz producers, are deservedly curious why a $50,000 Berzendorfer piano miked with a $1200 mike and using a $500,000 console, a $60,000 tape machine, and a monitor system costing $9,000 should need EQ to sound "right" to the engineer.

SELECTS AND ASSIGNS

A select switch is a switch with many inputs and one output. This kind of switch is found where a console circuit path can have several originating signals (eg. pre or post reverb send, mike or line input to the input module).

An assign switch is a switch that has one input (the signal) and many outputs (places where the engineer might want the signal to go. Eg. EQ out signal goes to buss assign switch, which can send the signal to one, a few, or all the busses.)

CONSOLE REVERB SYSTEMS

A console reverb system consists of:

The pre post switch, which makes the source of the reverb send from a particular channel either dependent or independent of the fader setting or motion. That is it taps the channel's audio before or after the fader (pre or post).

The reverb send level control at the channel.

The reverb send master, which sets the level of the console's overall reverb signal strength.

The reverb device (a Lexicon 224 for example) or reverb chamber (a room where reverberations occur; has a speaker and at least one mike).

A reverb return system -- usually just another input module, which can bring the reverb signal up to any of the recording busses or to the mix buss.

ELECTRICAL SIGNAL

Electrical signals come from the outputs of microphones or microphone like transducers. (A transducer is a device that converts one form of energy into another form). When operating properly, a transducer makes a signal in the new medium that exactly resembles the waveform of the energy in the original medium. A mike makes an electrical signal that nearly exactly resembles the air pressure/rarefaction phenomenon in air, which we call sound. Mechanical transducers lack the intelligence that the ear / brain transducer system has. When we are listening to someone in a crowded room, we can "tune out" the other persons who are nearly as loud as the person we are listening to. Mikes cannot do this. Mikes "hear" all sounds in the air in proportion to how loud they really are and cannot tell which sounds are the desired ones.

A good definition of electrical signal is: desired electrical waveforms.

The acoustical signals and leakage we spoke of earlier when transduced create electrical signals, which have the transduced acoustical signal mixed together with the transduced acoustical leakage and/or noises, if any were present and "on mike." Consoles that buzz, hum, crackle, etc., produce recording signals that contain electrical noises.

*** Please refer again to my table of signals and noises Illus. B3-1 ***

The set of definitions again:
Acoustic signals - musical sounds the engineer wishes to record
Acoustical leakage - musical sounds the engineer does not wish to record at least with the mike in question
Electrical signal - desired electrical waveforms
Electrical noise - unwanted electrical waveforms

ACOUSTICAL LEAKAGE AND NON-MUSICAL SOUNDS

There are many possible causes of unwanted acoustical energy. Some of these are:

Having a "soft floor" in your studio allows stomping noises from vigorous foot tapping musicians as well as the much more powerful kick drum thumps to travel through the floor and up nearby mike stands. When the mechanical energy thus transmitted gets to the mikes via their stands, the energy is transduced into low frequency pulses, which are recorded. These pulses may or may not be audible in the monitor system, depending on how full range the monitor system is. Thus it may be possible to be recording very low frequency thumps without hearing them. However, they may be visible on the VU meters from time to time. It is the relative motion of the air and mike that produces these acoustic sounds -- in addition, physical vibration of the mike produces mike outputs as well. If only one or two mikes are being used, then the problem may not be pronounced, but the more mikes you use, the more audible the thumps will be, since their energies are added together during mixing. Since thumps recorded together (during the same recording "pass") are in phase with each other, the more tracks you enter into the mix, the more you can hear them -- eventually becoming quite noticeable. Thumps should be avoided, since they will cause the disc cutter to have to reduce the amounts of bass he will be able to record on the finished record. If the disk cutter changes the EQ of your project he will, of course, also be changing your mix.

Recording in a studio where there is a noisy fan or air condtioning system operating may add low frequency rumblings to your recordings. If you record 10 vocal tracks simultaneously, each track having a different singer with a low frequency rumble in the room, this rumble will be in phase on each vocal track and will get louder the more vocal tracks you enter into the mix. Each singer only gets recorded on one track; the rumble gets 10 tracks. If this problem should show up during a mix session, put half of the "rumblized" tracks out of phase with the rest and juggle the phase switches until you "null out" the unwanted noises. Be very certain that you do not null out rumble or thumps in such a way as to reverse the phase switches on any stereo recording tracks.

If you position a quiet musical instrument (perhaps percussion) in front of a highly reflective surface (like the control room glass window), then as the percussionist plays, his body blocks off the room leakage from getting into his mike from the glass reflections. As he moves to another instrument however, the leakage reflected from the glass can pour into the mike. One way to avoid this is to have the percussionist wear a PZM as a pendant, which he would then carry with him to each percussion instrument he plays. The effect is that he is then always "on mike."

OPERATOR NOISES

Operator noise is caused primarily by improperly zeroed audio consoles or misuse of outboard equipment. Always cancel all console settings before beginning work. Some results of operator noises are:

Discovering that the snare reverb you have been hearing isn't from the monitor reverb system after all, but has actually been printed on the bass track along with the bass. (Cause: failure to see that buss reverb returns are off.)

You try to bring up some reverb for a mix at the session's end but hear no reverb return audio at all. (Unbeknown to you the reverb send pot on the bass mike, as well as the reverb send master, was turned up all the way during live recording so the chamber speaker was blown during the hours you were "exercising" but not listening to it).

Finding that one of the toms didn't get to the tape (due to a patch cord plugged into the wrong jack in the patch bay, such as the tape input jack).

Seeing very high bass levels during a tracking recording, you move the bass fader and thereby cause a loud crackle to be recorded on the bass track. (Caused by a dirty fader condition.)

NOISE PREVENTION

As important as noise reduction equipment, is the parallel science of noise prevention. Noise prevention starts with choosing the proper mikes, so that you don't put a quiet source with an insensitive mike. It involves seeing that ambient noise is low in your studio. (Ambient noise is the noise present when the studio is supposed to be "dead quiet.") Ambient noise is easy to study -- just put up the most sensitive mike in an empty studio, close the doors, and cautiously turn up the console all the way. If you only hear console noise, then your'e doing O.K. If you hear fans, clocks, rumbles, water running, motors, etc., then you should get rid of these noises.

After the mike choice, and making sure that you have good quality cables (so as not to get hums, radio interference, or cable crackle from cable motion), comes the all important operation of the mike input system. Use of the pads and trims will let you get the maximum quietness out of your console.

Next comes the proper setting of the console submasters and master faders. These must be set with wisdom. If they are set too high, excess console noise will result; too low and console distortion will occur.

Next comes the proper operation of the tape machine. There is no way to make quiet recordings unless your machine is properly set up. Machine set up is a very complicated study best left to a qualified maintenance man. If you don't know one, call a major studio and hire one to do the alignment. Always check that the machine is properly aligned. To do this first play your project's master tones (1 KHz, 10 KHz and 100 Hz). See that all the tones play back at zero. Now set the console into tone mode (this means a console mode where the console's oscillator is outputted to all the console's recording busses and check the record alignment on some blank tape to see that the machine reads back at zero levels. If the VU meters of the recorder read steadily at 0 level during record in repro mode then the machine is set up O.K.

If the machine is properly set up, then to continue with your program of noise prevention, you must make intelligent decisions as to the recording levels you will use for the different recorded sounds. Recording at too low a level makes for playbacks that are noisy with tape hiss; recording at too hi a level makes for soft or shattered sounding tapes due to tape distortion (called saturation.) Proper recording levels depend on tape speed, the distribution of the instrument's sound energy (if it has extreme highs or lows), and the percussive nature of the sound (whether or not it has percussive attack noises).

Now if the machine doesn't lose its alignment during the session or physically damage your tape (please refer to "Safety Copies" in Intermediate 7), you have only to protect the tape itself from partial erasure to continue practicing noise prevention. To do this, make sure you keep the tape at a reasonable temperature, and degauss and clean the heads carefully before working with it.

*** please see Intermediate 3 "Cleaning and Degaussing" ***

THE GAIN STRUCTURE DIAGRAM

*** For this next discussion, please refer to the Console Block Diagram Symbol Key and the Gain Structure Diagram - Illus. B5-1 and B5-2 ***

The gain structure diagram lets us see what the levels are through a console, and where the signal is low and high in level. First some more

Console Symbols for Line Drawings

Illus. B5-1

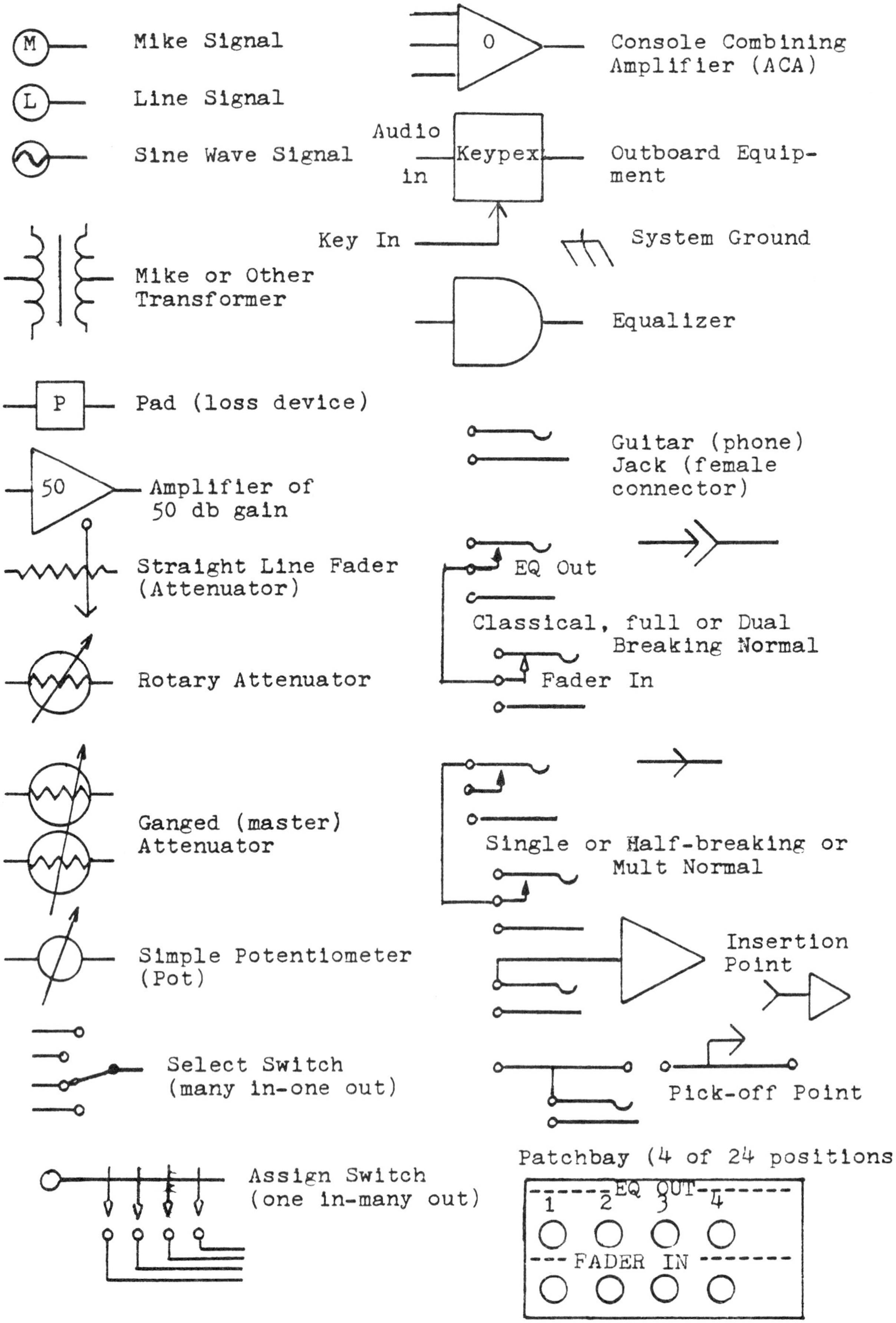

definitions:

Nominal level - the level at which the designer wants the device operated as shown by a signal reading of "0" on the VU meter of the device. This reading on the VU meter only is valid, however, if the operator has taken the trouble to set the device controls properly -- observed setting loss and other parameters.

Bel - a standard measuring unit for sound levels.

Decibel (dB) - one tenth of a bel; the audio industry's measuring standard (the bel is too large a unit to be really handy). Used for measuring the level or volume of audio signals whether acoustic or electric - one of the units that a VU meter is graduated in. The scientific definition of the dB is: 10 log (P2/P1) -- a power ratio. We will deal with the mathematical concepts of the dB after we have become more accustomed to using it as a concept.

Noise floor - the level in dB of where the noise resides in a device (about -56 dB from the console VU "0" for tape machines without Dolby units).

Signal to noise - the distance in dB between the noise floor and the nominal level.

Head room - the distance measured in dB between the nominal level and the higher level where the device begins to distort.

Line level - the signal strength at the output of a console.

Buss - a console combining system requiring a VU meter and a monitor position in the multitrack monitor system.

Assign switch - a switch with one input and many outputs.

Select switch - a switch with many inputs and one output.

*** Please refer to the gain structure diagram - Illus. B5-2 ***

The levels into the mike input of the console are usually around -40 to -50 dB (compared with the console VU meter). These mike signals (balanced signals) then go through a pad, which is activated to attenuate the mike levels if they are hot enough to cause distortion in the mike transformer.

The mike transformer helps to eliminate interference noises and passes the signal to the mike pre amp, where the gain is set by the engineer so that the signal at the output of the pre amp is at "0" VU (the same level as the output of the console). We will, for simplicity, ignore the fact that most American consoles are +4 "out" when the buss VU meter indicates 0 VU.

At the output of the mike pre, the signal is at the level where any kind of outboard equipment can be used. It is now at "line level."

The mike pre passes its line level signal to the mike line switch, which passes the signal to the inboard EQ.

The EQ passes the signal to the fader.

The fader passes the signal to the set of switches that assign the signal to any of the output busses of the console.

The signal then goes to the ACN or active combining network where it is combined with any other signals (from other input channels of the console) being sent to the same buss. If there are 32 input channels and 16 busses, then there have to be 16 buss ACNs, each one having 32 or more inputs. (A channel is the term for the console input strip that runs vertically and includes a fader, EQ, buss assign and reverb assign switches and the reverb send pots. (A pot is short for potentiometer, which is a simple rotary

GAIN STRUCTURE DIAGRAM   Illus. B5-2

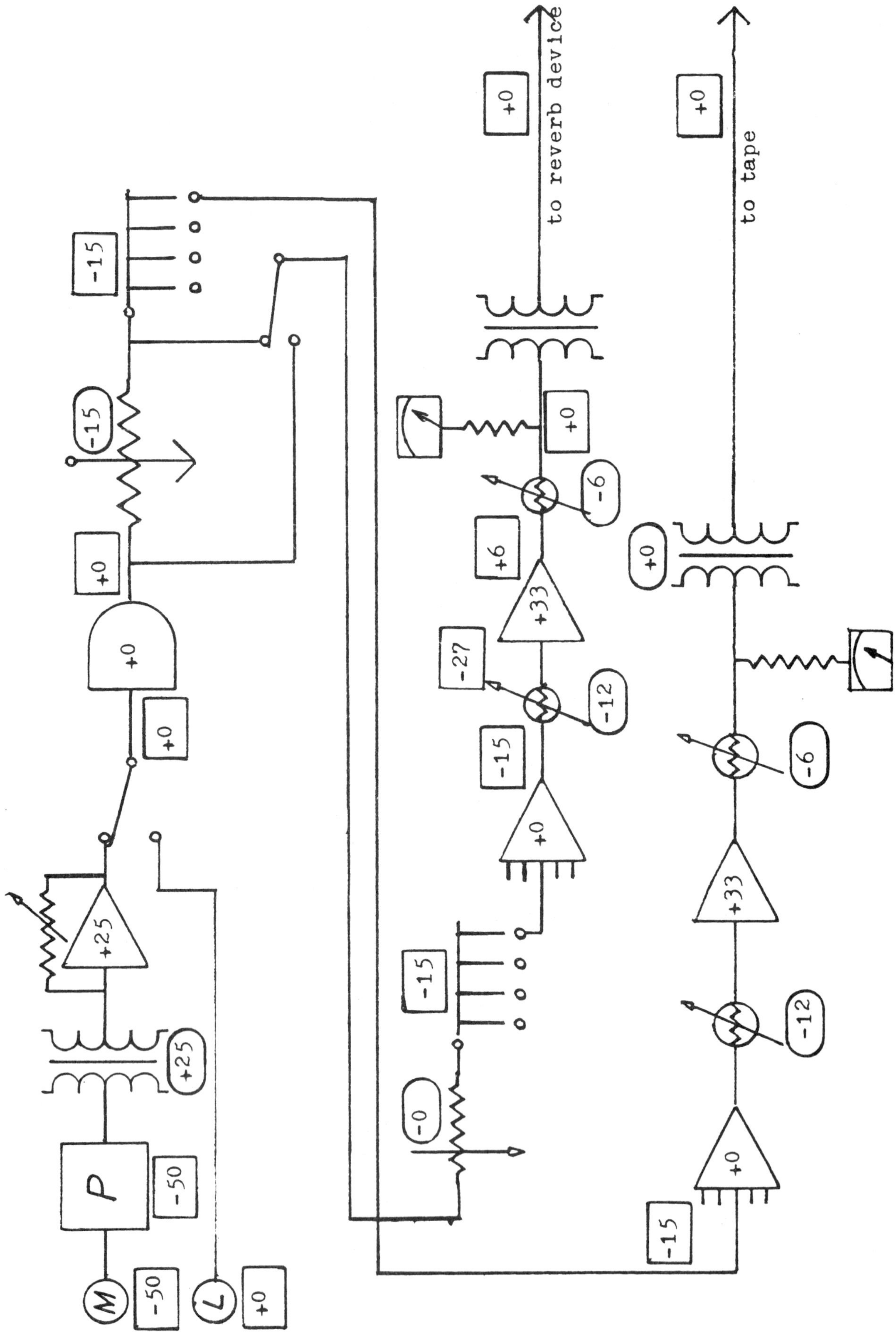

volume control. A pan pot is a panoramic potentiometer.)

The output of the ACN goes to a sub master, which is set lower than full volume (used with a setting loss) so that the entire mix on the buss can be turned up without having to move all the faders contributing to the mix.

After the sub-master is the program amp, which sets the gain back to the level necessary to go through the master fader (which may or may not have a loss) and still come out of the console at "0" on the VU meter.

At the final wire out of the console is the VU meter.

After the VU meter, the signal is shipped off to the tape machine via the console output wires (the line outputs).

The gain structure of the console shown in the supplied diagram goes as follows:

Mike level = -50 dB (50 dB below the console "0" VU indication).

The pad is not in use since the mike signal is quite low.

The transformer has a gain of 25 DB.

The mike pre amp is shown with a gain selected of 25 dB bringing the signal level to "0."

There is no EQ, so the output of the EQ is still "0." If EQ boost is used the output of the EQ will be higher; if cut EQ is used, the output of the EQ will be lower.

The signal then goes to the fader, where there is a setting loss desired of about 15 DB. (If no setting loss were supplied, then the mixer would only be able to fade signals and not "boost" them.) The signal at the fader input is at 0 and the loss is 15 dB (in this example) so the signal at the fader output is at -15 DB. This is the level that goes into the buss assign switches.

Switches have no gain or loss, they either make connecton and allow the signal to pass or they break the connection and allow no signal to pass.

After the buss assign switches, the signal goes to the properly numbered combining network. If you push buss assign switch #5, the 5th combining network gets the signal. There is no loss in modern combining networks, although some units may be connected for some gain (maybe 10 dB). The signal is not supposed to be changed by being combined with other signals, but personally I feel that combining networks do change the sound. For all critical applications, air is the best mixer. A blend will always sound better when mixed naturally in air.

Following the mix network is the sub master. The sub master is set with a loss so that it may be turned up in a session, allowing a buss mix to be adjusted in level with the use of only one level control. Our drawing has the sub masters with a setting loss of 12 dB (-12 DB). This makes the sub master output minus 27 (-27).

Next the signal is boosted to 6 dB above "0" level by the program amp so that the loss of 6 dB in the master fader will bring the level back down to "0" again. The program amp prepares the signal to drive the VU meter and the input to the tape machine through a long cable if necessary. In our drawing the program amp has a gain of 33 DB. Since the -15 dB out of the fader loses 12 dB in the submaster and 6 dB in the master, then the program amp must have 15 + 12 + 6 = 33 dB of gain.

Following the program amp is the master attenuator. The master attenuator is "ganged" (meaning more than one buss gets faded by the one control) so that multi-buss fades of equal volume per each included buss are possible. Our sub master has a loss of 6 dB, bringing the level at the master output to "0" VU.

Now the signal powers the VU meter and goes to the tape machine. Usually (especially for remote use) the output of the buss (the line out) goes through an output transformer so that the console wires do not really connect to the outside world except through a magnetic barrier.

THE DECIBEL

The decibel is simply one tenth of a bel and is named after Alexander Graham Bell. The bel is such a huge unit that all common measurements would be in very small numbers. So the decibel is used. One bel louder is about twice as loud -- the same increase in dB is 10 dB; 10 dB = 1 bel; 10 bel = 100 dB.

In ordinary usage, the dB means:

1. As a unit measure of volts there is a decibel equivalent for each possible voltage. The voltage output of a mike can be stated in volts or in dB as they are equally usable.

2. As a unit of measuring audio power (watts). Each 3 dB means twice or half as many watts, depending, of course, if you're going up or down in volume. 100 watts is 3 dB more watts than 50 watts. 50,000 watts is 3 dB more than 25,000 watts.

3. As a way of measuring relationships between two levels. Each 6 dB is twice or half as many volts as before. 30 volts is 6 dB more than 15 volts. 20 volts is 6 dB less than 40 volts. Since ohms (resistance), volts (electrical push or pressure), and amps (electrical activity or amount of electrons passing a given point in a given time) are all directly related, each 6 dB means twice or half as many (volts, amps, ohms) as before. If a mike signal coming into the console is measured at .01 volts and the console outputs this signal at 1 volt, then 1/2 the voltage would mean a .005 volt change at the mike stage but a .5 volt change at the console's output. At each stage of the console, there could be a different voltage level, making a comparison of voltage readings quite cumbersome. Using dB, the change is always the same amount (in our example 6 dB) no matter what the voltage reading: .005 to .01 volt = 6 dB and .5 to 1 volt = 6 dB.

4. Each 10 dB is approximately twice or half as loud.

PHASE ADDITION AND SUBTRACTION

*** Please refer to Illus. B5-3 ***

In the diagram supplied please note that the top and middle waveform together form the phase combination waveform to the right.

In the first case, waveform A has an amplitude of one unit (amplitude means height of the waveform). They are shown in phase. The voltage maximums of both waveforms occur at the same time as well as the voltage minimums of each wave. A definition of a waveform is: a frozen picture of voltage across time. The time we are dealing with for audio waves is very short; for a 1000 Hz tone, the time for one complete waveform is only 1/1000 second. If the maximums and minimums of two waveforms line up in time, the

PHASE ADDITION AND SUBTRACTION Illus. B5-3

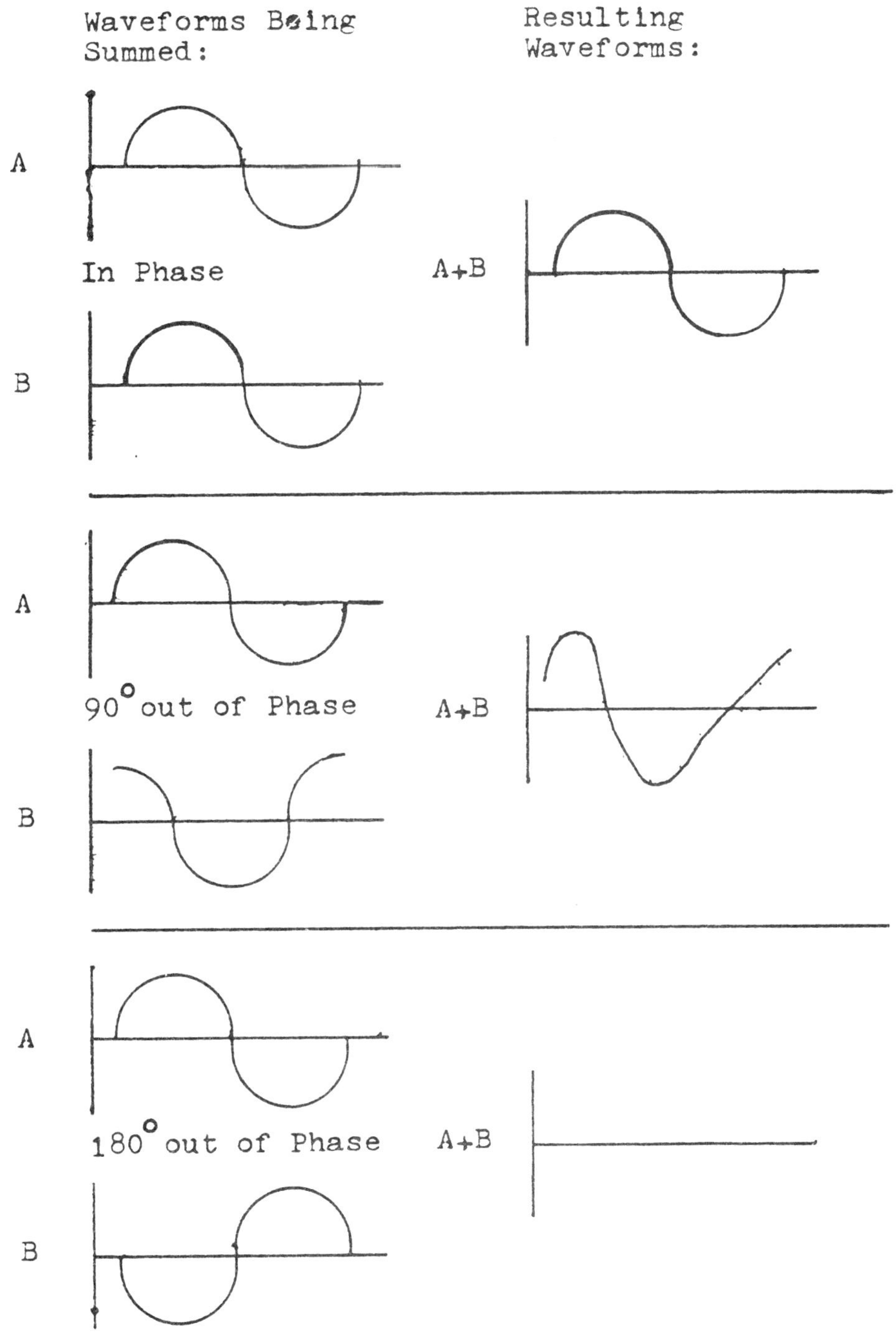

waves are in phase. In the first diagram, waveform A plus waveform B produce the right waveform, which has an amplitude of plus and minus (+/-) 3.

The third drawing shows two waveforms that are out of phase with one another. When waveform A is at +1, waveform B is at -1. When waveform A is at 0, waveform B is at 0. At all times waveform B is at the opposite polarity from waveform A and the resulting wave is one of zero amplitude.

In practical applications, we do not add the phase relationships in this way; we compare mike signals coming from two or more mikes that are in the same sound field (the mikes can "hear" the same instruments). If the addition of two mike signals to the same buss causes a loss of a major portion of the sound quality, then the mikes are out of phase. If the sound changes only slightly, and the volume increases greatly, then the signals are in phase. If you encounter two signals that are precisely out of phase when you add them together, you get a near perfect cancellation of the signal.

TRACKING

Review the explanation of "Getting Levels and EQ" section included in Basic 1. This stage of the session is best performed with an able assistant who can get and maintain good cue balances while you are getting the levels and EQ straightened out.

Console channels are operated in the mike input mode for each input channel to which you have plugged a mike; all other channels should be switched to line in mode.

The monitor system is operated in program mode.

Monitoring is by use of the multitrack monitor system (as opposed to through the faders).

The console VU meters are "on" the busses for all recordings.

The console VU meters are "on" the tape playbacks for all playbacks.

The cue system and the monitor system is set to the program mode (monitors the console busses). The tape machine is operated in reproduce mode at all times so the recordist (tape machine operator) can supervise the tape machine VU meters and compare the levels to those indicating on the console.

The engineer operates the faders, levels, EQ, and bussing, and "commands" the session.

The second engineer helps to operates the monitor system and the cue system, and keeps a complete and separate track sheet for each track of each tune recorded.

The recordist operates the tape machine, makes edits, watches for signals not showing up on meters where they should be showing up on and for tape path difficulties, uses a tape strobe from time to time to check that the tape speed is O.K. and keeps a box legend take sheet for all takes including tape reel number, take number, take time, and other recordist's code data. The recordist keeps tape locator times for auto locater operation later, and makes sure that the function buttons of the multi channel recorder are always set correctly, eg. not set on READY (to record) for a playback, etc.

To do a playback, switch the console's master multitrack monitor system switch into play (sometimes called tape) mode.

OVERDUBBING

In this mode (during Basic engineering) "bring up" the tape playbacks through the monitor system for monitoring, not through line in to the modules as during mixing.

The engineer operates the overdub mike input channels and keeps the monitor mix perfected.

The second operates the cue mix and keeps the track sheets up to date including the connotation o/d (overdub) and the date of the overdub.

The recordist operates the tape machine and keeps the proper tracks on ready, overdub, sync or play (safe) as required.

The monitor system setting of the overdubbed track should not be moved during the entire overdub, but rather all volume changes should be made with the recording faders. This is so that the overdub will "follow" the volume excursions of the track as it was recorded.

To overdub on track 12, the following settings should be used:

1. The monitor system is in sync mode with monitor 12 in non-sync, which means that monitor position 12 will be in program mode (to hear buss 12 while recording).

2. The overdub mikes are brought up through the console using the channels in mike position as required.

3. The recorder is in sync mode with track 12 in ready/input mode.

4. As part of the overdub is completed, punch in and out operations are accomplished by pushing the master sync function on the console monitor system (all monitor channels in sync), and using the tape deck automatic input mode for listening to the console during "punched in" periods.

5. All playbacks should be performed with the tracks all in safe and reproduce mode.

STACKING

Stacking is the mode of operation where several tracks already recorded are re-recorded onto another track so as to free up the contributing tracks for further recording work. To be a stack, the contributing tracks can not occur at the same time. Stacking does not involve mixing. An example might be where a rhythm and blues tune has special "feel" guitar parts that occur in different parts of the song. The parts do not overlap and since more tracks are needed for vocals, stacking is required:

1. Turn down the contributing tracks' playback volumes in the monitor system.

2. Turn up the stacking operation's recipient track on the multichannel monitor system.

3. Get a level on the different parts of the stack, being sure that no monitor system adjustments are needed during the course of the song. Mix volumes are now set with the channel faders of the console so that the overdubs stay in the mix without constant monitor system adjustment. This technique greatly reduces the mix session work load and allows the mixer to be more creative.

4. After the stack is perfected, record it onto the proper track (with the recorder in sync mode of course).

5. Make sure that the stack is properly recorded and then erase the contributing tracks completely from the beginning to the end of the song.

COMBINING TRACKS

This operation is a mix and requires the presence of the producer, the artist, or both.

In this operation, two or more existing tracks are blended together and re-recorded on yet another track so that the contributing tracks can be erased and reused. Combining tracks is a mix. The contributing tracks are happening at the same time (active together). It is imperative that the mix existing in the monitor system at the time of the combine be as perfect as possible especially in the case of the combine track. If the combine track is "featured" in the mix, any blend of the contributing tracks will do. When the combine track is "in perspective," then quite different mixes of the contributing tracks are usually required. This difference in the mix is caused primarily by the aural phenomenon of "masking." (Masking is when some sounds are hidden from hearing due to the presence of other sounds in the mix having the same frequency range.) For example, if a vocal combine of 6 tracks were made and mixed "way up front" in the monitors, then any reasonable mix would do; if the "blend" was moved back into perspective, as it would be in a final mix, then the presence of the bass, low piano, and low guitar sounds would make the low voices disappear. If the voices were mixed in perspective, then the low voices would seem predominant when the voice track was played alone, but in good balance when the vocal combine was played with the rest of the mix.

Once the combine is O.K.'d by the producer, then when any of the contributing tracks are erased, the mix becomes final. A different mix cannot be achieved without re-recording the contributing tracks.

To Combine Tracks:

1. Set the recorder in sync mode
2. Turn down all monitor pots (volume controls) of the contributing tracks
3. Bring the contributers up through the proper line input modules into a mix bussed to the recipient's numbered buss
4. Adjust the monitor system so that the recipient track is in perspective with all the other tracks in the mix and so there is enough gain available in the monitors, so that the recipient's monitor pot does not have to be turned up anywhere during the running of the combine
5. Transfer the mix to the recipient track using sel-sync, of course. Operate the combine faders to keep the contributers in a good mix and in good perspective.
6. Check that the new recording is all right technically and musically
7. Erase all contributing tracks for future re-recording.

PING-PONGING

Ping-ponging is the transferring of audio from one track of a recorder back to another track while adding new audio (from mikes) to the transfer. It is this form of recording that is primarily used at home with a two channel recorder:

After recording a guitar part on track one, another guitar part (perhaps lead fill guitar) is recorded on track two in a mix, with the playback of the originally recorded guitar part. If the mix and the performances are O.K., then track one can be erased and the two guitars, recorded in a mix on track two, can be used along with new audio, which can be recorded on track one and so forth.

Ping-ponged tracks are a mix and when the contributing track(s) are erased, the mix becomes final. Therefore the producer must be present.

Steps For Performing Ping-ponging:

1. Turn down all contributing tracks volume knobs at the monitor panel and listen in sync mode to all the rest of the tracks, except the track to be recorded, which is monitored in input mode (or buss).
2. Bring the contributing tracks up through the line input modules and buss them to the proper buss.
3. With all other tracks of the recorder in sync mode, put the track

to receive the ping-pong on record and repro mode.

4. Get a mix of the playbacks and the live inputs.

5. When the "run-downs" begin to sound O.K., record the ping-pongs. (A run-down is a pass when preparing to record, but before the record button is pressed.)

6. Continue to record until the mix and live performance sound perfect.

7. Check the new ping-pong track in reproduce along with the remainder of the tracks in as close as possible to a perfected mix (as close as possible to the kind of balance which the final mix will have).

Be aware that some recorders have difficulty when ping-ponging, stacking or combining from one track to another track that is directly adjacent to it. It is therefore a good idea to always leave a buffer track between the contributer track(s) and the recipient track(s) to avoid possible squealing while transferring. If ping-ponging is to occur during a four track session, always record track one, then track two, and ping-pong to track four adding live performances, thereafter erasing one and two and keeping the mix on track four. Next overdub on track 3, and ping-pong to track 1 and so forth.

GENERATION LOSS

Whenever you record and then playback a tape recording, you incurr a generation loss. Whenever you transfer tracks from one track to another track of the same machine, and use the resulting track, you suffer a generation loss. Each generation loss beyond the normal two (recording and playing the multitrack, and recording and playing the two track) is clearly audible in a good control room. It is therefore always recommended that you have more tracks available on the machine than you need. Record SMPTE time code on the last track (60 Hz sync tone then should be recorded on track 23) for use in syncing up your first multitrack machine to another one. If at all possible, record the time code live with the tracks, rather than striping the tape with it later as an "time code overdub," so to speak. A 60 Hz sync tone MUST be recorded with the original music to be effective in controlling possible machine speed problems.

MIXING

Multitrack playbacks are brought up on the console input channels as line inputs. The level and EQ of the various previously recorded tracks and overdubs are carefully blended and arrayed across stereo by the panoramic potentiometers, called pan pots for short. This final EQ should already be carefully tried and true due to the many rough mixes you've already done and given to the artist and producer, and the comments you've gotten back from the band during the course of the overdubbing sessions. At the mixing stage all clicks and pops must be edited out. Some slight "improvement" edits may be made to tighten up the band's returns from musical pauses and breaks; down beats at the starts of the songs can be tightened up as well. Several different versions -- a mono single mix, cassette duplicating mix, a mix without the lead vocal (for appearances on T.V. where the artist sings live using the recorded version of the band's track instead of trusting the T.V. audio man to get as good a mix as you now have), optical sound track release mix and a compressed mix -- should be mixed (if needed) while you have the mix well under control, including final EQ, effects, the fader cues and the muting and editing tricks worked out. The phase of the various tracks must be carefully watched so as to make a mix that sounds equally good in mono. The physical and electrical alignment of the two track used for mastering must be carefully checked by the engineering staff. Do not change two track machines during the mixing of an album project if you can help it. Following each final mix Keep careful notes so as to facilitate the resetting of the console to the present mixdown settings in case some major or minor changes are decided upon after close scrutiny thereby requiring another mix. The mixer should be fresh, fully rested, and not in Fletcher Munson fatigue when doing final mixes. A good mix includes:

1. Powerful and solid lows
2. Proper use of the very powerful mid range areas
3. Clear and clean highs
4. Proper but not overburdening effects
5 Dimension - some sense of depth: sounds behind the speaker plane so to speak
6. Motion - movement of the instruments using the pans to heighten the music
7. At least one true stereo track (hopefully something "up front" and really featured in the mix) perhaps a real stereo recording of a room full of string players
8. Some acoustic information -- not just straight digital delays and reverb.

A very helpful mixing technique (or for complicated overdubbing for that matter) is the use of dummy busses. A dummy buss is a recording buss that is used solely for the purpose of keeping an eye on the VU meters -- the output of the buss is not recorded.

When mixing a poorly recorded set of background vocals, for example, if the various BG VOC tracks are jumping up and down in level and making the mix very difficult to do, assign the first vocal track to buss one (in addition to the stereo buss where it's going anyway), and the other difficult tracks to consecutive numbered busses in the same order as they appear in the mix from left to right. Now when one of the vocal tracks jumps out, you will hear it leave the blend but also see it on the proper VU meter of the assigned dummy buss.

During difficult overdub mix situations, assign the channels to be mixed to a recording buss, perhaps buss 22, but also assign the overdub recording channels to dummy busses 1, 2, 3, etc., where you won't be recording them. Now you can see a "mix picture" in addition to the more important activity of listening with extreme care to the blend. You may find that a certain VU picture correlates with a really good sounding blend, and this discovery will help you get the blend in the future when you overdub the same instruments using the dummy buss technique again. This technique works extremely well when combining many overdubs of synthesizer harmonies, where you cannot always tell which track is which voice; sometimes you can get to know the fader's sounds very quickly by watching the dummy busses.

PATCHING MICROPHONES

Having a complete mike input list with no errors is absolutely necessary. Only with complete confidence in this list can you completely repatch the mike input section of the console and come up with a completely new mike input layout across the faders that will work.

On the surface cross patching does not present any serious problems, but some problems may creep in if you are not fully aware of what's going on inside the patch bay. For one thing, when doing a session with mikes cross patched and difficulties come up, you must give the mike in number, and not the fader number to your second engineer, who would otherwise unplug the wrong mike, by relating to the fader number, which would be wrong. Remember that there are 48 volts of DC at the mike out patch bay (this patch point is sometimes called mike in ???), and patching around using this bay with a mike fader up either the original mike fader or the destination mike fader is highly illegal and can make such a loud noise that the control room speakers and all the headphones in use could be blown, as well as your ears.

*** Please refer to Illus. B5-1 -- Console Symbols for Line Drawings ***

The circles refer to sources of signals; l=line signal and m=mike level signal.

A select switch has many inputs and one output. A select switch selects one out of several possible input sources to go to the input of the next console device.

Transformers are found at the input and output stages of most modern consoles and tape machines. They serve to get rid of noises like hum and interference from radio signals and other disturbances.

Assign switches, which feed one or many possible circuit paths, are found at the outputs of console systems. After the console faders, an assign switch determines which buss the fader's signal will be contributed to.

A pad is a form of attenuator (volume loss device that does not change the sound with different settings of volume). A pad is usually made with attenuation in steps of five or ten dB.

Triangles are the audio amplifiers. Amplifiers have a fixed or variable gain. Variable gain mike pre-amps for example are shown with a resistor symbol over the top, connecting the input to the output wire -- this means that the gain is variable. Remember 10 dB is approximately twice as loud.

Straight line attenuators are shown as the resistor symbol with an arrow pointing straight down. A straight line attenuator is the console fader.

Rotary attenuators are primarily used to conserve vertical panel space, such as reverb send level controls, and in the monitor panel to control the volumes of the playbacks of the tape tracks. Some console manufacturers use straight line faders for monitor volume as well as for input channels, but these faders are very small.

Ganged master faders are for fading more than one buss at the same time (as in a stereo fade).

A simple pot (rotary potentiometer) is sometimes found in a console system, and is used only when electronically safe (not in a record circuit), and is used to save money.

The combining amplifer, currently disguised under the names ACA (active combining amplifier) and ACN (active combining network) is for mixing two or more sounds together into one circuit. An eight buss console has eight buss ACA's and a 20 input console has ACA's with at least twenty inputs to accept one signal from each of the twenty console channels.

Outboard equipment is shown as a rectangular box with the type of unit written inside the box.

Console equalizers are shown as boxes with curved right ends.

Console patching points are shown as twin arrowheads or single arrowheads pointing "down stream." This means that there is a patch point where you can "break" the normal flow of audio and either change the path, or insert one or more outboard equipments into the circuit.

GROUND LOOP SOLUTIONS FOR DIRECT BOXES

*** Please refer to Illus. B7-1 -- the Direct Box Diagram ***

The symbol with the kinked wire at the upper left corner of the diagram is the symbol for a guitar type jack (female guitar connector). The upper one is for plugging the guitar into; the lower one is for plugging the direct box into the instrument amp, where the instrument was to have gone.

The line dropping to the bottom of the diagram and then moving to the right is the ground system wire of the guitar and the guitar amp.

The little switch at the lower right labeled "GND SWITCH" is known as the ground lift switch. It is shown here in the ground lifted position. If the switch is switched to the non-ground lift position, then the quiet, pure console ground system will be hooked up to the dubious ground of the guitar system. Using the direct box in non-ground lift mode is dangerous and can result in damage to the console and electrical shocks to the performers.

The two identical figures shown back to back to the right of the guitar jacks are input capacitors.

Next to the input capacitors is the pad input switch with its associated resistors of 4,700 ohms, 15,000 ohms, and 2,200 ohms. When the pad switch on the outside of the direct box is pushed to the ON position, the instrument signal no longer rides straight to the transformer, but has to go through the two horizontal resistors, and this reduces the volume. The vertical pad resistor also reduces the volume, and operates in conjunction with the filter switch and capacitor, to provide a high frequency roll off. Since capacitors pass high frequencies and block low frequencies, using the direct box with the filter switch closed will short highs to ground, reducing them, but not lows (since the cap blocks lows). In this way, highs can be rolled off.

Finally the main event: the transformer. The direct box transformer is used to transfer impedances from those that the guitar supplies to the impedance that the console needs.

Impedance can be thought of as gear ratios in cars. For instance, when a car is in high gear it has the potential to cover lots of ground (high speed), but has very little power to pull a load up a hill. When the car is in low gear, it moves very slowly, but can pull a great load even up a hill. High impedance is like a car in high gear: there are a great number of

JE-DB-D
DIRECT BOX TRANSFORMER

Illus. B7-1

Tech Data

jensen transformers
By REICHENBACH ENGINEERING

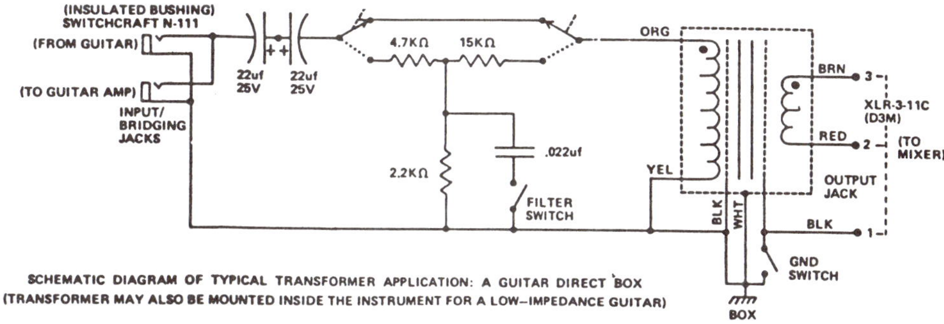

SCHEMATIC DIAGRAM OF TYPICAL TRANSFORMER APPLICATION: A GUITAR DIRECT BOX
(TRANSFORMER MAY ALSO BE MOUNTED INSIDE THE INSTRUMENT FOR A LOW-IMPEDANCE GUITAR)

GENERAL CHARACTERISTICS

Turns Ratio	11.55:1
Impedance Ratio	20K/150
Primary Source Impedance	20K ohms (typical guitar pickup)
Secondary Load Impedance	1K ohms (typical mic pre input)
Faraday Shield	Two shields, separate leads
Magnetic Shield	30dB, separate lead
Max. Input Level at 20Hz	+14.5dBV (Re: 0.775V)

PHYSICAL CHARACTERISTICS

Package	Mu-metal can
Termination	Wire leads
Dimensions	1-1/8" diameter, 1-1/16" high
Mounting	2 holes, 0.7" center-to-center, self-tapping screws supplied.

TYPICAL PERFORMANCE

Voltage Loss	21.3dB
Input Impedance	108K ohms @ 1kHz 80K ohms @ 10kHz
Secondary Source Impedance	181 ohms @ 1kHz 186 ohms @ 10kHz
Total Harmonic Distortion (Below Saturation)	0.45% max. @ 20Hz 0.28% max. @ 30Hz 0.14% max. @ 50Hz 0.017% @ 1kHz
Input Level @1% Saturation (dBV Re: 0.775 volts)	+9dBV @ 20Hz +14dBV @ 30Hz +20dBV @ 50Hz
Frequency Response (Re: 1kHz)	−0.75dB @ 20Hz −0.9dB @ 20kHz (No resonance peak) −3dB @ 40kHz
Phase Response	−28° @ 20kHz
Rise Time	6.6uS (10%-90%)
Overshoot	<1%
Common-Mode Rejection Ratio	>85dB @ 1kHz >65dB @ 10kHz

PAD CIRCUIT

For cases (1) & (2), the pad is switched out of the circuit when the signal source is a guitar pickup. For cases (3) & (4), the 10 dB pad is switched in to prevent overload of the transformer from higher output levels, i.e., from electronic instruments or guitar amplifier speaker outputs.

The pad presents a load to the output of the electronic instrument of about 6800 ohms and a source impedance to the direct-box transformer of about 16K ohms (approximating the guitar pickup). The latter is required to prevent transient distortion in the microphone preamplifier as described in requirement (a) under "impedance matching characteristics".

Electronic instruments may have some DC voltage present at the output which would cause low frequency losses and distortion in the direct-box transformer. The two electrolytic capacitors in series with the input signal block this DC voltage to maintain optimum low frequency performance.

FILTER CIRCUIT

The output signal of a guitar amplifier may exhibit severe clipping from amplifier overload. However, the high frequency roll-off characteristic of the typical guitar speaker attenuates the resulting distortion products which are upper harmonics of the fundamental notes.

Since the direct-box transformer does not roll-off these upper harmonics, the clipping or distortion will be heard in the recording. The filter circuit can be switched in to simulate the high frequency roll-off characteristic of the guitar speaker, more closely approximating the speaker's sound. The filter response is 6 dB per octave starting at about 4 kHz to 5 kHz. The value of the filter capacitor can be changed for more or less high frequency attenuation.

GUITAR CABLE LOSSES

The capacitance of a coiled guitar cable can be in the range of 1000 pF or greater. If the source impedance of the guitar pickup is in the range of 20K ohms, the resulting high frequency roll-off starts at about 8 kHz. There will be even more high frequency losses if the cord is longer or if the pickup impedance is higher.

This loss of the upper end of the frequency spectrum will affect the "brilliance" of the instrument. All of this suggests that the shorter the cable length between the high impedance pickup and the direct-box, the better the sound.

electrons moving around, but they have almost no power to push against a load (the mike input circuit of the console provides quite a load). Low impedance sources (like professional mike outputs) have fewer electrons moving around, but the electrons that are moving have a great amount of push.

The direct box transformer converts the high impedance signal of the guitar into the low impedance signal the console needs.

Direct boxes are mainly used to capture the instrument's electrical signal before it goes into the instrument amp/speaker combination, which is often noisy and of lo fidelity. In the instrument amp chances are that the pure instrument signal will distort, and most of the highs will be lost.

At high impedance high frequencies are reduced just by running through long cables. It is best, therefore, to have the shortest possible cables between the instrument, the instrument amplifier system and the direct box. This is why some professional studio musicians put a direct box inside their guitars and have their amplifiers modified to low impedance inputs.

The connection of the direct box for the audio signal is pretty straight forward: the instrument goes into the box and the direct box outputs go to the console mike input XLR and to the instrument amp, where the instrument used to go (uses a phone plug cable also known as a short guitar cord).

The functioning of the grounding switch on the instrument amp and the ground lift switch of the direct box are tricky to understand and to operate correctly.

Basically there are two types of instrument amps as far as the direct box is concerned: those with three wire AC cords and those with two wire AC cords.

*** Please refer to Illus. B7-2 -- Direct Boxes and Instrument Amplifiers ***

The problem is further compounded by the fact that, in the rear of the amp (sometimes in the front with the power switch), is an amplifier ground switch. In general the three wire amps have a three position switch, and the two wire AC type amps have a two position switch. The three wire (newer) instrument amp will be considered first.

DIRECT BOX USE FOR THREE WIRE AC AMPS

Plug the AC cord of the amp into the wall and hook up the direct box with the ground lift switch in the lift position. Make sure that the amplifier ground lift switch is in the mid position. Have the artist play a little and get a recording level. If the signal is noise free when he stops playing then you are done. If not, then try switching the amp's ground switch to either of the two outside positions. If all three positions of the amp's ground switch still give a humming or buzzing signal (either at the amp or in the control room), then get a three to two AC ground lifting adaptor and plug the amp's AC into the wall through the adaptor. Return the ground switch on the amp to the center position. Put the ground lift switch on the direct box in the non-lift position, which means that the console's ground system is now supplying the ground for the guitar amp. This should do the

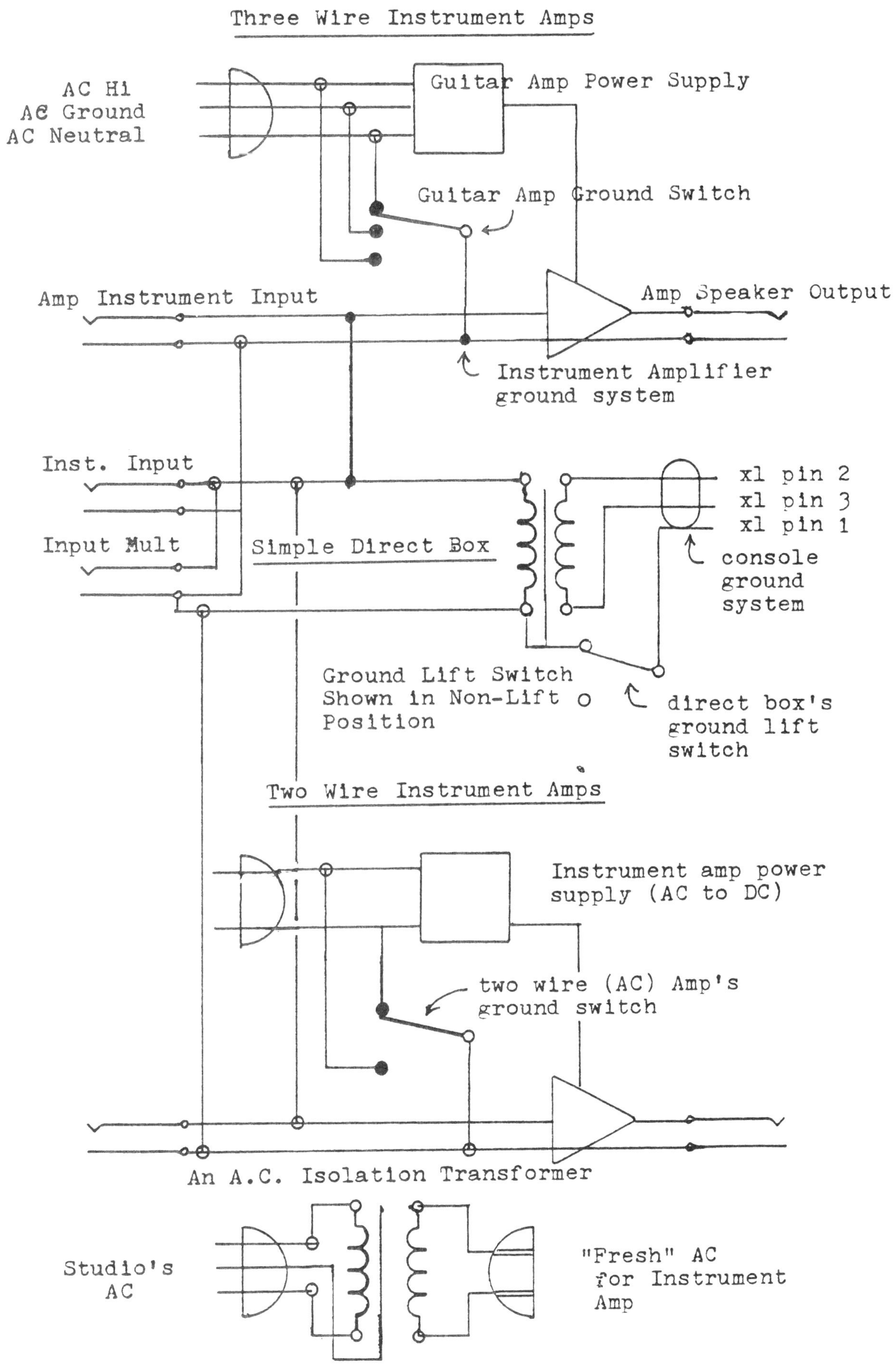
Three Wire Instrument Amps
AC Hi
AC Ground
AC Neutral
Guitar Amp Power Supply
Guitar Amp Ground Switch
Amp Instrument Input
Amp Speaker Output
Instrument Amplifier ground system
Inst. Input
Input Mult
Simple Direct Box
xl pin 2
xl pin 3
xl pin 1
console ground system
Ground Lift Switch Shown in Non-Lift Position
direct box's ground lift switch
Two Wire Instrument Amps
Instrument amp power supply (AC to DC)
two wire (AC) Amp's ground switch
An A.C. Isolation Transformer
Studio's AC
"Fresh" AC for Instrument Amp

trick. Starting off with this hookup is not recommended since now the console's ground system includes the dubious guitar amp circuit -- something the console designer might frown upon. Do not switch the instrument amp's ground switch to either side of center with the direct box's ground lift switch not lifted, as this will hook up one side of the guitar amp's AC plug to the mike input ground of the console.

If these measures fail to get rid of the noise, get an AC isolation transformer and hook up the amp to it. Put the amp's ground switch in the middle position and put the direct box ground lift switch in the non-lift position. Now the amp is no longer connected to the AC of the building, only to the new AC created by the isolation box, and the console is supplying the ground system to the amp.

When you finally get the noise out, check using your volt/ohm meter, that there is no AC voltage between the guitar amp chassis and the microphone body the player will use to sing into. If there is any AC reading, change the hookup combination to one which gets rid of the voltage and is without noise. If you must go with an AC potential condition, warn the player and put a foam wind screen on the mike. Use a plastic mike stand insulating adaptor, which allows the player to grab the mike stand below the plastic stand insert without shock danger. Bear in mind: if the foam wind screen gets damp from breath spray, the shock potential to the singer's lips will return.

If the electric direct box system is still noisy after all these efforts, then the instrument or amplifier system is malfunctioning and the player will have to get another.

DIRECT BOX HOOKUP FOR TWO WIRE AMPS

With a two wire amp having a ground switch, one or the other prong of the AC cord is attached to the guitar amp's metal surfaces. Never use a direct box that does not have a ground lift switch on a two wire amp (or a three wire amp without a middle position on the amp's ground switch).

Connect the direct box with the ground lift switch in the lift position, and try either position of the instrument amp's ground switch to get a setting that is both quiet in the control room and at the amp itself. If the instrument amp does not have a ground switch, manually reversing the AC plug in the wall does the same function. If the system is still noisy, get an AC isolation transformer and plug the instrument amp into it. Put the direct box into non-lift mode and see if quietness results. If not, then the amp is useless for recording.

When you have finally gotten the noise out, check, using your volt/ohm meter, that there is no AC voltage between the guitar amp and the microphone body the player will use for singing. If there is, find a hookup combination that gets rid of both the voltage and the noise. If you must go with an AC potential condition, warn the player and put a foam wind screen on the mike. Use a plastic mike stand adaptor, which allows the player to grab the mike stand below the plastic stand insert without shock danger.

PHASE CHECKING

If you record a true stereo piano on your multitrack machine with the left or right side out of phase with the other, then the piano will be audible when used in a stereo mix but will partially disappear in a mono mix.

Always check phase when making stereo recordings, or when more than one mike is used to pickup the same acoustic source (when more than one mike is in the same sound field).

To Check Phase....

1. Bring each mike up and set the level and EQ.

2. Test any two mikes to see where the volume levels are equal on the fader templates.

3. Bring the two mikes up in equal volume to the same buss and see that the mid range and highs and lows all remain in proper relative proportion to one another. If they do not (the sound changes), then one of the two mikes is out of phase. Switch mikes if you can and send both of them to the shop to be phase tested. Otherwise use the phase reversal switch on the console or a phase reversing patch cord, which should be colored red in the mike out/pad in patch bay, to put the phase right.

4. Check other mikes that are going to be combined with the first two.

5. Return buss assignments to normal.

Always connect a phase scope or meter to the monitor system left and right outputs so that phase anomalies will be more obvious.

CLEANING TRACKS

Tracks are cleaned to prevent artists and producers from wasting their time "recording from the top" all the way to the tune's end while overdubbing on a previously recorded track. If the current overdub is only a short one, perhaps in the middle of the tune, just erasing this portion of the track will leave remnants of the old recording ahead of and behind the new overdub. These remnants will greatly interfere with monitoring since they will "pop" in and out. The proper way to handle this is to clear ALL tracks that have the entry "ERASE" written on the track sheet when overdubbing or playing back the tape where you expect to go "all the way through." Cleaning tracks is also done to get rid of unwanted coughs, laughs, pops, groans, and other sounds recorded along with desirable vocals and other recordings. Sometimes a singer will "come in" before the proper time, having been fooled by two identical musical phrases in the song. This missed cue will have to be cleaned.

During mixing, when the faders are set to proper positions, having to move them down to "duck" noises that weren't cleaned is to increase the mixer's work load unnecessarily. Fader settings are a form of audio information, which in some cases takes hours to achieve: never move a fader unless you have to. Instead use mute switches, or better yet, clean tracks.

*** See also "Rocket Punch Outs" in Intermediate 3 ***

Y CORDS AND MULTS

A "Y" cord (or box) is a device that has two or more input (female) connectors and one output (male) connector. The purpose is to take several mike outputs and plug them all into only one mike input channel of the console.

Properly speaking, Y cords are illegal and should never be used to combine signals. The only proper way to combine signals is to use a console combining amp. In a dire emergency (like insufficient inputs on the studio console for an upcoming session) Y cords may produce a better "sound" because you will be able to use more mikes to cover the instruments.

Y cords are used to combine dynamic mikes rather than condenser mikes.

Mults are devices (can be cable variety or made up in small boxes) that have one input and several outputs. The input connector is a female and the several outputs are the male connectors.

The main use of mults is to connect the outputs of a two track to the inputs of other copy machines. This is why mults are sometimes called machine mults, as opposed to mike Y's, which are also called mults: mike mults. A handy stereo machine mult has a left and right input and maybe five left and right outputs (clearly marked of course). This way, 5 stereo cassettes or quarter track copies can be made at once from a two track playback. When wiring Y's or mults, simply wire all the XLR pin 1's together, pin 2's together, etc.

*** Please refer to Illus. B8-1 - Tape Recorder Block Diagram ***

RECORDER BLOCK DIAGRAM

Starting at the left side of the page, the three semicircles labeled P, R, and E are the play, record and erase heads.

The circle with the L is the line input connector (coming from the console buss out); this is the audio input to the machine.

The circle with the sideways "S" is a hi frequency oscillator within the machine.

Beginning at the top left, this amplifier symbol is the reproduce amp for developing a strong signal from the playback head.

The sync amp is for playback during sel sync operations. (Really there isn't a separate amp for sync playback, but this drawing style is easier to understand for now.)

The bias amp is for boosting the internal hi frequency signal to the level required for erasing the tape at the erase head.

The line amp is for preparing the signal to the strength required to operate the VU meter and feed the lines going to the console for monitor purposes or mixing.

The record amp is a combining amp that mixes the bias with the audio signal and prepares the signal to be used by the record head.

The volume controls...

The repro level is for adjusting the playback level so that the tape machine will indicate 0 on the VU meter when a zero level tape is being played.

The sync level control is for adjusting the level when the machine is in sync mode.

The erase level control is for adjusting the proper amount of hi frequency tone so as to properly erase the tape.

The bias level is the control that causes the tape recording to be made with the least amount of noise and distortion and with the strongest signal.

The record level is the control that causes the tape to be recorded at the proper volume. For example: a vocal reading -3 dB on the console once recorded, must play back at -3 dB for the machine to be considered operating properly.

The input volume control is set so that when the tape is not rolling and the machine is in input mode, the machine VU shows the same volume that would be recorded if the machine were in record.

The repro EQ is set so that during playback of the reproduce alignment tone tape, the proper amount of lows and highs are played back from the tape.

The record EQ is to set the amount of high frequencies so that they will be recorded in proper proportion to the midrange and lower frequencies. There is usually no low frequency record EQ so the amounts of lows being recorded are not adjustable.

Now the switches...

The play/sync switch determines which head will play the tape, either the record or the play head.

TAPE RECORDER BLOCK DIAGRAM   Illus. B8-1

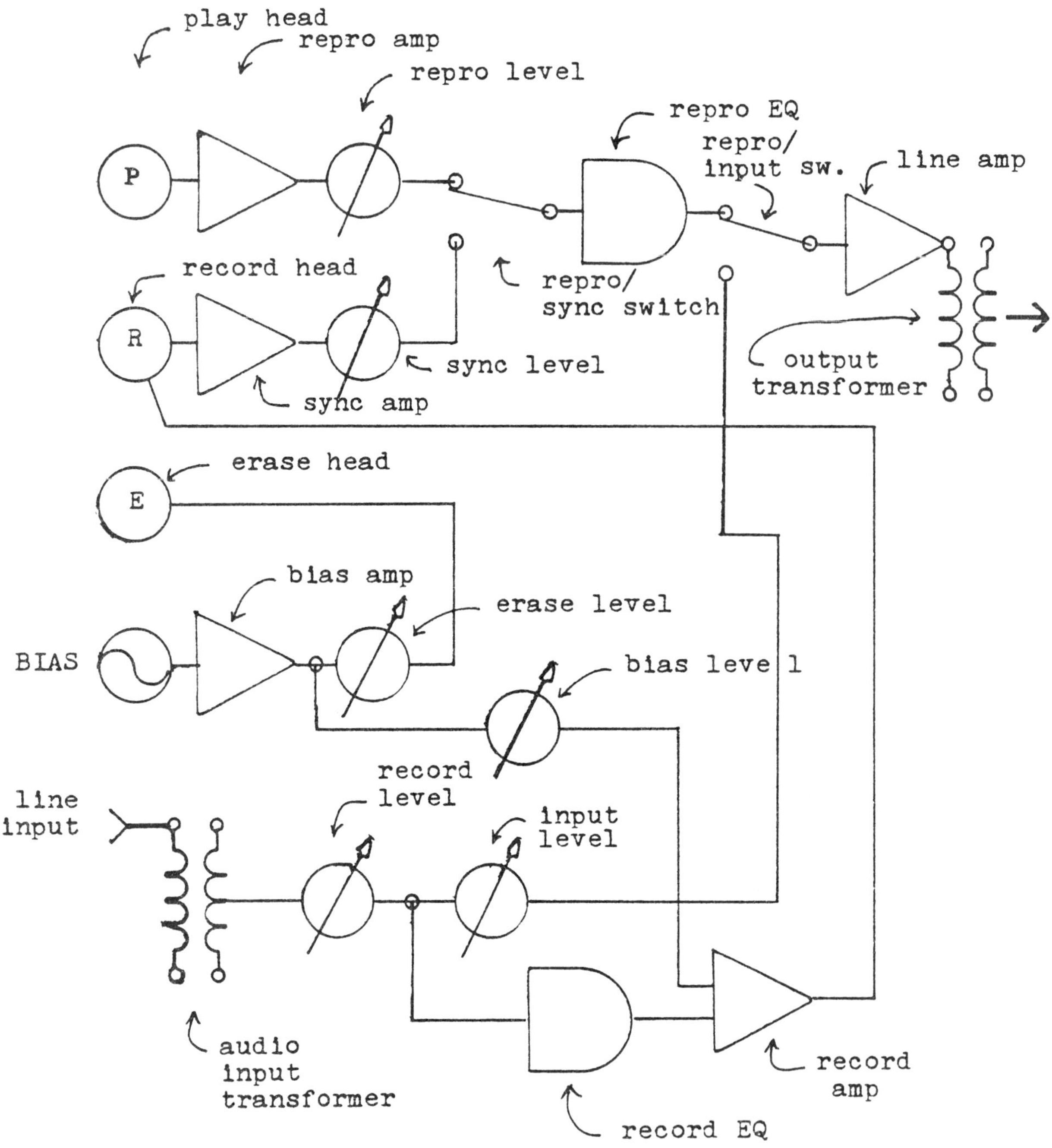

The play/input switch determines whether the machine will playback tape, or if the signals at the recorder's input will be used for the machines output. In the input position of the play/input switch, any signal presented to the input of the machine is observed on the machine's VU meters, as well as being audible at the output of the machine as though the signal were already recorded.

During recording, the strong bias signal from the internal sine wave source is mixed together with the weaker audio signal and the mix of the two is presented to the record head for recording.

The play EQ boosts bass slightly and reduces the amounts of high frequencies dramatically. It is the reduction of highs from a tape playback that reduces the tape hiss, which would otherwise prevent quiet playbacks.

The record EQ does just the opposite -- reduces bass slightly, and boosts highs dramatically.

*** What the tape machine does to the audio as it prepares it to be recorded, and how a recorder plays tape, is all discussed later in more detail ***

BIAS

*** Please refer to Illus. B8-2 - The Ampex Paper on Bias ***

To understand magnetic recording you must understand magnetic inertia. Magnetic inertia is similar to pushing a car -- its hard to get started, but once it begins to roll keeping it moving is easy.

As you apply more and more magnetic energy to a tape the tape responds with three types of results:

First comes the inertia area where large increases in magnetizing energy cause little or no resulting magnetization increase on the tape.

Then with more magnetizing energy comes a region of magnetic activity, when even slightly more magnetizing force will produce more magnetization on the tape. This area is known as the linear region and is where all tape recording signals reside.

Above the linear region is the region of diminishing returns known as tape saturation. Above this volume (magnetically speaking) increasing the console signal stength will cause little or no increase in the strength of the recorded signal on the tape. When you record into this region, tape saturation results. Tape saturation is one kind of tape distortion.

Recording below the linear region causes distortion and over bright recordings; recording above the linear region causes distortion and dull recordings.

On the professional recorder the bias control sets which region the machine will record in.

When setting up a tape recorder in the studio, bias must be checked. To not check bias is to risk making distorted or noisy recordings.

Bias and the audio signal are mixed so that the strong bias frequency tone (supplied by the machine) overcomes the tape's inertia so that the weaker audio signals (supplied by the engineer) are recorded above the inertia area and in the linear region. The bias is recorded too, but the playback head cannot play back this very high frequency. The bias can be sensed however by special equipment, which can tell if a tape has ever been

Bias

Substances that can be magnetized have one thing in common: the electrical currents that produce the magnetism are far less effective at low levels than they are at higher levels. As current increases from zero, very little magnetism results until a point is reached that in effect overcomes inertia. This slow start is generally referred to as hysteresis loss and will vary with each magnetizable substance. Magnetic tape suffers from these losses and its magnetizing pattern is similar to this:

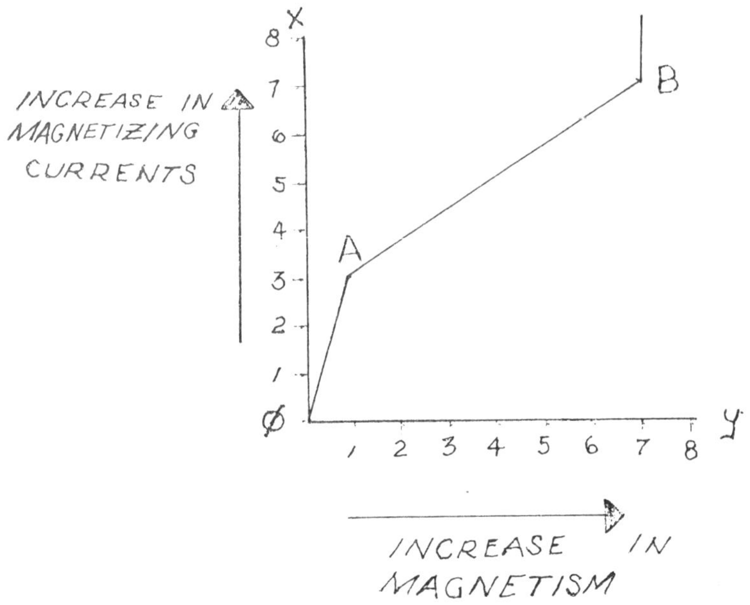

As current increases from Ø to 3 on the x axis we see that magnetism has increased only from Ø to 1 on the y axis. From Ø to A is the area of losses. From point A to point B, we see that a relatively small increase in magnetizing currents brings about a similar increase in magnetism. Most important, A-B is linear, each incremental increase on the x axis results in a correspondingly equal increase on the y axis, ie, each increase of 1 on the x axis above point A results in an increase of 1.5 on the y axis.

Point B is called the saturation point, the point where the tape is magnetized as much as it can be magnetized and any additional increases in current will produce NO increase in magnetism.

The magnetizing current for magnetic tape is the recording current. The amplitude of a recording current is a direct reflection of the amplitude of the sound (music) that produced it. The strength (amplitude) of the magnetic (recorded) pattern on tape is therefore a direct reflection of the amplitude of the sound that produced it and when played back it should reproduce the same amplitude of sound.

What would be the results if we recorded music on the entire magnetic tape plot, Ø-A-B? Imagine a musician playing a long sustained note and periodically increasing the volume of the note at points r,s,t, and u. The note is reflected as a recording current on the x axis and a reproduced sound on the y axis.

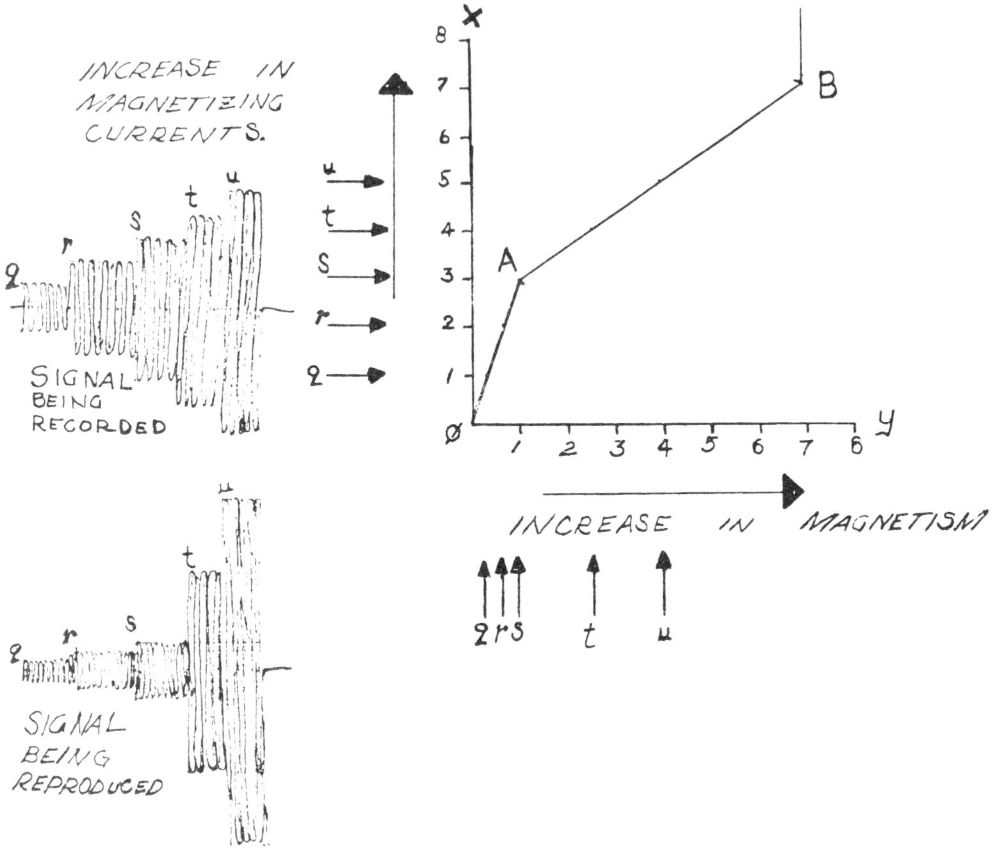

Notice that r, s, and t on the y axis are too low in amplitude then suddenly at u the volume becomes tremendous. Point s and point A are the point where inertia is overcome. Above point A (s) each incremental increase in recording current results in a correspondingly equal increase in reproduce current.

Recording currents do not have to start from Ø. They can be made to begin at point A by adding another current to begin at Ø and magnetize the tape enough to overcome inertia. This additional current is called bias. It is 3 or 4 times greater in amplitude than the recording current in order to overcome inertia and is very high in frequency to prevent its being reproduced on play back.

Bias and record currents are mixed in the record head. Figure 1 is a photo of 100 kHz bias as seen on an oscilloscope, figure 2 is a 5 kHz recording signal and figure 3 is the result as they are mixed in the record head.

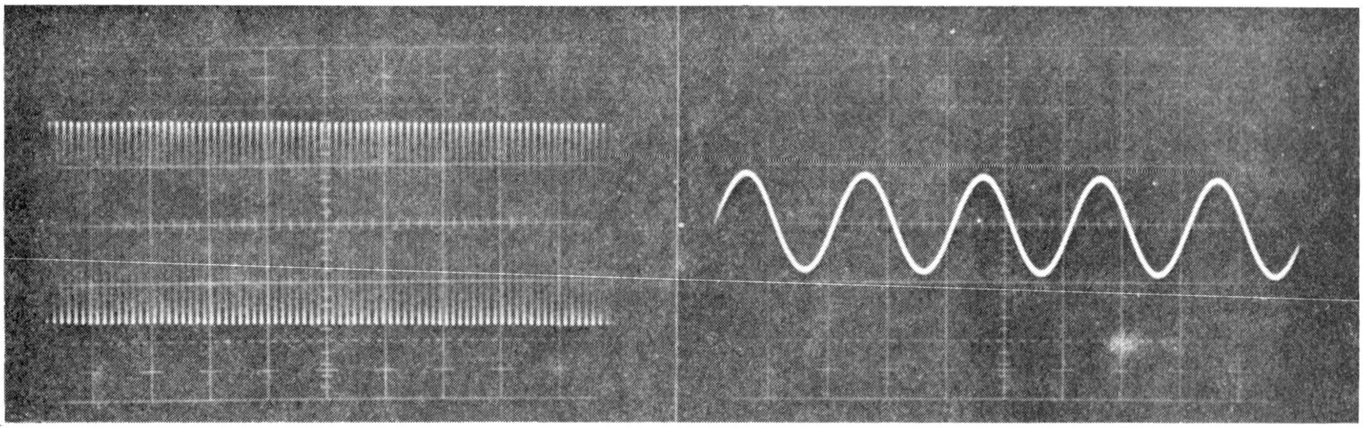

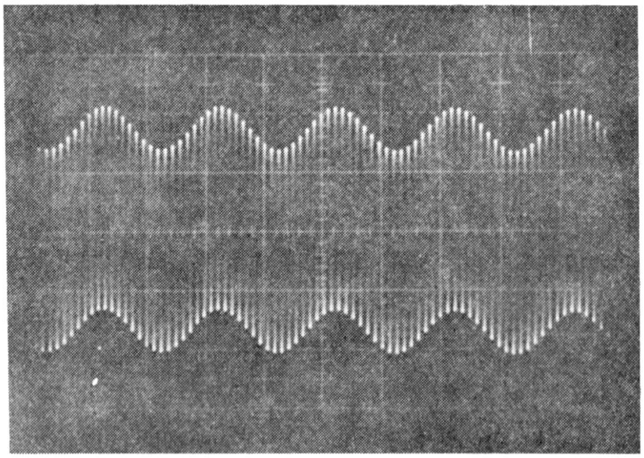

Summarizing, we see that bias is a magnetizing current applied continuously during the recording process to overcome "inertia". Recording signals then added will go onto "prepared" tape.

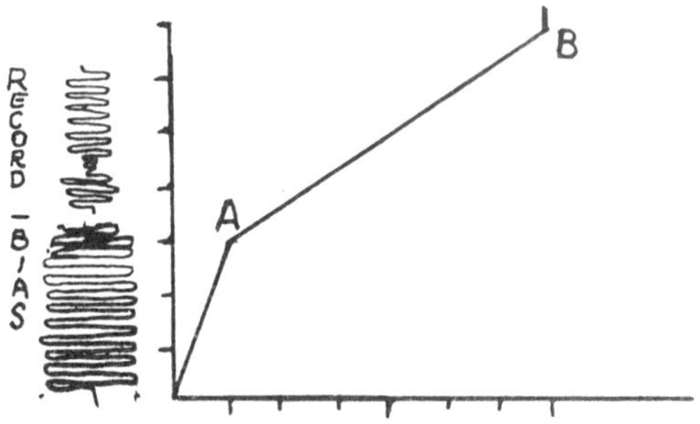

Equalization: Pro-Audio Gen. 2

Equalization is simply the process of restoring a sound to its original state after the recording and reproducing of that sound. "Restoring" would indicate that something is lost in the record-reproduce process and something is.

Ideally, if we recorded all the frequencies that humans can hear, (roughly 20 Hz - 18 kHz) using the same recording energy level for all these frequencies, we would get back (reproduce) all these frequencies with the same results. Graphically it would look like this.

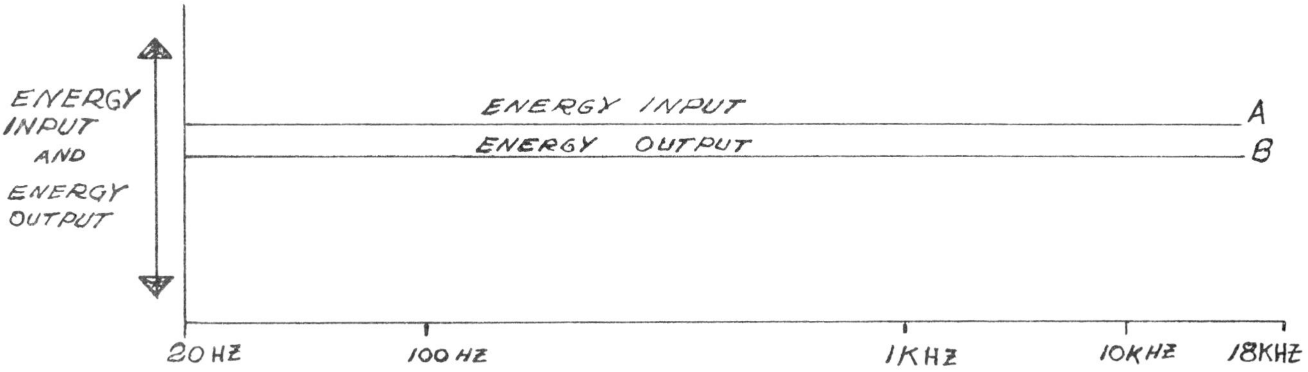

The energy ratio, input to output, remains the same (linear) for all frequencies and is called a linear or flat response. Notice that there is a loss between input energy and output energy, but the loss is linear or flat for all frequencies. Equalizing, or correcting for this linear loss would simply be a matter of turning up the amplifier enough to give back the loss. Such would be the ideal situation.

The record and reproduce process differs considerably from the ideal. Many things involved in the record-reproduce process won't cooperate completely, the most uncooperative being the magnetic tape itself. Energy losses occur that are non-linear, the higher frequencies losing more than the lower frequencies. A graph showing actual record-reproduce losses would be more like this:

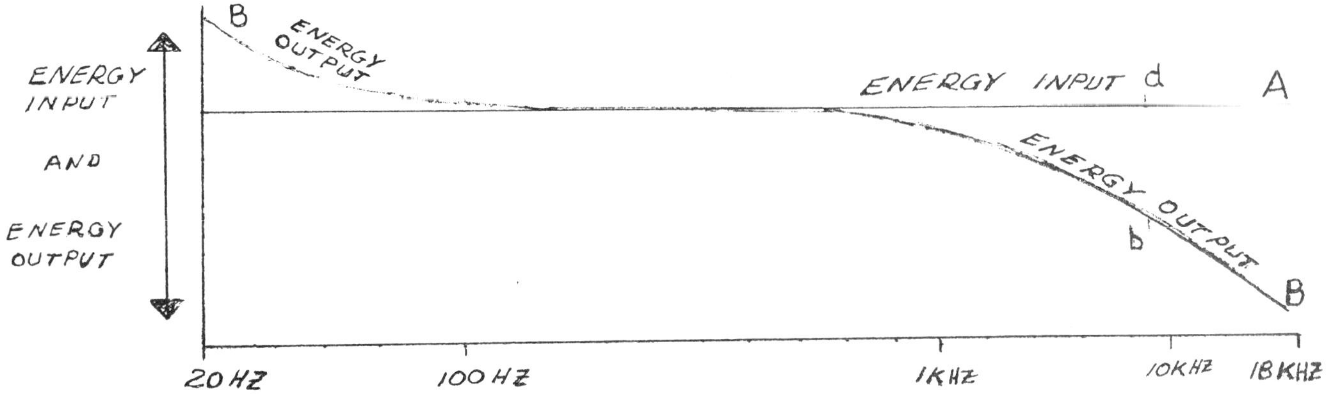

A high frequency note recorded at energy level and frequency coordinates "a" would be reproduced at energy level and frequency coordinates "b" greatly reduced in volume, and would probably be "lost" in accompanying notes.

Notice the plot B (losses) shows a small "hump" <u>above</u> plot A in the low frequency area. This indicates that the record-reproduce process actually increases the volume of low frequencies. Equalization for low frequencies, therefore, is a <u>reduction</u> in volume.

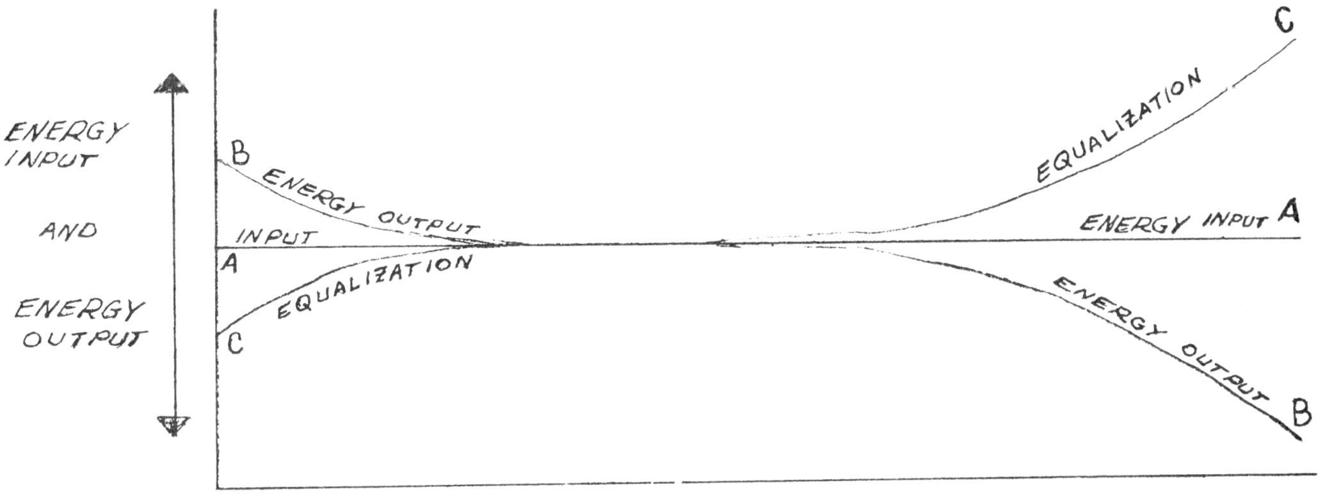

Equalization increases the energy level (volume) to compensate for energy losses at the various frequencies. For example, if we know that we have a loss of 5db at 4500 Hz we will turn up the volume at 4500 Hz an additional 5db. Added algebraically, -5db loss + 5db equalization = 0db change from the original. Graphically the equalization plot (C) will be a mirror image of the loss plot (B), and the algebraic sum of any two points that are one above the other will equal the desired plot (A), the energy input.

Subjectivity played a small part in determining the equalization plot; ie. the likes and dislikes of humans. The National Association of Broadcasters (NAB) established the equalization curve (plot) that is used throughout the U.S., Canada and Mexico. Europe has its own equalization standard, established by the Consulting Committee for International Radio (CCIR). It varies slightly from the NAB standard in that both the low and the high frequencies appear a little louder than when recorded-reproduced using the NAB standard.'

Tape speed plays an important part in the equalization process. Losses will change as tape speed changes and equalization must change to offset the losses. Professional audio recorders/reproducers have a separate equalization circuit for each tape speed. The following tables show the amount of equalization required to meet the NAB standard for reproduce amplifiers.

AUDIO
CONVERSION FACTORS

English to Metric and Metric to English conversions applicable to Magnetic Recorders.

TO CONVERT	INTO	MULTIPLY BY
Inches	Centimeters	2.540
Inches	Meters	2.540×10^{-2}
Inches	Millimeters	25.4
Ounces	Grams	28.349527
Pounds	Grams	453.5924
Pounds	Kilograms	.4536
Centimeters	Inches	.3937
Meters	Inches	39.37
Millimeters	Inches	.03937
Grams	Ounces	.03527
Grams	Pounds	2.2046×10^{-3}
Kilograms	Pounds	2.2046
°F - 32	Temperature (°C)	.5556

STANDARD TAPE SPEEDS

Inches per second	1.875	4.7625	Centimeters per second
	3.75	9.525	
	7.5	19.05	
	15.	38.1	
	30.	76.2	
	60.	1.524	Meters per second
	120.	3.048	
	240.	6.096	

erased or edited (remember the Nixon tapes?) because the continuity, character and frequency of the bias will vary.

THE HYSTERESIS CURVE

The hysteresis curve is a method of visualizing the resistance to initial magnetizing efforts, the linear region, and the point of diminishing returns (tape saturation).

Using Ampex's bias explanation (Illus. B8-2) note that on the graph on page 1, the vertical direction indicates the input to the tape -- the "magnetising currents," which are applied from very small amounts to very large amounts in efforts to magnetize the tape. The horizontal axis is the resulting magnetic result on the tape (recording made).

As the recording signal is increased from the 0 point to point A, the amount of applied signal goes up by 3 units (vertical axis), but the resulting magnetic result only goes up 1 unit (both units being the same increments). If the audio were coming out of the console 3 units high only a 1 unit high recording would be made. The recording, in addition to being lower in level than expected, would also be distorted.

As the applied magnetic force to the tape is further increased, the vertical part of the graph goes from 3 units to 7 units (4 units of signal strength). A signal applied at this strength results in a horizontal signal strength of nearly the same number of units. The important point here is that as the recording signal is moved smoothly from point A to point B, a gradual amount of increase in the resulting recorded signal strength occurs on the horizontal axis. This is the linear region where a given change in the applied recording signal will result in an exactly similar amount of change in the recorded signal on the tape.

After the applied recording signal goes above the vertical axis point of 7 units, the tape becomes saturated and can no longer accept any more signal strength. No matter how much stronger the vertical recording signal becomes, the horizontal recorded result gets no stronger.

This is a somewhat simplified diagram of hysteresis, but it explains the phenomenon adequately for now.

DISTORTION OF TAPE MACHINES

*** Please refer to the Recorder Block Diagram Illus. B8-1 ***

Let's follow the signal from the input of the tape recorder to the final output of the recorder during playback of a tape...

From the circled L (the line input of the tape machine) the signal enters the input transformer. This transformer is for isolating the actual signal wires of the machine from the outside world. The reasons for having this transformer are:

1. To provide for rejection of noises, which may be radiated into the input wires from CB radios, amplifier hums, or noisy light dimmers.

2. To allow either balanced or unbalanced signals to be used as inputs to the machine.

3. To prevent ground loops.

Since there is a level control directly after the transformer, it is possible to use input levels that will distort the transformer, and then you would only be able to turn the transformer's distorted sound down with the pot; however, you would not be affecting the distortion amount at all. Transformer distortion is not at all like the sound of solid state distortion. Of the two distortion types, having the tube / transformer type rather than solid state (clipping) is the lesser of the two evils.

Supposing that the input audio from the mike is clean, and the transformer is not distorted, then the next kind of distortion (during recording) might be the record EQ. The record EQ can distort due to too hot a level on a signal with too heavy a high frequency content. Monitoring playback while making multitrack recordings is the only way to prevent accidentally making distorted recordings, recordings missing some tracks or recording on bad tape, and realize any of these problems immediately as they happen.

Next in line is the record amp, where the EQ'd signal is added to the bias waveform to be sent to the record head. If the record amp is distorting, a horribly distorted recording or no recording will occur. If the sine wave coming from the bias oscillator is even slightly distorted (not a pure sine wave), then a very peculiar type of distortion will occur. Audio distortion due to a distorted bias waveform causes musical distortion of a harmonic nature and is sometimes difficult to hear.

During playback the reproduce amp can distort. If it does, this is likely to sound very much like console distortion except that there is a large amount of high frequency boost on the audio in its recorded form. Changing reproduce cards will clear up the distortion; if not, the record circuit was doing the distortion and the distortion is "on tape."

If the input signal to be recorded is "hot" enough, the transformer will distort making for a harsher signal. If there are lots of highs in the sound, the record EQ will distort the highs in its attempt to boost highs to the degree necessary to make quiet recordings. With transformer distortion, record EQ distortion, and the following record amp distortion, the signal comes out of the record head too hot to be recorded on the linear region of the tape. The tape may distort. With the maximum level coming from the tape during playback, the playback chain may distort. Remember that the VU meter cannot show the full level of bright transient sounds due to the great mass of the pointer compared with the speed of the electron motion the VU is attempting to indicate.

MAGNETIC FIELDS AND ACCIDENTAL TAPE ERASURE

Magnetic fields occur in record heads, erase heads, magnetic tape (that's been recorded), ribbon mikes, tape head degaussers, tape degaussers, airport X-ray machines and an occasional magnetized razor blade used to edit tape.

The varying magnetic field at the record head exactly resembles the varying voltage at the output of the console (except for the presence of the bias signal).

The erase head produces a very high energy high frequency magnetic field, which is above the saturation point of the tape. The erase signal is so strong that it completely replaces the previously recorded signal. Erasing over virgin tape adds about 3 dB of noise to the tape so don't "punch in" on a virgin track at the top of the tune for an overdub that begins in the middle of the song. By not recording on all tracks at the basic track session, you can make the playback of the beginning of the song much quieter. If you delay recording on each track until just before its musical part begins, you will reduce the tape hiss during playbacks at the beginning of the tune.

The magnetism left on magnetic tape is in the form of very small bar magnets lined up with their north and south poles pointing up and down the length of the tape. With virgin tape, the little magnets (the size of iron molecules) begin with no particular alignment -- some face down, some face up the tape. After being recorded, nearly all the little iron molecules on the tape are lined up facing north and south in the same pattern as the audio waveform would look on an oscilloscope; all the magnets pointed north pointed down the length of the tape (towards the takeup reel) for the parts of the signal above the middle line of the oscilloscope (positive going waveforms caused by pressure waves), and all the little magnets point up the tape (towards the supply reel) for parts of the signal that are below the middle line on the oscilloscope (negative going waveforms caused by rarefaction at the mike).

Ribbon mikes are the only kind one should expect to have any magnetic potential in the studio. It is possible for a ribbon mike to partly erase about 1/5 of a reel of 1/4 inch tape by being set down on top of the reel. Ribbon mikes have the strongest magnets of all mikes.

Head degaussers are used to completely magnetize recording heads by their strong alternating magnetic field (north/south/north/south 60 times each second). The strongest magnetic field comes from the head degausser when it is turned on or off. Never use a degausser that has a damaged AC plug or wire. Never unplug a head degausser while it is next to the tape head.

By holding the degausser against the recorder's heads, the alternating magnetic field causes the tape head's iron molecules to first be magnetized north then south and so forth. As the degausser is pulled away some of the metal molecules in the record (or play or erase) head fall out of the "dance of the molecules" and come to rest aimed in a random direction somewhere between north and south. The directions the molecules face depend on where the degausser field is when the particular head molecule falls out of the "dance." The molecules of metal in the head hold on to their magnetic orientation due to molecular "glue." The molecules with the strongest molecular glue fall out of the dance first (when the degausser is moved to a distance of about 1/2 inch). Molecules with the weakest glue fall out of the dance last, at a distance of about 8 inches. The distance really depends on the strength of the degausser.

Keep recorded tape at least three feet away from the nearest degausser and even farther when the degausser is first turned on or off.

Tape degaussers are usually large, bulky devices. They put out a tremendous alternating magnetic field. There are usually two points on the surface of the degausser that are "hot" magnetically. You can find these hot magnetic center areas by fishing around with a screwdriver. The instant the tape degausser is turned off, it puts out a really tremendous magnetic field pulse. Magnetising screwdrivers (to recover a dropped screw) is quite easy if you switch off a tape or head degausser with the screwdriver blade touching the degausser.

When deguassing tape remember to rotate the tape so that all of the tape gets to the hot area. Continue rotating the tape as you slowly draw the tape to a distance of about 1.5 feet, then quickly remove the tape from the proximity of the degausser and turn off the beast (because they can overheat).

Airport x-ray machines cannot erase tapes with their x-rays, but the equipment that powers the x-ray machine produces powerful magnetic fields that do erase tape.

Tapes have been erased at airports.

Some international producers, who travel by air with their master tapes (to record strings in England perhaps), wrap their tapes in steel foil. Steel foil affords slight protection only. Mu metal is the only anti-magnetic metal able to protect against magnetic fields. Mu metal is a very soft metal that can lose its protective qualities by: being heated, being subjected to a strong electric current, or being struck by a sharp blow.

When flying with magnetic tape, always carry on the reels personally and clearly mark the reels "magnetic tape do not x-ray." Remember, though, that tape put in the baggage compartment is cooled to -40 degrees Farenheit or lower due to indirect exposure to the outside air temperature at tremendous altitudes at which jets fly. Always make careful safety copies of you multitrack master tapes before you fly with them.

TAPE HANDLING AND STORAGE

When pulled from storage, magnetic tape should be found tails out (the end of the tape is on the outside of the reel) and very smoothly wound, especially 1/4 inch tape. The tape should be carefully rewound, watching for possible sticky spots due to a splice having "leaked." If a splice does leak, then the tape will want to stop instantly where the two layers of tape are glued together by the leakage. For this reason, tape that has been stored for a very long time should be wound off its storage reel with great care. Use the minimum deck tape tension, or possibly spool the tape directly from reel to reel using the much lesser play tensions (put the deck in play mode with a splicing tape dispenser holding the end-of-tape arm from swinging out) and use your hand as a clutch to prevent the tape from getting too much speed. If some splicing tape glue did leak, clean it up using some new splicing tape as a sticky "mop."

After rewinding the tape it can then be used for whatever purpose intended.

After use, the tape should be wound to the top, and then spooled or played to the tail, but not fast forwarded. (Spooling is a special fast play speed where the tape lifters are energized to prevent head wear.) An ideal spooling speed is about 60 ips.

The tape should then have its end (tail) fixed to the reel flange of the reel by a piece of blue tape. (Blue means tails out -- red means heads out.)

The tape can now be stored.

The important thing about tape storage of is that the tape have a constant tension wind, and the tape pack is flat with no pieces of tape standing above or below the main body of the pack. If any of the wraps stick out a little, edge damage will result, even if the tape is stored for a short time.

As tape is wound onto its storage reel in fast forward mode, air is trapped under each layer of tape. This trapped air lets each individual wrap settle at its own position and a poor wrap is thus produced. Since the tape is slightly stretched as it is wound, those pieces of tape not supported by being wound directly over the wraps of tape below contract back to their original un-stretched length. When they contract, they move towards the center of the reel, where the circumference is smaller. Now one edge of the loose wrap of tape is one length, trapped by the layers of tape over and under it, and the other edge of the same wrap of tape is shorter, from shrinking towards the center of the reel. When this kind of damaged tape is played back, the damaged edge will not properly come into contact with the playback head, and most of the hi frequency audio is lost from these track(s).

It is very dangerous to work with tape that is on hubs. (Hubs are reels without the flanges attached.) The same air flotation phenomenon that causes the tape to ride up and down, causing bad packs and edge damage, can also cause the tape to ride up and over, forcing the tape to "dump" (slip over the top of the pack tearing the tape).

It is safe to fast wind a hub onto a reel but not the other way around.

1/4 inch tape on hubs is used because: (1) buying tape from the tape dealer on hubs is less expensive than buying tape on reels, and most stereo tape recordings go out of the studio on plastic 7 inch reels or cassettes anyway and (2) tape weighs less on hubs making tape storage easier. The only 1/4 inch tape output from a studio that requires reels are masters or long tapes that have to be shipped.

WAVEFORMS, FREQUENCY, WAVELENGTH, AND AMPLITUDE

A waveform is an instantaneous picture of audio voltage across time. (The expression of some quantity across another is supposed to bring into your mind's eye a graph of the first quantity in the vertical and the second quantity in the horizontal.) Since waveforms change very quickly the term instantaneous is used to infer a picture of a brief time period. In the graph a central line is used to represent the zero volt line; waveform excursions above this line are positive volts, under the zero line are negative volts.

The height of the waveform is its amplitude or volume. With no volts reading on the vertical axis of the graph, even with a strong power amp and a good speaker, there would be no audible sound from our audio system. Starting with the tiniest amount of signal however, any increase in the waveform amplitude will produce a noticeable increase in the volume.

The frequency of the waveform is how many complete cycles occur in a second. The term cycles per second (cy/sec)was "improved" by abbreviating it to Hz which is short for hertz. K is the abbreviation for thousand, therefore, 1 KHz is one thousand cycles per second.

A complete cycle is when the waveform completes an excursion from the very bottom to the top and back to the same point at the very bottom, or begins at the center (or any other point), and completes one cycle back to the same place on the graph.

All acoustic mediums -- wood, metal, air, water, glass -- have a different speed at which sound travels through them. The faster the speed at which a medium transmits a sound, the more spread out the waveforms become. This causes a complete cycle to have different lengths in different mediums and means that the same sound can have different wavelengths.

To calculate the wavelength of a certain frequency waveform you must know the speed of sound in the given medium. The speed of sound in air is approximately 1130 feet per second (ft/sec.) The wavelength formula is:

wavelength = velocity in the medium / frequency

So for a 1 KHz frequency in air, the wavelength is:

wavelength = vel / freq
=1130 ft/sec / 1000 cy/sec

* the two label denominators (bottoms of the fractions) being sec cancel each other, leaving ft/cy left as the label:

= 1.130 ft/cy = the wavelength of 1 KHz in air (one complete wave each 1.13 feet)

the wavelength of 10,000 cy/sec (10 KHz) is:

wavelength = vel / freq
= 1130 ft/sec / 10,000 cy/sec
= .113 ft/cycle (1.356 inches)

The velocity of sound in a mike wire is the speed of light or 186,000 miles per second. The wavelength of a 1 KHz tone in a wire is therefore:

wavelength = vel / freq
= 186,000 miles/sec / 1000 cy/sec
= 186 miles/cy (one complete wave each 186 miles)

* Again the label denominators are the same and cancel, leaving miles per cycle as the answer label.

WAVE PROPAGATION

Acoustic sounds come from most sources as outward radiating spheres of air pressure disturbances. Sound changes the air pressure from the normal value to air pressures above or below the normal value in relationship with the frequency, waveform and amplitude of the sound. A 1 KHz tone in air raises and lowers the air pressure 1000 times each second. Frequencies above 20,000 cycles per second (Hz) are too fast for our rather massive hearing apparatus, so we are unable to hear them. They are above the range of our hearing. Dogs can hear well above 20 KHz and porpoises can hear far above this.

The tried and true analogy of sound behaving like ripples in a pond is not satisfactory for this discussion. Sound radiates in spheres like rapidly expanding balloons one right inside the other in a constant array as long as the sound source continues to radiate the sound.

If the observer moves beyond about 30 feet from the source, the size of the sound spheres begins to get so large that they do not act like spheres any longer. If sound energy were visible, the sound would look like huge flat walls, compared to the size of the observer. This type of sound energy behaves differently and is called far field sound. This brings about the concept of near field and far field sound.

For near field sound, as you double the distance from a microphone to the source, the intensity goes down by a factor of four times as much. As you half the distance from the source to the mike, the intensity goes up by a factor of four times as much.

Intensity is the pushing power of the actual sound energy in the air, as measured on a 1 cubic centimeter surface exposed to the sound energy. Since the area of the surface of a sphere is 4 x PI x the radius squared, you can understand that the energy of a sphere of sound would decrease 4 times if the distance of the measurement point was moved twice as far away: the distance being doubled means that the measurement radius is doubled so squaring the change in the radius yields a sphere 4 times as large. After all, the 4 and the figure pi remain unchanged so it's only the radius that changed. If the surface supported by the sound energy got 4 times bigger when it moved to twice the distance, where it was measured at first, then the energy is spread out over 4 times as much surface: our 1 square centimeter measuring area will receive only 1/4 as much sound energy as before.

This relationship of double the distance = 1/4 the intensity is called the inverse square law:

The change in the intensity = 1/change in the distance squared

In the near field, each time the mike distance is doubled or halved, the dB reading is changed by 6db.

Past the near field the sound behaves differently and does not reduce so rapidly per increase in distance. Since it is rare to put a mike over 30 feet away from a source, we shall only consider the near field.

As the spheres of sound travel through the air they remain relatively undistorted. When they strike a flat barrier however, like a wall or ceiling (called a boundary in acoustics), they "reappear" headed in a new direction as would be predicted by billiard ball like reflections, with no spin on the ball. The curvature of the sound spere is now away from the boundary as if the boundary were the source of the sound.

REFLECTORS, DIFFUSERS AND ABSORBERS

Reflectors, diffusers and absorbers are room treatments that are used to vary the way sound energy is reflected or not reflected off of boundaries (surfaces) in a room. Reflectors bounce sound off the boundary. Absorbers retain the sound energy so that less than the expected amounts of acoustic energy reflect off the surface. Diffusers are curved surfaces that bounce the sound at differing angles, according to the particular point at which the sound hits the diffuser. A diffuser operates the same way a curved rail (curving towards the inside of the table) on a billiard table would. Each and every point on the curve will reflect a ball (sound wave) in a slightly different direction.

Reflectors can serve to make spaces (rooms) sound warm. The human ear/brain combination is remarkably able to tell the nature of the surface from which a sound is being reflected. Wood is an often used reflector in the studio, since it has "warmth" and makes certain instruments sound good. For example, good quality reflected sound comes from recording a string section near wood walls and/or near a wood floor. A very dangerous (and not so "warm") sounding reflecting surface, which most all studios have, is the glass window separating the studio from the control room. The sound coming from the glass is very apt to "leak" into the mikes placed nearby and add a "harshness" to the sound.

Absorbers can be anything from a fake wall made of "transparent" (to sound) porous burlap concealing cork and fiberglass treatment, to a simple drape or even a lead lined velvet drape. Of all absorbing materials lead and dead air trapped between two immovable (heavily constructed) surfaces are the best. Sound is absorbed by a surface that is able to be moved by the sound energy. When the sound energy moves the absorbent surface, some of the energy in the acoustic sound is absorbed. The physicist says that the energy goes into heat energy in the absorber, which gets "hotter" as the sound's reflections get softer.

Diffusers are mainly used where there are very large, flat building surfaces, which would otherwise set up "standing waves." (Standing waves are what occur when two good reflective surfaces are opposite one another in a room permitting sound to bounce freely back and forth.) The large recording studios almost always use diffusers, since a sound hitting a large, flat wall will send a strong reflection directly to the opposite large, flat wall, which sends a reflection back. Soon there are a series of reflections flying back and forth in the room. This is not a usable acoustic environment for a recording studio. Having a diffuser allows only a small amount of the sound hitting the first wall to be reflected "straight out" on a course for the opposite wall. The rest of the sound energy gets bounced off in all directions thereby thwarting the standing wave.

ENCLOSING SOUND - SMALL ROOMS

If you make a recording in a small room and you don't use close miking techniques, you will have a small sounding recording. Vocal booths and other small rooms should only be used in the worst of situations, unless close miking techniques will produce a reasonable sounding recording. (Close miking means recording with the mike so close that the sound of the room is not audible.) It is sometimes possible to make a vocal recorded in a small room sound like the performance happened in a large hall -- but not if there is a large quantity of room sound included with the voice.

ACOUSTICAL SPECTRUMS AND DIRECTIVITY

The term envelope means a volume picture across time. Imagine a graph with a dot starting at the left and moving across the graph at a steady rate, moving up with more volume and down with less -- the graph the dot leaves behind is an envelope.

*** Please refer to Illus. I1-1 -- The Envelope Diagram ***

Envelope number one is of an electric bass. The note begins with a burst of high frequency energy as the pick leaves the string with an intense clicking noise. This noise is louder than the actual tone part of the note, but it is very short-lived. As this clicking sound dies, the actual round part of the bass note (the tone) starts and begins to slowly ring down in volume until perhaps one minute passes before the note has faded away .

Claves produce a sharp ping noise with almost no tone part to the sound. The sound transient lasts less than a second.

An organ note has a mundane envelope: there is no sound until a key is pushed, then the sound suddenly appears at full volume. The volume remains unchanged until the key is released, when the sound suddenly stops.

*** Please refer to Illus. A5-1 and the explanation in the text on Attack, Decay, Sustain and Release ***

ENVELOPE SHAPE - WHICH MIKE?

The more drastic the rise time of the sound the more transient or percussive it is. Very fast sounds like clicks, snaps, claves, pings, etc., need very fast mikes to accurately track the sound waves. The fastest mikes are condensers, due to the extreme lightness of the capsule element, therefore the crisper the sound you want (on crisp sounds) the more you should use condensers. The smoother the envelope the more you can get away with using dynamics or ribbons.

ENVELOPES Illus. I1-1

1. Bass

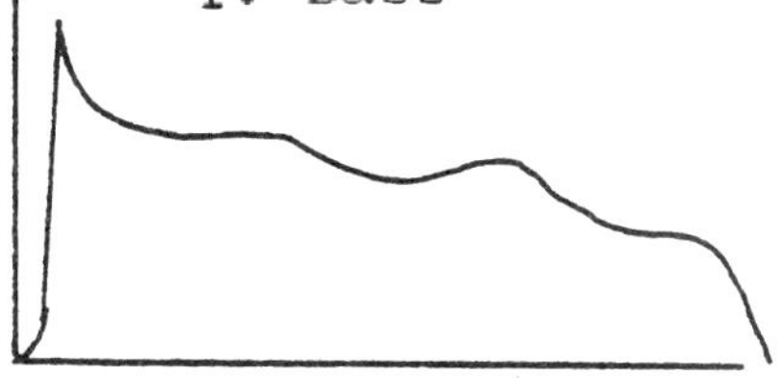

2. Claves

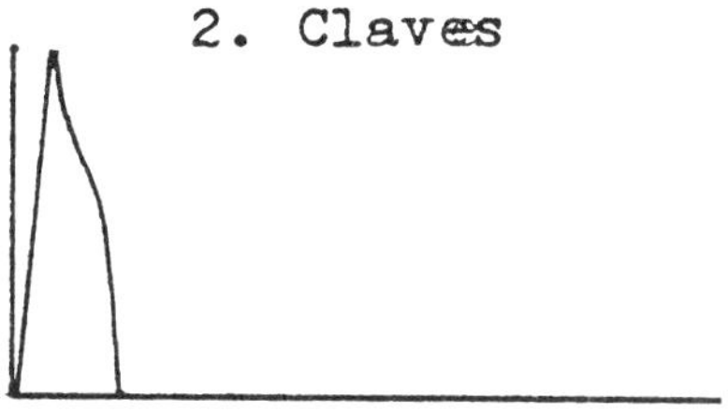

3. Organ Note

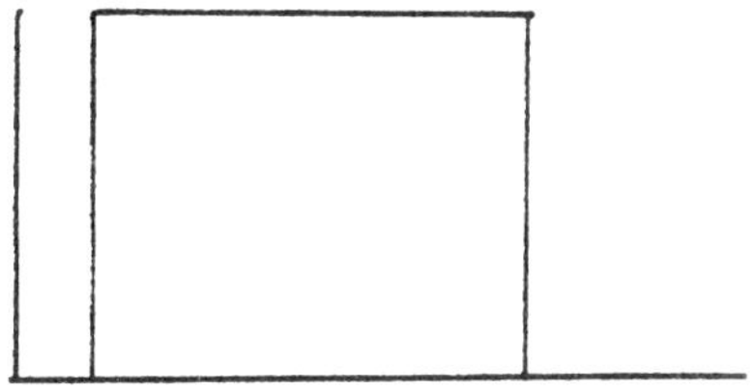

4. Organ Note with Tremelo

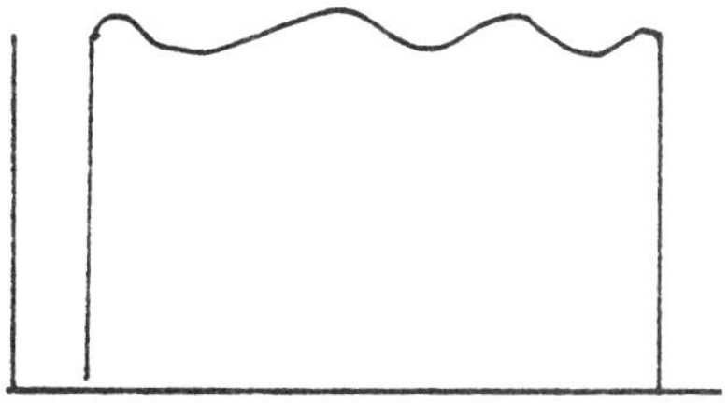

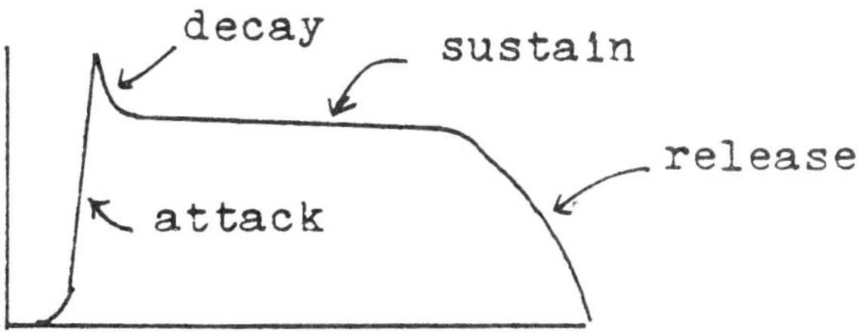

THE CONSOLE BLOCK DIAGRAM

*** Please refer to Illus. I2-1 -- the Console Block Diagram ***

The two signal sources at the upper right are the two most common input signal sources in the studio -- the mike and line input. The mike input arrives at vastly different levels due to the use of different mikes, different miking distances, and the volumes of the miked acoustic source. The line level signal always comes in at the same volume (level) -- at "zero" or line level. For the line input signal, there is direct access to the "line level" part of the console: the EQ, fader, etc. For the mike input circuit however, the circuit must also include: a pad (a variable loss circuit), a mike input transformer (to provide electrical isolation and perform common mode rejection to eliminate noise), and the mike preamplifier. The gain combinations of the mike pre amp and the pad are usually combined into one switch. The pad is not active at low level mike signals and the mike amp is at high gain, so the low level signal arrives at the EQ at the same level as the line input -- at zero level. As the mike is replaced with a more sensitive one (with higher gain) or as a louder instrument is used, the gain of the mike pre must be reduced so as not to pin the VU meter. (Remember that the fader is supposed to remain at the ideal fader setting loss.) As the signal of the mike gets higher, the transformer is threatened with being overloaded, so a 10 dB pad is included and the switch position at which this occurs skips reducing the gain of the mike pre. The next lower gain position would perhaps add another 10 dB of pad, again leaving the gain of the mike pre the same. Now the mike input channel is at its lowest gain (for mike input use).

Next in the circuit is the mike/line switch. This switch only selects whether the mike amp or the line input shall go through to the EQ.

The EQ follows the mike/line switch. The EQ may have a small amount of gain (perhaps 10 db) or loss, or (most usually) will have no level shift at all. These gains and losses are with no EQ dialed in. When you boost 10 dB of highs, 10 dB of mid range and 10 dB of bass, the EQ would then have a "gain," which is due to the boosting it is doing. This means that the levels at the output of the EQ will be higher, making the levels "down stream" of the EQ higher. The EQ is supposed to operate with its output at a precise level, if you use boost EQ, then the output levels of the EQ will be too high. You should reduce the gain of the EQ at its input, but there are no controls for this, so you must reduce the gain at the mike pre. This means that the mike pre will be running a little lower than it could but the EQ will not be in danger of distorting.

After the EQ comes the fader. The fader is an attenuator that does not change the sound at different volume levels. The input and output of the fader are tapped off for reverb sends (improperly called echo sends in the U.S.). The reverb system will be discussed later. For now, we will follow the main console recording buss.

Having set or controlled the level with the fader, the audio signal goes into the buss system. The buss system begins at the buss assign switches (the stack of 16 or 24 switches, usually located at the top of each input channel, which are used to send the mike or line input signal to any one of 16 or 24 busses.)

CONSOLE BLOCK DIAGRAM Illus. I2-1

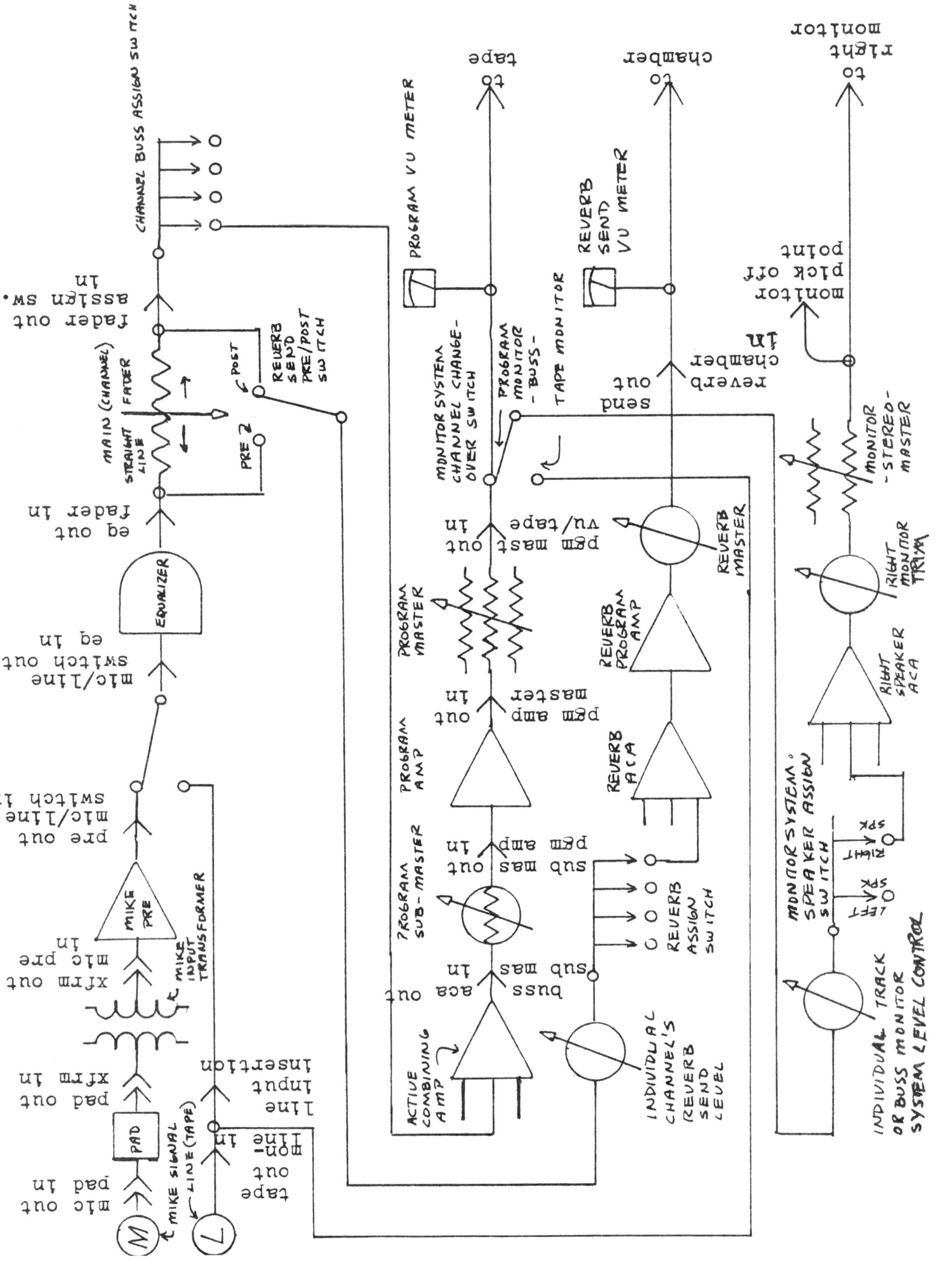

After the signal passes into the buss assign switches, the signal does not "come out" again until after it is combined with the other signals sent to the same buss (if any) at the active combining amplifier (ACN). The ACN combines signals together elctrically: not at all the same as when sounds mix together in air.

For each input to the buss system, the input modules, reverb returns, etc, there has to be an ACN input wire. 32 inputs, plus 8 reverb returns = 40 input wires to each ACN. For each buss there has to be an ACN. There is an ACN for all places where the console does mixing: to each speaker from the monitor panel, to each reverb send, to the solo from modules, to the solo from monitor system.

There are really a lot of goodies packed inside today's consoles, but when we examine the console as a one in/one out device, we make the understanding much easier. Once understood, it is a simple task to expand the understanding to a system with more channels.

Anyway, the ACN feeds the submaster for that buss. The submaster goes to the program amp (obsolete in some designs of today). The program amp sends its signal to the program master, along with the signals from several other program amps, so that several busses can be faded together with one hand.

The output of the program master feeds three things: the VU meter, the program part of the monitor system, and the tape input wire, which runs back to the tape machines.

This completes the single wire system of the input and bussing systems of the console. Now we have only the reverb and monitor systems to consider.

The reverb system is very much like the input system, except that the reverb send source is always a line level signal coming from pre or post the faders (depending upon the setting of the pre/post switch). From the console channels, the reverb send comes from the pre/post switch. In the pre mode, the fader setting or movements have nothing to do with the reverb send level. You can have full reverb send level even with the fader closed (fully attenuated). In the post position, the reverb send is after the fader. In post, the fader's settings and movements also control the reverb send.

Once the reverb send has been set to the desired level by the pot, the signal goes to the reveb assign switches.

Reverb assign switches determine where the particular send will go. Most consoles have at least four sends. As more interesting and inexpensive special effect and delay equipment becomes available, additional sends will be needed. With today's mixing techniques and the variety of interesting sounding outboard equipment available, consoles can easily use up 8 sends quite quickly.

The reverb send signal, after going out of the input module, feeds into the reverb send ACA. The ACA combines all the reverb sends from whichever modules are being sent to its particular number, and sends the signal to the reverb send program amp obsolete in some console designs.

The reverb send program amp feeds the reverb send master. (There should be one separate master for each reverb send system.)

The output of the reverb send master feeds the VU meter for the reverb send, and also the reverb send wire leading to the reverb device or chamber.

Since the reverb send system combines signals, has VU meters, and can be monitored (by patching reverb send out to tape in and monitoring tape), the reverb systems of most consoles can be used as busses (by our definition) for recording.

The reverb returns and special effect inputs during live recording and mixing can be considered as "coming back" on spare faders since reverb (or special effect) return channels should be just like the input modules having mike/line switches, EQ, sends, and faders. This is so they can be used as input modules for very large recording sessions.

*** Please see Illus. I2-2 -- The Monitor System Block Diagram ***

The monitor system is a dual input system consisting of buss (program) and tape (playback). The program (PGM) monitor signal is derived from the buss master output. The tape (sometimes called sync) signal comes from the console line in patch point.

Most monitor systems allow you to monitor individually channel by channel either tape or program. This is done in some consoles by a sync switch (which means listening to tape) or by an overdub switch (listening to program), which is located on each monitor module and hopefully on each cue module as well.

In addition to the individual channel by channel switches, there should be a master monitor changeover switch that lets the operator listen to all tape or all console.

American consoles use the sync terminology, meaning that when the master monitor switch is in the program mode you can overhaul one channel at a time to tape by pushing sync on the proper monitor module. With some sync switches pushed, if you press the tape monitor master switch, then all monitor goes to tape anyway regardless of the sync switch positions.

English consoles like Neve use the terminology "OD" for overdub. In these systems, the console is left in tape mode on the master monitor switch, and the channel to be recorded has its OD button pushed. This converts the channel's monitor source to program. Pressing master program converts all monitor system sources to program, regardless of the OD switches positions.

The monitor position switch (shown completely un- automated), having selected tape or program, feeds its signal to the monitor channel volume control for that playback (or buss output) position. This volume control adjusts the loudness in the speakers of the instrument recorded (or being recorded) on that numbered track (or buss).

The monitor channel volume control feeds the monitor speaker assign switches. These switches determine which of the four (or more recently, two) speakers in the control room the sound will come from.

From the left or right speaker assign switches the audio goes to the left or right ACA to be combined with other audio also selected to the left (or right) speakers.

The monitor ACA's feed the left or right monitor trim so that the individual speakers can be adjusted to sound as though they are at the same volume. These controls should be at full volume. If the speakers do not sound the same or if one is louder than the other, one or even both speakers may have been damaged by a previous mixer using the room. Speakers shouldn't be damaged by normal use; usually the monitors get damaged by somebody "hot switching" a mike -- unplugging or unpatching a hot mike being monitored at

MONITOR SYSTEM BLOCK DIAGRAM Illus. I2-2

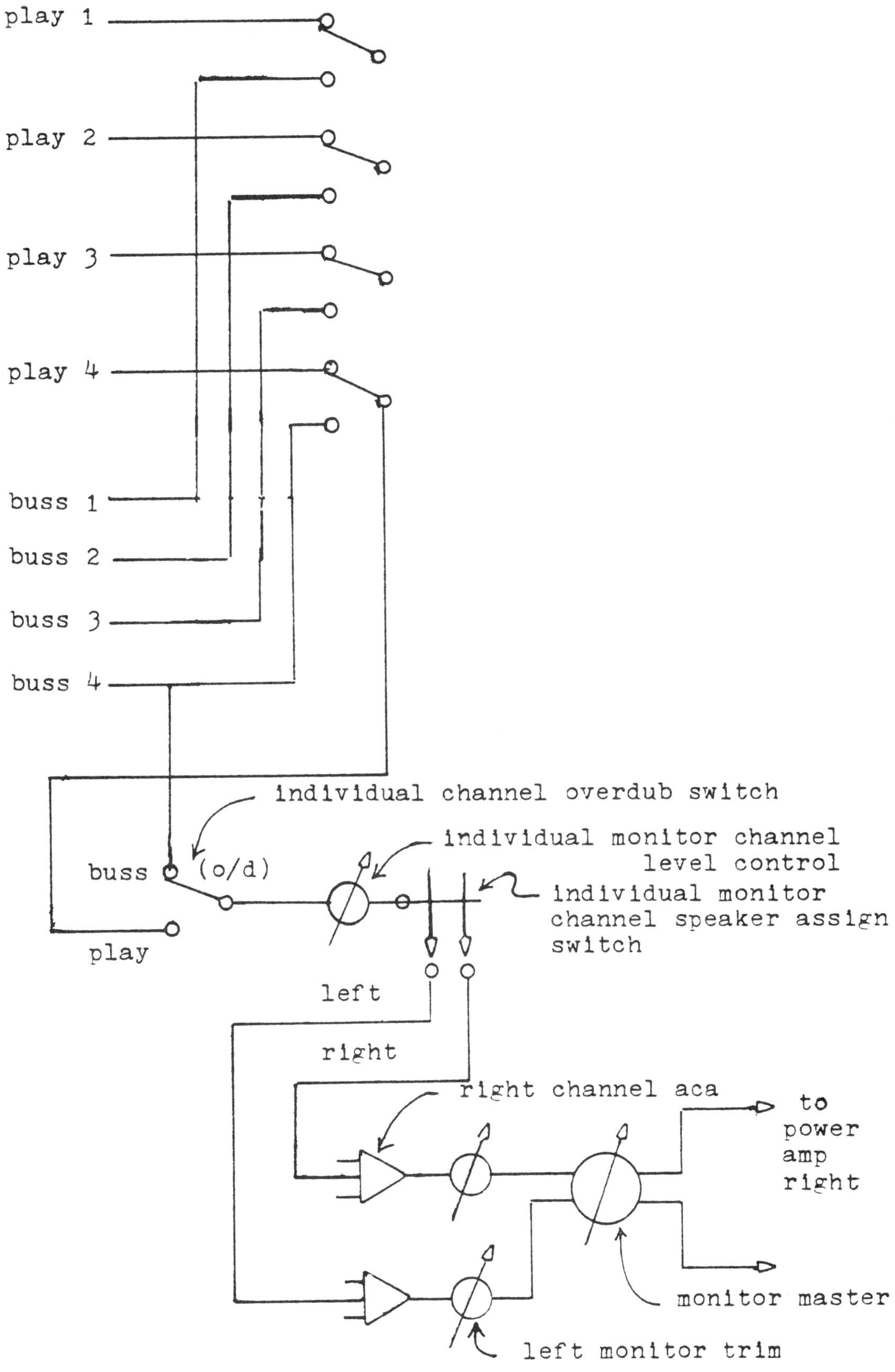

loud volumes at the time -- or by feedback, which may not only damage the speakers, but the ears of those in attendance.

Next in line is the monitor master. This is the stereo (or quad) volume control for the monitor level. This should be set with reference to a sound pressure level meter to the "proper loudness" and all session work including mixing should be done at this same sound pressure level. A good level for background music is 80 DB. For music intended to be played loud 85-95 dB on the sound pressure level meter is better.

High volume playbacks will be requested by the musicians present. Since they have been playing very loudly in the studio and are in a higher compressed state of hearing than you, they must therefore hear the playback at loud volumes. By all means, insert your ear plugs and turn up the monitor master or leave the room for the duration of the playback but give them the treat they desire.

There's nothing wrong with exercising the speakers to jazz, please or impress people who don't know any better or who are impressed by such things, but there IS something wrong with the mixer exercising his ear drums while he is supposed to be tracking or mixing.

SELECTS AND ASSIGNS

*** Please refer to Illus. B5-1 and I2-1 ***

A select switch has many inputs and one output. They are used at the inputs to console devices to decide which input will be selected. For example the mike/line switch is a select and might select off, line 1, line 2, mic 1 or mic 2.

The assign switch is used at the outputs of console devices to determine which paths the signal will take. An example of an assign switch is the buss assign or the reverb assign, which might assign the reverb send pot to chamber 1, chamber 2, an EMT or the Lexicon 224.

PATCHING

There are five basic types of patches. Patches are points where the signal "comes up" to the patch bay and therefore can have its "normal" pathway through the console changed by the user.

1. The classical or dual (or double breaking) normal
2. The newer, half breaking or mult normal
3. The invisible normal
4. The insertion point
5. The pickoff point

Patch bays are female telephone connectors in dual rows, usually twenty four long, where console components (rather than telephone lines of course) are available to be patched. In the U.S., the convention in use is to have the output of the previous console component above and the input of the following component below (outputs on the top, inputs on the bottom). For some unexplainable reason (at least in American consoles) the "format" is broken for outboard equipment and tape machine patch points: their patch bay inputs are on the top and the outputs are below. European console patch bays stick to the format: all outputs on top and all inputs on the bottom including support gear and tape machines.

In the example (Illus. B5-1) the top row of the bay is EQ OUT, the bottom is FADER IN.

Since you want the signals to flow to and away from the patch bay whether you have any patches in the bay or not, there have to be normals to complete the signal path of the various circuits. If normals were not used, you would have to patch each circuit by hand and hundreds of patch cords would be needed to do any session.

The differences between the types of patch points arises from the way they make their normal.

The dual breaking or classical normal is the oldest. If you follow the signal in from the upper left side of the drawing, you'll see that the signal flows into the patch from the EQ and feeds the "feeler" of the EQ OUT patch point. If there is no patch cord present to push the feeler up, the normal touches the signal feeler and the signal flows down to the patch below. It flows into the lower patch through the normal, and if no patch cord is present, the normal passes the audio to the feeler of the FADER IN jack. The signal then flows out of the patch bay and into the fader.

If you were to patch into the upper patchpoint to get the EQ OUT audio, then the feeler would move away from the normal and the signal path to the fader would be lost. The signal would only be present on the other end of the patch cord. If you patched this signal to a legal and usable circuit then there would be a continuation of the audio; if not, then the signal flow would be lost.

Meanwhile, the fader still listens to a dead wire since the EQ OUT patching broke the normal. You could patch an active audio signal, perhaps the output of another active console equalizer, into the normal broken FADER IN and give the fader something to do, otherwise the fader is cut off from the console.

Notice the symbol that lies under the different patching diagrams; these are the way these patching circuits appear in our simplified block diagrams throughout the book. The symbol for the full or double breaking normal type is the double right facing arrowhead drawn on the single line.

The most widely used patch type today is the mult normal or half breaking normal. This is the second type shown and operates in this fashion:

The signal coming in from the output of the EQ enters the patch bay at the EQ OUT jack. The normal in this bay is not used. Instead the normal in the next bay (FADER IN) is hooked up to the main wire of the EQ OUT jack. For this reason, when patching into the EQ OUT bay, the signal is not stopped on its way to the fader. There is created an instant mult or double output for the EQ in question. When patching into the fader, however, the signal from the EQ in the bay above is blocked from going to the fader. Instead only the signal send via the patch cord is inputted to the fader.

Invisible normals come into play when there is relay switching of multitrack outputs in conjunction with a patchbay. Consider the case of a studio having two 24 track machines that feed a 24 channel relay switching system, which in turn feeds the console's line inputs. In this case there would be two patch rows: one labeled 24 TRACK OUT #1 and another labeled 24 TRACK OUT #2. These signals would also be going to the switcher system. The switcher system would INVISIBLY feed the normal of a bay called CONSOLE MONITOR/LINE IN. There would be no bay called RELAY SWITCHING SYSTEM OUT. The set of relay switch tape outputs would feed the normals of the CONSOLE LINE IN bay without ever coming up to a patchable row -- thus the invisible normal bay. You do not lose any patching options: you can still patch the machine play outputs and patch a new input to the console's MONITOR/LINE INPUT.

Insertion points are not normalled patch points but they do occur in patch bays. These are extra inputs to console devices (usually the ACA's), which are in addition to the usual inputs to the devices. So if there was a wild or auxiliary input position in the patch bay for each ACA, then when all the input modules were assigned to a buss, you could still get another input to the buss via the WILD or "AUX ACA IN" patch point. No signal paths are broken with the insertion point patches.

The pickoff point is the last type of patch shown. In this particular type of patch, no signals are broken and none are made. There is no real mixer function at all to these patch points, rather they serve as maintenance "look-see portholes", so that the maintenance engineer can see if the signals are healthy at the points in the circuit equipped with pickoff points without doing anything more than plugging in a patch cord that is connected to the test equipment.

*** Please refer to Illus. I2-1 ***

In our diagram are displayed the two primary types of patching -- full and half breaking normals.

Starting at the beginning: the first patch -- the mike input is the double breaking MIKE OUT/PAD IN patch bay. This bay is a double bay like all the others in the console, except pickoff point bays and insertion point bays. The top row is the MIKE OUTputs from the mike input panels in the studio. The next lower bay is the PAD IN patch point.

Next comes the PAD OUT/MIKE TRANSFORMER IN bay. This bay serves as the input to the mike pads. The mike signal comes out of the pad with the amount of loss in level dialed in by the recording engineer.

The TRANSFORMER OUT/MIKE PRE IN bay follows. This is the bay that has the transformer output wires above and the mike pre-amplifier input wires below. This is the last point where the signal levels are at low level prior to undergoing the mike preamp volume boost. Because of this tremendous gain available in the mike pre amp, it is best not to change patches anywhere "above" (ahead of) the MIKE PRE IN position without muting the input channels involved in the patch (both sending patch and receiving patch). In fact, don't even touch the patches while the input mike input modules are hot: scratchy connections in the patch bay can cause noises loud enough to blow up your monitor speakers.

The MIKE PRE OUT on modern consoles is usually not patchable until after the mike/line switch. (Our drawing is an old fashioned complete patch bay including all patches.) The MIKE PRE AMP OUT bay is usually where mult normals begin.

Next comes the MIKE LINE SWITCH OUT/EQ IN patch bay. This bay is again a double row bay usually of 24 positions of which the upper row is the output of the mike/line switch, which could either be the playback of a tape or the output of a mike preamp, depending on the setting of the mike/line switch. The lower of the two bays is the input to the particular channel's EQ. Each of these patch bays are numbered to identify the modules to which the patch points belong.

Next comes the EQ OUT/FADER IN patch bay. This bay offers the EQ output. Having an EQ IN and an EQ OUT patch point makes the console have patchable EQ's. Also included is the fader input. The fader output is usually patchable.

After the FADER OUT/BUSS ASSIGN SWITCHES IN bay, the signal disappears for a while.

The signal reappears only after being combined by the ACA to which it was assigned. These combined signals appear at the ACA OUT/SUBMASTER IN patch bay.

Next comes the submaster and the buss amp (also called program amp). The bay associated with it is the PGM AMP OUT/MASTER IN bay. Sometimes this bay is not included due to the absence of a program amp in the console.

The final bay for most consoles is the BUSS OUT/VU-TAPE IN bay. In this bay, if you patch into the program out, you get the buss mix (having been combined through the ACA, and having passed through the sub, the program amp and the master). If you patch into the VU/TAPE IN point, you get a wire to the tape machine and the proper numbered VU meter on the console to watch the levels with.

In addition to the mike input patches, there are two additional patches for the line input (from the tape machine playbacks).

The first line input position on the patch bay is the TAPE OUT/MON-LINE POSITION IN. This is the actual playback wire from the machine and the input to the monitor system tape monitor mode, and the line side of the mike line switch. This allows repatching of the tape playbacks to the console line inputs (for mixing) so that the instrument layout across the faders will fit into the mixer's usual way of doing a mix. This requirement for repatching will become clearer later. The other use of the LINE IN patch point is to get to the monitor system input for the purpose of convenient monitor system listening inputs.

The LINE INPUT INSERTION is provided so that you can have separate line inputs to the EQ/fader/buss system for mixing, which can include signal procesing equipment, but still have the tape playback (monitor) system, which goes directly from the tape to the speakers without effects equipment being involved.

All patches upstream from the MIKE PRE OUT are at mike level and outboard equipment may not be used. All patches after the MIKE PRE OUT are at "line" level. All line level signals may be patched anywhere within the console or to outboard equipment without fear of hum, buzz, or noise being introduced, as long as professional quality equipment is used with common sense.

There are three reasons to do patching:

1. For emergencies
2. For adding to the console's basic equipment
3. For conveniently changing recorded tracks or mike input

arrangements to familiar locations on the faders

When emergencies come up, like when a mike preamp suddenly develops a frying noise, it is necessary to quickly repatch the mike to another module that is operating properly. This would be done in the console by patching the MIKE OUT to another PAD IN. In a studio not having mike out patching, we would re-plug the mike into another XLR position in the studio at the mike plug-in panel.

If, for example, during a mix you perfect the vocal changes and mark them on the vocal fader and then try to add some EQ, only to find that the EQ in that module isn't working, you could "borrow" the EQ from module 30, and by writing "EQ from channel 30" on the module you were using, save the fader markings and keep the familiarity you have already established.

When special EQ is desired, such as parametric EQ on the snare, then patching outside equipment into the console signal flow is done. There are hundreds of patchable outboard equipments available for patching into line level patchpoints in the console.

There are some basic patching don'ts:

1. Don't patch line level signals into mike pre amps
2. Don't patch the output of a device back into its input
3. Don't patch the input of two devices together
4. Don't patch the outputs of two devices together
5. Don't patch MIKE OUT until you have already patched to the MIKE PRE IN of your choice
6 Don't pull the MIKE PRE IN patch first -- pull the MIKE OUT patch first instead

There are some do's to remember as well:

1. Pull patches left overnight. You can leave them in the patch holes, but not completely inserted. This helps keep foreign substances from settling on the contact areas of the patch bay normals. If cigarette smoke or other insulating contaminants do settle on the contact points, the normals will fail, making you have to manually patch the signals from one device to another in the console.
2. Clean the patch cord ends with Brasso so that they shine. Be sure to leave no Brasso residue.
3. Pull patch cords by the plastic handles, NOT by the wire jacket.

CONSOLE SOLOS

There are two types of solos:

1. Monitor solos
2. Muting or destructive solos

Monitor solos operate by muting the monitor system entirely and then replacing the monitor feed with the audio at the point where you pushed the solo button (eg. fader solo #8). The monitor consists of one or a mix of however many solos you push until all the solos are disengaged.

Muting solos kill the monitors also but do so by killing the audio paths in all the console channels, except for the one you have soloed. If the console has program solos in the input modules, and you push one during a take, then all the tracks on the multitrack machine will go dead except for the track being sent to by the module you soloed. This would, of course, ruin the take. This is why some persons call these solos destructive.

Things to remember about solos:

1. Use solos only to get the sounds, not after the tape is being recorded. After recording starts, let the producer listen to the music.

2. All solos should have an associated light, which shows that they are activated.

3. There should be a redundant (backup) solo lamp somewhere on the console in case a solo is pushed and its lamp is burned out. In this case, there would be no monitor system function at all and without the redundant lamp, you would have no idea of why the console had "gone dead."

SUBMASTER/MASTER AMD VCA/GROUP FORMATS

In the "old days" there were submasters and masters. Today the trend is for no recording buss submasters or masters at all. Only mixing busses -- quad, mono, and stereo -- now come equipped with masters. The theory is that the gain structure of a console is best suited by having all the gain "up front" in the mike preamp, and having no gain stages after the mike preamp at all. The reason this change in design is possible is because in the old days the combining was done after the buss assign switches by use of resistors. These resistors caused a rather large level loss necessitating a "gain make up stage" directly beyond the passive combining network (PCN). The signal after the PCN was too low to properly pass through the long wire "runs" to the output of the console, too low for the VU meter to register and too low for the tape machine to record.

Today's consoles use active combining networks. There is no loss: the signal emerges from combining at the same strength as it went in.

In this course of study I will mainly deal with old style consoles since they have the most components you will encounter and modern consoles can then be considered to have "missing" controls. Operating an "oldie" will require knowing the meaning of all the patch points and associated controls necessary to their correct operation.

*** Please refer to Illus. I2-3 -- VCA and group VCA ***

VCA means voltage controlled amplifier. There are also voltage controlled filters, EQ's, faders, masters, oscillators, pitch shifters, to name only a few. The reason behind the voltage control is so that a control voltage can control audio volumes without the audio having to be sent here and there to be controlled. In VCA consoles, the fader is actually a "light dimmer" and controls not the audio, but a small voltage running to the voltage sensitive input of a high quality VCA. Three things happen as soon as this design is implemented:

1. The small crackles of noisy faders will no longer add crackles to the audio

2. Connecting the control voltage from group master # 1 to faders one and two will allow the volume of channels one and two to be controlled by the motion of the one group master.

3. The audio does not have to go from the EQ to the fader and then back into the console to the group master, it can be "faded" right at the input module's VCA.

The VCA does not care where the voltage comes from. In a system where the presence of 5 volts means a full volume fader setting and 0 volts means a completely faded fader setting, the presence of 5 volts at the voltage input to any VCA would open it up for the audio. This means that there can be other sources for the fader VCA voltages besides the group masters -- one other source is control voltage generated by a mixing computer: console automation. On older automated consoles, the unevenness of the tracking of the VCA's one to the other caused funny things to happen during fades: one or two instruments, whose VCA's resisted being faded the most, would be "featured" against the mixer's wishes during fades, since all the other tracks would be faded faster by the more "willing" VCA's.

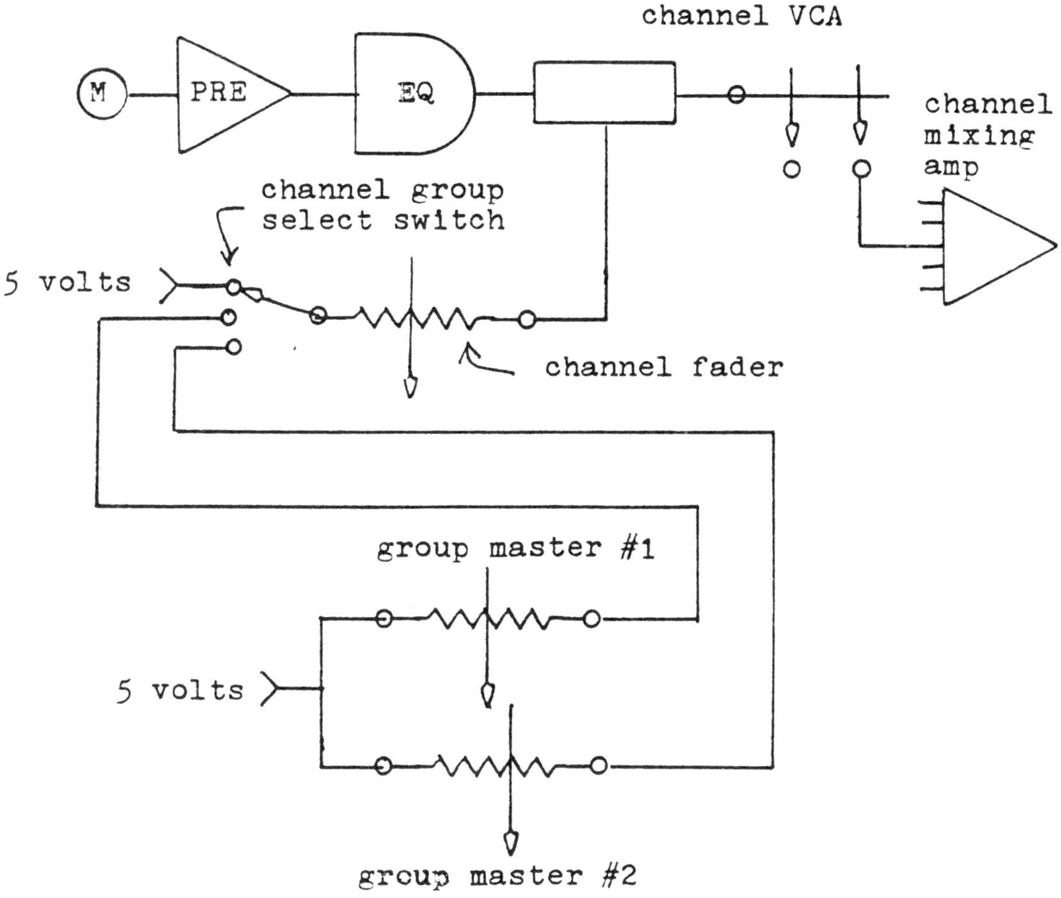
channel VCA
M
PRE
EQ
channel
mixing
amp
channel group
select switch
5 volts
channel fader
group master #1
5 volts
group master #2

PHASE SCOPES AND METERS

A phase scope is a simple, patchable oscilloscope that has a moving point of light, which indicates the phase relationships between any two console signals, usually the left and right mixing busses. If there is no signal, the scope shows only a small dot in the center of the screen. In the presence of a left speaker signal a horizontal line appears across the center of the screen proportional in length to the level of the signal. A right mixdown signal generates a vertical line in the center of the screen with length proportional to the level of the signal. If the two mixing busses have any common signal, like a bass in the middle, then the phase between the two busses can be seen by examining the shape of the pattern of the tracings on the oscilloscope. A straight line from the lower left to the upper right indicates a completely in phase relationship. A line from the upper left to the lower right indicates a completely out of phase relationship. Any other picture indicates a phase relationship between in and out of phase, with nearly circular patterns meaning 90 degrees out of phase. Strong out of phase displays caused by low frequency instruments must be avoided. This is because hi energy, low frequency out of phase program material will cause "lifts" in the master record when it is cut. This means that the groove gets shallower and disappears off the record, leaving the record player needle no clear path to follow, so it jumps to the groove next door. This causes a record to skip.

Phase meters are VU like meters that normally point straight up to show no phase relationships present. Pointing full scale to the left usually means fully out of phase; pointing fully to the right means fully in phase relationships present. Using phase scopes gives more information than using phase meters.

Some studios use oscilloscopes with a circular screen so that they can twist their oscilloscope cathode ray tubes (CRT) 45 degrees to clockwise. This makes a left signal appear as a line from upper left to lower right and a right signal go from lower left to upper right. The advantage of doing this is that the oscilloscope picture then exactly resembles the activity of a phono stylus, which would be tracing the groove made by the mix now in progress. A strong vertical component is to be avoided especially if caused by punchy low frequencies. Very high speed, hi frequency motion in the phase scope display is to be avoided as well. This kind of display can be caused by two mikes, separated by a distance of a foot or more, picking up hi frequency information such as overhead drum sounds. To prevent this hi frequency phase anomaly from being a problem, locate the drum overheads with the capsules nearly touching, positioned at about a 45 degree angle.

DEGAUSSING AND CLEANING THE HEADS

Before playing any important tape, or certainly before playing any master or standards tone tape, degausse and clean the heads. Do this as follows:

1. Turn off the tape machine.
2. Plug in a head demagnetizer at a good distance from the tape deck, as well as from any recorded tapes.
3. Degausse the heads by first lightly touching the degausser tip to the top of the head and then very slowly drawing the deausser down the head stack; when you get the tip to the bottom, draw the degausser straight away, moving very slowly at first and then more rapidly as the distance reaches 3 inches or so.
4. Do step three for all three heads (or more if there are more heads) and then turn on the tape machine again
5. Swab off the heads with a cotton swab soaked in xylene or denatured alcohol, whichever your studio uses. Note: be careful!! Xylene can melt some types of recorder head material.
6. Swab off all the tape path components including rollers, tape lifters, the capstan, and the pinch roller. Only use denatured alcohol on the rubber pinch roller.
7. To really clean the heads (I recommend you do this once each morning) soak a Kimwipe (a lintless tissue) in your solvent and lay the soaked Kimwipe against the head so that the gap area is adhered to. Then degausse the head with the Kimwipe stuck to it. If any magnetic particles are "hiding" in the gap, they will be shaken out into the Kimwipe and carried away. This is the only way to clean a set of heads that has been "lapped" or re-machined due to wear, since this process may leave magnetic metal particles in the gap. A head with particles in the gap will play back tape with noise, distortion and intermittent garbling of the sound.

SET UP PROCEDURE

Playback alignment - source: Ampex reproduce alignment tape

Tone	Machine adjustment
1 KHz	playback (PB) level
10 KHz	pb head azimuth
10 KHz	pb hi freq EQ
100 Hz	pb lo freq EQ
1 KHz	re-check PB level (if off repeat from the top, skipping azimuth)

Following down the playback alignment listing, a reproduce alignment tape with only two frequencies 1 KHz and 10 KHZ is needed. This is because it is very difficult to accurately set the low frequency playback EQ to low frequency tones found on a full track reproduce alignment tape. (A full track tape is one which is recorded edge to edge.) The low frequency tones always read too high causing you to set the low freq playback EQ too low. So for the purposes of this book, we will set the low freq play EQ after setting up the recorder for record and use the tones we record ourselves (100 HZ) as reference tones for setting up the low freq play EQ. This procedure is reasonably secure from error since their are no adjustments for low freq record EQ available on most machines.

Record Alignment - source: console tones

Tone	Machine adjustment
1 KHz	bias
1 KHz	record level
10 KHz	record head azimuth
10 KHz	hi frequency record EQ
100 Hz	lo freq playback EQ
1 KHz	re-check record level. If off, return to 1 KHz record level step and repeat, skipping azimuth.
1 KHz	set input monitor calibrate (also known as CAL or INPUT)

There are two major types of alignment procedures -- those for continuing sessions, which already have a set of project tones recorded, and first sessions needing the creation of master project tones.

These two major types are further broken down into those sessions requiring use of the Dolby noise reduction system (or other noise reduction), and those that do not require Dolby units.

All possible combinations are covered in the alignment notes in Intermediate 7. Memorize both of the above listings before you get to Intermediate 7.

If you look ahead do not be dismayed by the seeming complexity of the alignment procedures, they are not really very complicated to do. The reasons for doing the alignments the way I present them are difficult to understand though, and would require a thorough knowledge of Dolby units and recorders. This is why I deal with setups as procedures. If you cannot get the desired results by following the procedures I give, then something's wrong with the equipment or its hookup.

When aligning machines, listen for flutter and watch for tape shedding on the guides. Look at the heads to see if there is abnormal head wear. Check to see if the record cables are attached to the proper busses. See if the playbacks are appearing at the proper monitor system outputs. Always check tape machine terminations.

As this is not supposed to be a maintenance book, I will not go into the finer points of setting up recorder heads for zenith and the other adjustments. These techniques are best learned from a qualified maintenance person. We will discuss head alignment further, however, in Intermediate 7.

EDITING TAPE

*** Please refer to Illus. I3-2 -- Joe Tall's "Tape Editing Facts" ***

There are two basic forms of editing, narration and music. Narration deals with making a recorded tape play exactly like the script from which it

What Is Tape Editing?

Tape editing means simply to cut out from a recorded tape those parts that are **not** wanted and to do the work so professionally that the cut (edited out) material will not be missed. In a well-edited piece of tape there should not be the slightest hint that editing has been performed-no clicks, parts of sounds or anything but the smoothly played back recording, whether it is speech, music or information bits.

The only kind of tape that can be edited is one that consists of **one track** of sound (or other information). You may consider this **"one track"** to include any number of tracks **all recorded in the same direction.** Thus you may edit tape recorded on **one track only** on a two- or four-track recording machine; also any tape with from 1 to 24 tracks **all recorded in the same direction.** You **cannot** edit two- or four- or eight-track tapes where some tracks are recorded in one direction and some recorded in the opposite direction. The only time you **can** edit multiple reversed-track tapes is when you have two **identical** tapes recorded on the same machine and you want to interchange parts of one tape with parts of the other; that operation may be safely performed.

Tape recorded in cartridge form or on cassettes should normally not be edited, simply because they contain tracks of sound in opposite directions. However, broken tapes in cartridges or cassettes may sometimes be spliced together.

Tape Splicing Materials

Tape splicing means to join together two broken, or cut, ends of tape. The preferred method of splicing includes the use of a tape splicing block or jig like the **Editall Block** and the proper grade of pressure sensitive splicing tape. Pressure sensitive splicing tape has been developed to the point where you may feel perfectly safe in using a spliced tape over and over again for recording fresh material (after it has been erased, of course). A properly made splice will leave no space between the spliced ends and allow no adhesive to creep out from under the splice. Splicing tape should be chosen for the work it is intended to perform. For instance, where the tape used is comparatively thick 2 mil. tape (including the oxide coating) one should use a splice that will not loosen in case the tape **and the splice** is subjected to maximum tension. Thus you will find that the polyester-base adhesive tape and Editabs for use with Editall Blocks are matched in strength to the recording tape they are to be used with. Where the splice requires maximum strength the material used (not counting the adhesive thickness) is one and one-half thousandths of an inch (mils) thick. Where the splice requires minimum thickness, as in videotape splicing, the splicing tape is only a quarter of a thousandth of an inch thick!

Good adhesives for splicing tape must have several basic good qualities to be completely trustworthy in splicing. The first of these may be defined as high shear strength or ability of the splice to resist being pulled apart. Of course, any splice, no matter how strong, can be pulled apart if the shearing force causes the recording tape itself to stretch; then the splice will just lift off and presto!—no splice. However, if the splice resists shear to the point where the recording tape itself **begins** to stretch, it is more than strong enough for its job. The shear strength of good splicing tape **increases** with age; the splice becomes much stronger a few hours after application. It does not weaken with age, provided that the tape is made of good quality polyester. I do not recommend or use acetate splicing tape for long storage; it is all right to use for only comparatively short periods of time. With age, under some conditions of tape storage, acetate splices have been known to become absorbed into the tape base with the result that splices mysteriously disappear. Not good for tape libraries!

The second desirable quality for a splicing tape is dryness of the adhesive. In actual practice, the adhesive should be applied in precisely the correct thickness to do its job. Too much adhesive will feel very sticky; it will adhere well and seem fine; but in actual use excess adhesive will be forced out from under the splice by the tremendous pressure exerted by tightly wrapped tape and cause one round of tape to stick to the next round. This sticking sometimes results in oxide being stripped off the recorded tape; sometimes it will cause a tape reel to move jerkily or even not at all. Besides all the foregoing, we must realize that the "stickiness" of the adhesive is closely related to its shear strength. Generally speaking, the **least** adhesive that will give good adhesion will give the best shear strength. A splice with poor shear strength that **"creeps"** and exposes adhesive to the head surfaces may not only gum up the heads but part unexpectedly with disastrous results. Incidentally, I would like to stress that tape rewinding force should always be just sufficient to wrap the tape smoothly. Too high winding tension results in sticky splices and distorted tape; it is wise to have this checked and adjusted to proper tension.

The shear strength of a splice and its "stick-ability" have a direct relation to the length of splicing tape that should be used. If the splicing tape is of poor quality its leading end may tend to lift off the recording tape when going around a sharp bend in the tape travel, like around a right-angle guide. In a case like that you would have to use a longer splice so enough would be left to hold the two tape ends together. With high quality splicing tape like that furnished by Editall, seven-tenths of an inch (.7 inch) is sufficient to splice tape and will not begin to creep until a shear strength of over four pounds is reached. A one and one-half inch length will exceed the shear strength of the strongest available tape by several pounds. A splice of from ¾" to 1½" is recommended, well rubbed down by fingernail or orange-stick to obtain the best possible adhesion.

The width of the splicing tape should always be slightly less than the width of the recording tape. This provides a safety factor in case, due to excessive heat, humidity or pressure, there should be a slight exudation of adhesive at the two edges of the splice. This ooze is often due to the way adhesive is coated and slit and cannot always be avoided in manufacture. The above width "safety factor" will keep you out of trouble. I began using 7/32" wide splicing tape just as soon as 3M began making it upon my recommendation. It is the proper width to use for quarter-inch tape splicing, and, as mentioned previously, preferably made with polyester base material.

Splice-edge adhesion (due to the slight ooze of adhesive mentioned above) led to another expedient in tape splicing during the early 1950's. I found that if the splices were pre-cut to length at a sharp angle they tended to stick less layer-to-layer. If there was any layer-to-layer adhesion, the tapes would still strip away from each other easily without causing the tape drive to jerk or "wow." Therefore, before editing and splicing a sequence on tape, I cut many of these splicing strips with scissors and stick them, ready for use, to the edge of the aluminum base of the tape recorder. I noticed very soon, however, that every splice applied created a faint noise at the splice, due, no doubt, to a chemical reaction of the adhesive to the aluminum. Thereafter I stuck these splices to a piece of clear wood or a clear piece of plastic and the noise was eliminated. For some years I cast about for a method of making a "prepared" splice that could be temporarily stored on some inert material and that could be rapidly used. What was finally invented, in 1965, was the Editab. It is simply a precut (actually diecut) splicing tab, adhering to a die-cut guide which keeps the tab's edges about two hundredths of an inch away from the edges of the recording tape, all temporarily adhered to a release paper. By stripping the Editab off the release paper from one end (colored red), applying it in place on the tape ends, then stripping off the guide from the other end (transparent) a perfect splice is most rapidly effected.

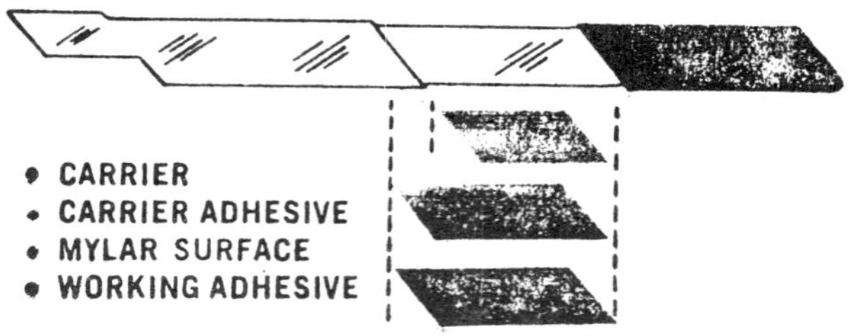

There is another advantage, for the blind, built into Editabs. A fraction of an inch from the transparent end of the Editab the carrier angles away from the main body sharply. If, in making a splice on an Editall quarter-inch block the editor fits this angle exactly at the Editall's edge the splicing tab will fall precisely over the cut ends of tape in the groove of the Editall Block, at its diagonal cutting slot.

The Editall® Splicing Block

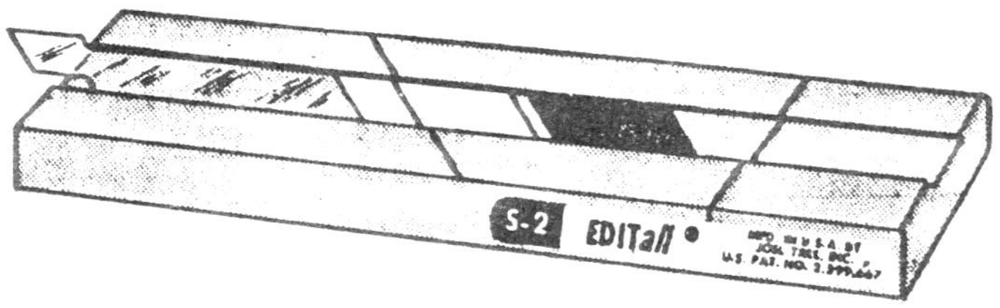

EDITALL S-2 BLOCK WITH CX-1 EDITAB

All Editall Blocks (except the Model KA-2, where the guiding principle does not hold) are made with slightly curved tape-holding grooves dovetailed at the top so that the tape cannot slip out. The tape is forced into the groove by slight pressure and is held in the groove by its edges butting against the dovetail in the Editall's sides. The tape can be slid within the groove by more or less force, depending upon the thickness of the tape. For example, 2 mil. tape will, in the quarter-inch size, fit quite tightly in the S-3 Editall groove and require about four ounces of pull before it will slide in the groove. One mil. tape, on the other hand, will still hold sufficiently for accurate cutting and splicing but will move in the groove with much less applied force. Tape that is accurately slit to specifications will perform admirably; stretched tape will not hold and oversize tape, naturally, will not get into the groove at all. The fact that the tape, whose tendency is to remain flat, is bent slightly in the groove, causes it to exert enough pressure against the ridged sides to prevent its moving during the cutting operation. Because of this it is possible to assemble in the groove many bits of tape for simultaneous splicing, an operation performed frequently in phonetics demonstrations, music concrete and some electronic music assembly.

After the section of tape to be discarded has been cut out, the cut pieces of tape are placed in the groove, and the two ends, which have been cut (in the diagonal slot across the groove of the block) at a precise 45-degree angle with a razor, are butted together firmly. To obtain the best possible splice, push the tape ends together until there is a slight bulge where they butt. Then a length of splicing tape 1½ in. long and 7/32 in. wide should be cut with both ends at acute angles; this piece should be placed with its center over the butted tape ends in the grooved block.

The splice should be smoothed down firmly with the fingernail, until the splice takes on the color of the tape. If using Editabs, remove an Editab (CX-1 type for normal ¼" tape splicing) from the release paper by its red end, center the splicing tab over the cut ends of tape in the Editall Block, smooth down firmly with a fingernail and remove the whole spliced tape from the block groove. The removal should be accomplished by grasping the spliced tape by **both hands** (at the ends of the block) pulling it taut and then lifting it **upward** out of the groove. **Do not attempt to peel the tape out of the block!** All you would accomplish by this will be to practically destroy the tape by ruffling its edges severely! When using Editabs, remove the Editab guide from the finished splice by peeling it off **from the transparent end** after the spliced tape has been removed from the block.

The cutting slot in all Editall Blocks is machined to accommodate a standard single-edged razor blade which is nine-thousandths of an inch thick (.009"). Stainless steel blades are furnished with Editall Kits and are practically non-magnetic. However, even stainless steel can be magnetized in a strong magnetic field so it is advisable to make certain, before editing, that your blades have not become magnetized. If so, every time you cut the tape you will leave a bit of magnetic energy on it which plays back as a noise click. Tools like blades or scissors can be demagnetized by the use of a bulk-eraser or any small degausser. Use a slicing motion in cutting tape to get the cleanest possible cut without deforming the tape. **Read the instruction sheet** that comes with all Editall Blocks.

Tape cutting slots in all Editall Blocks are made at an angle to the tape. There is also a perpendicular slot but this is to be used only when splicing on tape leader or editing information tape that requires abrupt cutting. Any **audio** tape should be cut at an angle; the specific angle of cut depends upon the number of tracks recorded (in the same direction, remember?) and the tape speed. The reasons why recorded tape should be cut at an angle are many and to explicate them properly would take possibly fifty pages of print this size. In short, there are two main reasons for diagonal cutting for sound tapes:

1. When tape is played back it is actually the transformation of magnetic marks on the tape (being "read" by the playback head gap) into magnetic energy that creates (or re-creates) the sound you hear. No matter how tightly you splice, some magnetized bits of iron oxide or other magnetized material may get into the joint between the two tape ends. If the tape is perpendicularly cut the play head will "hear" it as a heavily recorded noise. In addition, unless you cut tape by using an oscilloscope, you cannot hope to make sure that the recorded bias voltage (which remains part of every recorded tape) will not jump from one extreme to the other at the cut. Thus, if at one cut end the bias recorded is at maximum and at the other at minimum, the jump in voltage will play back as noise, albeit at high frequency.

2. The other principal reason for cutting diagonally lies in our hearing. To be heard as sound, not noise, any sound has to become established in our hearing for a fraction of a second. If we hear a sound that is shorter than this minimum in elapsed time from onset to end we hear noise! This minimum varies in time from one person to another; also in frequency. So, to be perfectly safe in editing, cut at a diagonal.

How Do You Find Out WHERE to Cut the Tape?

One of the principal reasons advanced for the growing popularity of tape recording in its early years was that tape can easily be edited. However, tape recorder manufacturers, except those making professional machines generally, still do not make it easy to edit tape. They make recording machines that are beautiful to look at and simple to use in every respect except editing. To edit tape two things are necessary: 1. an exact way of marking the tape to be cut so that the actual cut will correspond with that part of the tape precisely at the gap of the play head. 2. a way of leaving the playback amplifier "on" but with the brakes and drive motors "off." You cannot edit tape with any facility unless both these conditions have been met. I can show you how to measure accurately to establish a tape marking-point. You will have to see the manufacturer's agent or a tape recorder serviceman in order to be able to use your machine for editing with motor drive and brakes off.

Establishing A Tape-Marking Point

Were it not for several disadvantages one would remove all impediments, including head covers and shields, and mark tape for editing right at the point where the sound is reproduced—at the playback head gap. The reasons why you should not do so are these:

1. In order to see the marks on tape you have to mark it with a black or yellow grease pencil. Although these marks are easily removed from the head surface with Freon or even alcohol, some of the stuff might remain. If it should it could cause the head surface to wear unevenly, ruining the head for good reproduction.

2. Constant marking of the head surface could result in the head alignment being pushed off azimuth, or to one side, resulting again in inferior sound.

Because of the above reasons it is best to mark the tape for editing somewhere away from the tape head gap. The only way to do so accurately is to establish a tape marking point somewhere along the path of the tape as it goes from one reel past the heads to be wound up on the takeup reel. Then, by measuring the exact distance from this established tape marking point to the play head gap, and marking off the **same distance** on your Editall block (measuring from the center of the diagonal slot to your **right** on the Editall Block) the measuring job is done. Thereafter all you have to do is mark the tape **at the marking point on the tape recorder**, line up this mark with the mark you made on your Editall block and cut the tape at the diagonal slot. With ¼" tape you should make certain you have picked a tape marking point that **can** be marked off on the Editall Block you are using. For example, an S-2 block is four inches long overall, while an S-3 block is five and three-quarter inches long. The diagonal slot is one inch in from the left end of each block. That means that if you have an S-2 you will have to find a place to mark the tape on your recorder less than three inches to the right of the play head gap. On the S-3 the marking place would have to be within four and three-quarter inches to the right of the gap. Just to be sure the above marking process is clear let's go over it again.

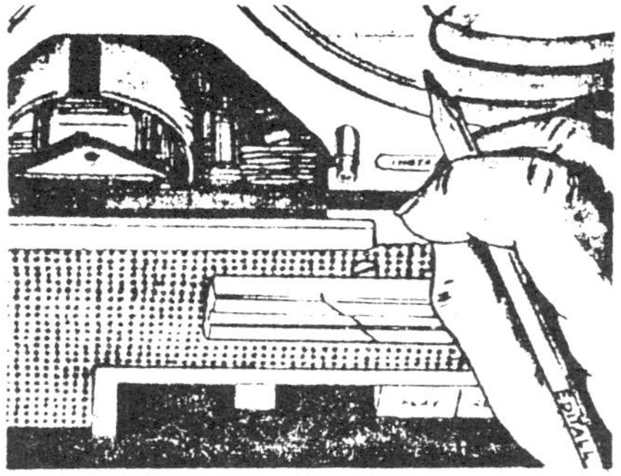

The drawing above puts the reader where he is looking over the shoulder of a man establishing a tape marking point. (The drawing is distorted but the action is clear!) The left-hand mark on the tape was made on the play gap, the right hand mark on the tape was made on a convenient marking place, the metal stud.

Since sound is reproduced at the magnetic gap of the playback head, you must know where this gap is in order to edit tape accurately. On professional machines it's easy: the head farthest to the right is the playback head and its gap is in the middle of it and may be seen with a good magnifying glass. On machines with two magnetic heads the same thing is again true—the right hand head is either play or record-play head. Where a machine has just one head, with two gaps, the right hand gap is the play gap. Thus, to establish a tape marking point, find a place to the right of this gap, within the distance available to you on your splicing block, where you can mark the tape. If there is no convenient place, like the stud in the above figure, cement a little block of wood to the tape deck in a convenient spot just about clearing the tape as it travels to the capstan and takeup reel. Put some kind of mark on this block of wood so that you can mark the tape exactly. Then you measure this distance between the gap and the marking point, or actually mark a piece of tape. Put this marked tape in your Editall Block, with the left hand mark bisecting the diagonal slot to the left. Mark your block permanently where the right hand mark falls. Thereafter all you have to do when editing is to mark the tape at this marking point, put the tape in the Editall Block with the mark you made coinciding with the mark on your block and cut the tape at the diagonal slot.

For those who are not gifted with sight there is an audible method of finding the play head gap. Record a few seconds of sound on tape. Take this recorded tape off the machine and place it in the Editall Block with the recorded sound right on the perpendicular slot. Cut this tape, discard the right hand portion and splice a piece of blank tape or leader tape onto it. Put this tape back on the recorder and turn the reels or pull the tape slowly by hand until you just hear the beginning of the sound. At this point the splice will be directly over the play head gap.

Don't move this tape yet! Mark the tape by pricking thru it with a knife or razor edge at your tape marking point, place it in the block with the splice bisecting the diagonal slot and mark the block where the tape has been pricked. If these marks are made so that they can be felt, a blind man can learn to edit tape very accurately, especially by using Editabs to splice with. Editabs have been designed with this in mind so that sightless persons can edit tape by sound and touch.

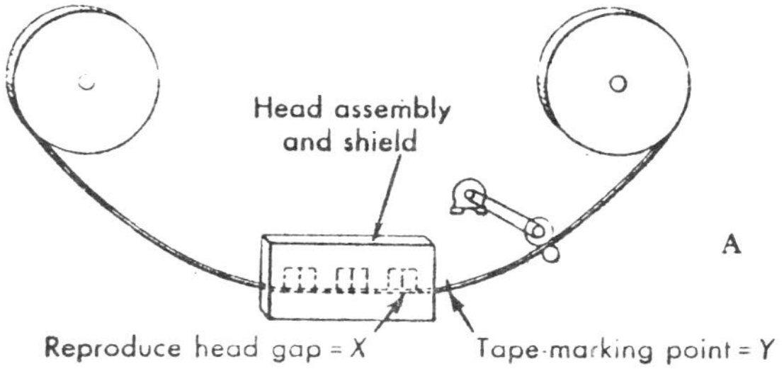

A

Diagram showing tape in correct position on recorder and ready to be marked. The distance between the reproduce head gap (X) and the tape-marking point (Y), once established, should not vary.

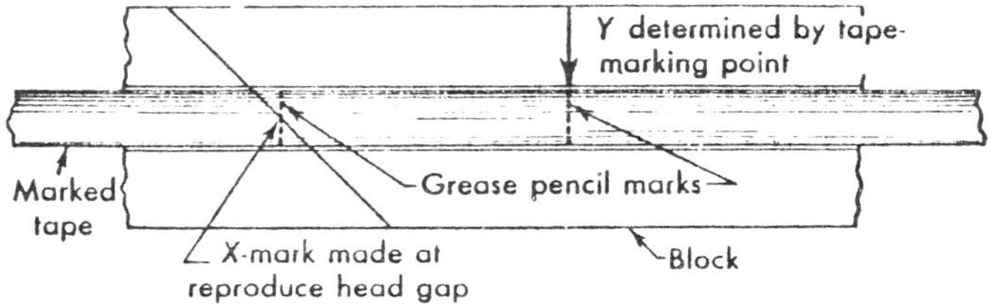

B

With X (the mark on the tape) placed at the center of the diagonal cut on the EdiTall block, Y indicates the point at which the EdiTall block should be marked.

Sound Recognition

The difference between a professional tape editor and a novice can be expressed in very few words: the professional **knows** where to cut; the novice does not. To learn where to cut takes much time and practice; it is impossible to acquire this ability overnight. A good working knowledge of sound, phonetics and music helps but practice in editing is the only key to success in producing well-edited tapes. The novice should begin his learning with practice in recognizing sounds played back at speeds much lower than the speeds at which the tape was originally recorded. Sound on any magnetic medium, including tape, is made up of magnetic marks recorded at a certain speed, for example, 7.5 inches per second. When we play them back at that speed, the reproduced sound is almost exactly the same as the original, allowing, of course, for any degradation caused by various kinds of distortion. (We still have not attained perfection in recording, regardless what the advertisements tell you!) If sound recorded at one speed is reproduced at a lower speed, the pitch of the sound is lower and the sound is dragged out in articulation. Thus, if a tone at the frequency of 1,000 cycles per second was recorded at the speed of 30 inches per second and played back at 15 ips (inches per second) you would hear it at a frequency of 500 cps (cycles per second). And so on down the scale. At 7.5 ips the sound would be 250 cps and at 3.75 ips 125 cps. It is obvious, that if we are to recognize sounds (for what they are) at abnormally low speeds (for example, the speeds at which we pull tape pass the playback head while looking for a place in which to cut) we will require considerable practice. We have to acquire a sort of unnatural ability to correlate that peculiar-sounding gurgle with the sound that "ng" makes at quarter speed. Never lose sight of the fact that you have to recognize a sound before you can cut in it, before it or after it! Machines have been built (sound spectrographs) with visual graphs of sound, but nothing yet invented, or likely to be invented, can take the place of a trained human brain as far as this kind of aural recognition is concerned.

Practice in acquiring the ability to recognize sounds at low speeds should begin with those easiest to recognize—the fricative sounds of f, v, s and z and with the hard sounds of t, b, p, d, and so on. At very low speeds there is great similarity between the sound of t and the sound of s, for example; the beginning tongue click constitutes almost the only differentiating sound. After you have gained the ability to recognize all the hard consonant sounds, begin on the vowel and diphthong sounds, those of a, e, i, o, u, and their various combinations. Learn what difference there is between sh and z sounds and other sounds that may appear almost the same. After a while, you will find that recognition of sounds will become automatic and you will have little difficulty in recognizing sounds at low editing speeds. It would help, also, if you learned to turn tape reels by hand at low but constant speeds.

Recognizing sound by listening in reverse

Very often, when listening to a compound sound—a "diphthong sound," for example, like "oe" in poem, it is not easy to separate the "o" sound from the "e" sound. However, it can be done rather easily by listening to the tape played in reverse. In normal listening we are most apt to supply actually missing sounds out of our imaginations! On one occasion, to demonstrate to a news broadcaster that he was actually omitting parts of words, I edited a sample of his speech leaving spaces where the part was omitted. He was convinced! But people accustomed to his speech never were convinced—their imaginations had supplied the missing parts of sounds. To get back to reverse listening: when you cannot edit a sound accurately because you cannot hear it accurately, listen in reverse. Then you will find the definite separation between sounds; you will be able to cut "oi" into the "ee" sound of the "i" and then the "aw" sound of the "o." In editing tape by using reverse listening, don't forget that the reverse is true; that if you are editing a speech of a man too lazy to enunciate all the sounds of words, don't try to clip his speech tightly. That would make his recorded and **edited** speech sound most unnatural!

An example of editing by reverse listening is the following: A correspondent in a recording pronounced the word "steamer" as "streamer." the "r" had to be cut out. The tape was recorded at 7.5 ips and that made the cutting even more difficult. After trying to hear the "r" cleanly in normal mode I gave up and reversed the tape. Cutting between the "ee" sound and the "r" sound in reverse was fairly easy. What was more difficult was to separate the softly enunciated "t" from the "r"—but that too was possible in reverse. It is a very valuable technique for difficult editing. When you practise it you will find that sounds like beginning d, f, k, and so on lose their normal hard character and become less definitely voiced than in normal playback. Note also that good loudspeakers at normal levels of sound and a quiet editing room are required if you are to hear any sound clearly without straining your hearing.

Speech Characteristics

Mood

After you have learned to recognize and edit speech sounds of all kinds and in every conceivable combination, you should be ready to undertake editing connected speech. In this category of tape editing there are many things to consider, primary among them what we call "mood." You've heard "mood music"—speech can be even more moody! It is one of the more fascinating aspects of tape editing when we are dealing with spoken human reactions and sensibilities. We learn to sense the "mood" of a series of words. We sense all the emotions in the way words are spoken. It is true the human voice can be the most superb musical instrument ever known. However, it is very easy to ruin a fine expression by careless editing; the mood of the edited version must match that of the original unedited expression. In condensing a speech by editing, one must not jump from a sentence that is spoken in one mood to another sentence—or part of one—that is in an entirely different mood. Even if the edited version makes perfect sense in all other respects, if the moods differ the whole thing will appear ridiculous.

Very few people, as we may notice by listening to recordings, remain in exactly the same mood for any length of time. Moods change, perhaps ever so slightly but still perceptibly, and although these changes are not striking in the original speech, they become immediately noticeable in a crudely edited version where the mood of spoken sounds has not been taken into account while editing. Since the primary reason for tape editing is to improve on the original while maintaining its sense in all respects, the editor must learn to match the mood of the edited to the original speech, thus preserving the real meaning and color of the speech.

Pace (tempo of speech)

Every speaker has his own pace in speaking; it may change according to mood but within each mood, generally, there will be a corresponding pace or tempo. Pace includes the spacing between words or speech sounds as well as that between the words themselves. We must allow for breathing sounds, for "ers" and "ahems" and all the countless vocal tricks that people play with their speech apparatus. There is a simple way to maintain a speaker's pace which is also an efficient way to edit speech. Imitate the speaker's tempo and then **say** the desired edited version to yourself, or out loud, **before** you cut the tape. If you can **say** it easily and in the proper tempo and rhythm, you can then cut the tape to **sound** the same way. Then, cut from sound to sound, always. **Do not** cut in the middle of so-called "quiet" tape unless you simply cannot avoid doing so. This rule is based upon the fact that when you cut from sound to sound you automatically maintain the speaker's natural pace.

For example, we will edit a simple sentence: "John, my oldest brother, is home in bed." We are to cut out the phrase "my oldest brother." How would you say what remains? The answer depends upon the context of the original sentence. If it was very important that the original "is" was accented or inflected, we must retain the complete "is" with its original meaning and accent. Then we would cut the sentence to read "John is home in bed," meaning, perhaps, that he is not roaming around with a gang of thieves! If the "is" is not accented, we would cut directly to the "s" of "is," making the edited version "John's home in bed." A tape editor must retain meaning and, at the same time, create a natural-sounding tape; if it sounds natural it is right, if not, it is completely wrong.

Editing with rhythmic background sound

When there is a repetitive sound in the background of the speech you are editing you must treat it exactly the same as a musical accompaniment and keep it in the same rhythm in the edited version as in the original. If you don't, the listener becomes immediately aware of the discontinuity of rhythm and the impact of your piece is lost. The ticking of a clock, the clicking of railroad car wheels on rail-ends, any repetitively patterned sound in the background should be retained in its original tempo. In some cases this can easily be done, even if it means transposing the rhythmic sounds from one spot to another in the edited tape. In other cases one must resort to rerecording the rhythmic sounds themselves over the edited tape to provide a final authentic version.

Speech habits

You will notice, while using the "say it to yourself" method, that a speaker has speech habits that result in elision of endings of words, or he may slur over certain sounds and not even pronounce others. These speech habits should be retained in editing or the person's character will apparently be changed. Very often, in the editing of radio program material especially, an attempt is made to eliminate coughs, "ahs" and "ers" completely. When this is done (very often it cannot be done because of the slurring together of the "er" and the following beginning vowel sound) the result is speech of an unworldly character, speech that simply cannot be at all natural. Clearly separate hemming and hawing can be cut and doing so, especially in a recorded political speech,

may be applauded as a distinct service to long-suffering listeners. But, in general, do strive for naturalness in editing at all times. Mechanical speech is not human, not natural and certainly not credible.

Inflection

Inflection in pronouncing words always gives them definite meaning which applies only in the context in which those words are used. To illustrate we can use the old story of a man accused in court of stealing a chicken. His denial, interpreted on paper as a confession, was an incredulous **"I stole a chicken?"** It is accent and inflection that give meaning to words, make them musical and worth listening to. Read a speech by a great modern orator, for example Winston Churchill (whose records are available to some extent) and then listen to the same speech. What a world of meaning those words evoke, merely by the way he intones, accents and inflects them! The same words, at various times, may be serious or laugh-provoking, ironic or guileless, bitter or gay. Inflection poses a problem in editing that is almost insurmountable. It is very difficult to use a heavily inflected word or phrase except in its natural context. The only exceptions to this rule are these: a speech may be cut abruptly after a heavily inflected word by: (1). a sudden noise (like a cough or a click). 2. sudden applause, 3. another speaker interrupting. For example let's take the following sentence; "On Wednesday next the ship will dock." We want to edit it to read "**on** Wednesday the ship will dock." Because of the way the original was worded, the last syllable of Wednesday is up in the air, because the speaker knew he was going to say "next" after it. If he had intended to say "the ship" after "Wednesday" he would have inflected "Wednesday" differently. Say it to yourself both ways and you will realize the truth of the above statement. People shape their lips in anticipation of speaking the word to follow. The more you study editing the more you will appreciate this fact. The safest way to edit speech complicated by inflections is to avoid cutting after an inflection **unless** the next word can be voiced naturally. If it **cannot** be voiced naturally, interrupt by the insertion of some short extraneous, but believable, sound like a cough or click. This insertion of slightly shocking (to the hearing only!) sounds will be discussed more fully later on.

Cutting within sound

In normal everyday editing we cut from the beginning sound of one word to the beginning sound of the next word that we plan to retain. For example, in this sentence: 'Editing according to the rules we are following is not difficult," the obvious way to eliminate the qualifying phrase "according to the rules we are following" would be to cut from just before "according" to just before "is." There is another way to edit this sentence; that is to cut in the middle of the "ng" sound in "editing" to the same sound precisely in the middle of the "ng" in "following." This "cutting within sound" is a professional refinement in editing both speech and music that can be very useful. The technique is used frequently, in many cases where a speaker garbles a word and corrects himself abruptly. Or it can be used to correct a bar or two of music by inserting the corrected piece recorded after the original was finished. In using normal editing technique to correct a mispronounced or garbled word that is correctly repeated immediately by the speaker we cannot produce a smooth edit. It will **sound** edited! We are left with a heavily accented word (the correction) but with no indication **why**. By cutting within sound we can edit from the "good" part of the mispronounced word to the unaccented part of the corrected word. We can't do this always, but we can in the majority of cases. An example of the method: "The President returned to Washlig—Washington—by train." If we cut from the middle of the "sh sound in "Washlig" to the same point precisely in "**Wash**ington," we have a perfectly natural "Washington" with normal accent and emphasis. Which is a good reason to use this cutting-within-sound technique wherever possible. An extra dividend comes with use of this method: by cutting within sound one avoids any possible addition of editing noise. Any noise created by the splice is drowned in the sound.

The limits of hearing and how to employ them in editing

Hearing sensitivity

The human ear, when it has not been desensitized by noise or extreme age, is an extremely sensitive apparatus. Most of us, living as we do in noisy environments, can never appreciate how sensitive it is. At its point of greatest sensitivity, at approximately 3,000 cycles per second, it is almost sensitive enough to detect the noises caused by the collision of particles of air as they bump in space. Our hearing declines in sensitivity at both the low- and the high-frequency ends of the audible spectrum, which is ordinarily estimated to extend from 20 cps to upward of 20,000 cps.

Hearing Fatigue—its avoidance

In editing recordings made against changing background sounds we have to make sure that we can clearly hear everything recorded on the tape. In order to do so we must appreciate that we are apparently most sensitive when we make no conscious effort to be so. It seems we see best when we glance at a thing only casually; we think best when we do not strain to think; and we hear best when we do not have to try too hard to hear! The hearing gets tired easily during a long spell of editing and, once it has become affected by hearing fatigue, it can cause us to make errors we would not have made under normal conditions. There are two ways to avoid this incapacitating fatigue: 1. stop and rest frequently. 2. Adapt your speaker amplifier to reproduce sound in a particular way.

Previous to this we described methods of recognizing sounds at low speeds. As you realize, this necessitates learning to recognize the differences between sounds heard at low frequencies. Should both the loudspeaker and your hearing be inefficient at low frequencies, as they generally are at low volume, you will have to strain to hear or else turn up the volume of sound considerably. Either way you will become rapidly fatigued. Besides, when the intensity of sound is raised almost to the point of hurting your hearing, your recognition of pitch changes. The cure, then, is to incorporate into the playback system a low frequency boost. That is half the battle. We still have to hear highs clearly at normal volume in order to tell the difference between a "t," "d" or "s" or similar sounds. This means that a slight high boost is necessary. Thus our loudspeakers and, in turn, our hearing (if room acoustics are average), are furnished with sound that is high at both ends and average in the middle. We have, in making hearing less fatiguing, compensated for deficiencies in both hearing and loudspeakers. It is best to have variable control so that you can find the exact amount of boost to prevent hearing fatigue. This writer, for example, found that only 3 db boost at 50 and 12,000 cps were needed to reduce fatigue onset almost 50 per cent. Remember that this boost is intended for use only in editing—do not employ it for any other purpose. Remember also that excessive bass boost can induce fatigue as quickly as lack of bass boost. If you balance the sound so that it is comfortable for you to listen to over long periods of time it is just right.

It is difficult to describe hearing fatigue because it may cause considerably different effects in different persons. Individual reactions may take varied paths. When my ears and their neural pathways leading to the brain get tired, several things happen: the physical components of hearing seem to tighten up and perception of sound, particularly pitch, becomes affected. Low-pitched sounds then appear flat and strangely distorted. High-pitched sounds appear shrill and ear-piercing. The resonant tones called "wolf tones" cause discomfort, almost pain! Strangest of all, we sometimes will not hear very short duration sounds! If any of these symptoms of overloaded hearing occur, don't try to continue to edit tape, because you will make serious errors in editing. Your hearing may be working at only 50% of its efficiency.

Auditory fatigue is of more than ordinary interest to the student of tape editing because it exaggerates the normal limitations of hearing, which we can take advantage of in editing. Some of the errors of perception present in high degree during auditory fatigue are also present, although less obviously, during normal listening. The point to be aware of, therefore, is this: If you know what these quirks of hearing are, you can use them to good advantage in editing.

Time factor in hearing

One useful quirk of hearing has to do with what we hear as noise and what we hear as musical sound. Sounds that rise gradually and fall away gradually are musical sounds to us and are not disturbing to listen to; sounds that begin or end too abruptly disturb our hearing and we hear these beginnings and endings as noise. It takes us a definite time, varying with individual perception, of course, in which to recognize a sound as musical in nature. If the time is too short for recognition to take place, we hear noise. If the time is **very** much shorter we hear nothing! To apply this observation in tape editing, we use a diagonal cut in splicing tape in order to make sound start and stop gradually. For example, in splicing tape at a speed of 15 ips, a 45-degree angle cut makes the splice 1/60 second long. Were we to cut sound at this same speed of 15 ips at a 90-degree angle there would be two possible reasons to hear noise: 1. as explained previously, the possible accumulation of iron dust in the space between the tape ends being read as noise by the play gap or the difference in voltage between the bias voltage on each tape end, and, 2. the too rapid onset or cessation of sound being heard as noise by the hearing system.

It is quite difficult, in comparing the theories of hearing regarding this point with personal experience, to come to a conclusion as to the specific length of time necessary for recognition of sound as being musical in nature and not noise. However, this writer has found that for all cases in his experience of over 20 years in tape editing a 45-degree angle cut is enough to eliminate noise. That is, at the speed of 15 ips, where the length of time is 1/60th of a second. At 7.5 ips that same splice would take 1/30th of a second, which is even better.

Hearing persistence

The persistence of the sensation of hearing is another of the facts of hearing that has led to the development of a very interesting technique in tape editing. Persistence of hearing is akin to persistence of vision, which has made motion pictures possible, but it persists nowhere near as long a time. We have just learned that it takes us a definite time to hear sound as musical and we know how to use that fact in editing tape. Now we can find out how to use persistence of hearing in tape editing.

Scientists have measured the time that sound persists in our hearing after the stimulus that caused that sound has disappeared. They found that it varied appreciably with the intensity of the stimulus attacking the nervous system of hearing. You yourself have probably experienced extreme persistence of hearing, for instance when suddenly shocked by an extremely loud burst of sound close by. However, persistence in everyday experience is not commonly recognized simply because it is a normal happening and we are accustomed to it. And just because it is a natural experience we must learn to use it naturally in editing tape.

With normal hearing, at sound intensities commonly encountered, the persistence effect lasts but a short time. This writer's experience indicates that it cannot be depended on for much longer than .03 to .06 seconds. This is perfectly normal and, in fact, desirable; music as we know it could not exist if persistence time were much longer. And how difficult to understand anyone if the sounds of his speech persisted in our hearing and blended together in an incomprehensible mixture! Persistence time varies with loudness and with the pitch of the sound. If the sound is more intense, persistence will be longer. The effect on our hearing of persistence may be used in editing, sometimes to avoid rerecording and still get a natural-sounding transition from one tape to another. For example, you have recorded a program and wish to eliminate the commercial or other messages. There is usually, in these programs, a managed round of applause, followed by the message, which in turn is followed by the resuming program. If you cut the tape immediately after an isolated handclap at fairly good level and join it to the first sound of the resuming program material, leaving no space between, the applause will seem to blend into the following sound. In fact, when this operation is properly performed, it will sound as though the applause continued for a split second **underneath** the following sound. This can be called a "persistence-blend." Of course, there are many conditions where it may be difficult to create a natural-sounding persistence-blend. All the factors we discussed previously—mood, pace, level, background—affect the creation of this kind of edited effect. It would not do, obviously, to cut from a laughing to a serious voice, or from an extremely rapid pace to a very slow one of the same speaker. Moods must agree; if laughter, for instance, is to be persistence-blended to a following voice, the latter must be in a jovial mood or be saying something amusing. A good deal of practice in making these blends will show you exactly what can be done by means of persistence-blending and what cannot be done.

Memory obliteration or hearing shock

In the same category as persistence-blending is an editing trick that is useful occasionally. It is useful when there is either no time for dubbing (rerecording) or no equipment available for that purpose, and when there is a noticeable difference between background sounds on two sections of tape that have to be joined together. The insertion between the two tapes of any slightly shocking sound that is in character—a cough, a mike noise, a door slamming—and that has been recorded in the same background as either of the two tapes, will cause the listener to forget the difference in backgrounds and accept the sequence as perfectly natural. The intervening noise has shocked the hearing system into forgetting temporarily everything before the shocking noise, so he does not notice a background change at all.

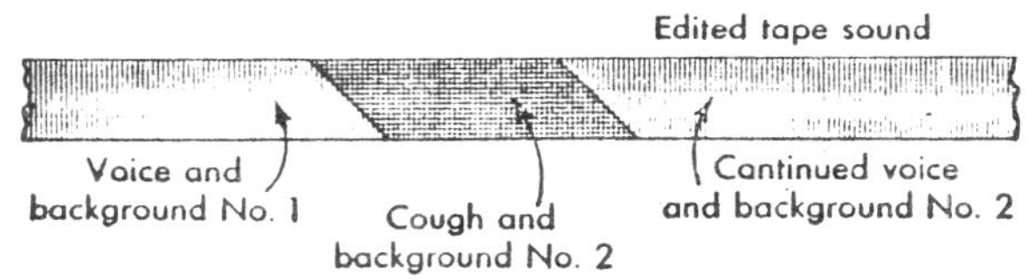

Attention of a listener to a playback is distracted from a change in the background sound when a different, slightly shocking sound, such as a cough, intervenes between the two different background sounds. Tape is edited in the manner shown above when *unnoticeable* shifts in background sound must be made.

In utilizing any of these techniques, do not consider the job well done unless you can listen to the sequence critically without noticing the transition. It must sound natural—as if it happened that way. An imaginative editor, for example, in joining heavy applause to a following voice in a persistence—blend, might cut part way into the first following word, on the assumption that if the sequence had actually happened the way it was edited, part of the first word would have been drowned out by the applause.

Effects of editing too closely

When different sounds are spliced too closely together, a "third" sound will be heard which may puzzle you until you realize what has happened. It may be caused partly by transient vibrations in the physical ear, partly by hearing persistence and partly by lack of recognition time. In any event, this garbled sound can always be eliminated by the insertion, at the splice, of a minimum of 1/60th of a second of tape from the same background of either of the two tape segments spliced together. If you get into the habit of saying to yourself what you intend to edit, you will not have to perform this inserting operation to remove a "third" sound.

Conclusion

The foregoing description of tape editing techniques should be sufficient, if studied and adhered to in practice, to get you started as a tape editor. Of course, there are many effects that cannot be created by tape editing alone. For example, there may be reverberation in a speech you are editing which might cause disturbing effects, like cutting at a reverberated word and having its echo in the background of another word where you don't want it! Situations like this require rerecording and a minimum of three tape recorders plus mixing and other equipment. They cannot be cured simply by editing, which is all this little book pretends to cover. So practice assiduously and, in time, perhaps you will become expert enough to put this book away and create your own book on tape editing!

was narrated, even though the original recording has mistakes and noises on it. Music editing is done for technical reasons, musical reasons, practical reasons, and for formatting a project so the songs will play in the correct order.

You must have a copy of the script to do narration editing.

The easiest and quickest way to record narration is with a two track. Set up the console with a 40 or 50 Hz oscillator tone coming up through one fader; the other faders are used as the narrator's mike inputs. Get a level on all the narrators on buss 1 for the left track of the two track (top track); get a 0 level setup on the tone fader to record on track two (right track). Mute the tone fader and begin.

As you record down the script, eventually someone will make a mistake. At this point un-mute the "tone key" fader for about two seconds, while the artist begins again at a point he or the producer chooses. Wherever the "pickup" re-starts, put a slash mark on your script. For each goof put another slash and record another slate tone on track 2. Four tries at the same sentence (the first three mistakes) and your script will have three slashes at the beginning of the sentence, one for the second try (having blown it the first time), one for the third and one for the successful fourth try.

This is what I call the silent slate method. Explain to the narrators that it is not necessary for you to say, "Take 2," or anything for that matter, nor should they wait for any signals, except from the producer perhaps, before they pickup from the point in the script they deem the best place. The narrator will usually pick the proper point to re-start at least as expertly as the producer.

After the recording is finished, begin editing by re-winding the tape to the beginning. (For editing you want an machine with an edit function where pressing a button puts the machine into play mode without the takeup reel activated -- this allows you to "dump" tape into a waste basket located to the right of the deck.) The editing goes like this:

1. Bring the two track up through the monitor system
2. Head leader the tones
3. Head leader the narration (play the leader and hold a red felt marker on the upper edge of the paper leader to make the program beginning stand out from any other leaders (by turning it red)
4. Now zoom through the recording, listening for a beep from track 2 (the sound of a 40 Hz tone at fast forward speed)
5. At each beep, refer to the script and find what word the narrator started his pickups at and back the tape up to that point
6. Cut the tape at the slashed word and enter the edit mode (on the 440 press edit and play together)
7. Spool the tape onto the floor or a small trash bin placed at the right edge of the machine, until the correct number of beeps passes (one beep per slash)
8. When the last beep passes, cut the tape at the beginning of the next reading at that same word or sound
9. Splice the head and tail pieces together
10. Now zoom through the tape in fast forward until you find the remainder of the slates and do all the necessary narration edits
11. Tail leader the work
12. Play the work from the top for timing (or use a tape timer) and write the running time on the box legend in the place provided (if this pass

is for a smooth wrap as well as a timing, don't stop or fast forward the tape anywhere; do the wrapping at play speed)

When you are preparing to cut different multitrack takes together and the tempos or levels are suspected of being too different, do an exploratory edit first. This is done by bringing the multitrack up through the faders into a reasonably good mix (or use the monitor system's mix). Next make a two track recording from about a minute ahead of the contemplated splice point to just past the splice point. Without changing any fader settings, put on the other multitrack take and record from the just ahead of the contemplated cut point onward for at least a minute. Now edit the TWO TRACK together at the cut point, where you planned to edit the multitrack tape together, and see what you have. If the two track edit sounds O.K., then cutting the multitrack master should work out O.K. too, at least with the type of mix you have going at the time you did the exploratory edit.

A mix done in pieces best illustrates the musical editing technique.

If the tempos sound all right, but some or all of the tape playbacks -- eg. vocals, guitars, etc. -- are switching track numbers on the multitrack machine, then the mix will have to be done in two or more pieces. Mixing in pieces is a relatively harmless project, unless you have large numbers and quantities of effects happening or huge amounts of reverb in the mix. The mix may have to be done in pieces because the instruments change faders the instant the splice goes through and no mixer is fast enough to re EQ, pan, level, reverb, and effect so many new playbacks that quickly. In some cases there are no multitrack edits causing track allocation changes, but the overdubs were done in such a way that the string tracks turn into horns or vocals and then into percussion as the tape progresses.

In all of these instances:

1. Review the dynamics of the tune to see where the real boogie or hot (loud) spot is. Set the levels of the rhythm section, including a rough mix of the lead tracks, for on-scale VU levels at this hot spot. This is so that you won't have to re-adjust the mix when you realize how hot the mix gets farther down the tune. You cannot change the mix master fader to keep the levels on scale because this would make the mix jump in volume at the edit point.
2. Return to the beginning of the song and, without changing the levels of the rhythm tracks greatly, get a final mix.
3. When the beginning is finished, mark the console using a grease pencil, which will indicate the beginning settings. Use different grease pencil colors for updating the mix in future parts of the song, so you can tell where the faders, EQ's, etc., were set at different points in the song, in case you have to start from there again for a re-do. Use different colors for each song section if you want.
4. Begin the two track mix, using the faders to try to "hang on to it" or "run with it" until the mix finally gets out of hand and you have to stop the 24 track and the 2 track.
5. Carefully review the 2 track mix you just finished and decide if it's a "keeper." If it is, re-mark the console in a new color, which you will use as the fader's "jumping off" settings for mix piece #2.
6. Set up the mix for the "out of hand" area, where the tracks

change instruments for example.

7. Re-start the two track, mixing from the cut point in the song, remembering to always start well ahead of the actual cut point to allow the tape players and recorders to get fully up to speed.

8. When you get what sounds like an O.K. piece #2, splice the finished pieces together to see how they sound, especially as regards the splice area. You must re-do the present piece if it isn't satisfactory.

9. Mix all pieces in this way until the end of the song has been reached right down to the end or fade.

10. Now cross your fingers and play the completed mix for the members of the band, who are probably worn out from exhaustion from this particularly harrowing mixing technique -- at least from their point of view.

Remember that if vocals have strong effects like reverb switch tracks, it is best to re-patch the vocal playbacks into the fader where the vocalist started out to preserve pan and effects amounts, rather than trying to patch all the effects around from module to module as you mix the different sections of the tune.

Always try to cut just ahead of transient noises or points where there are "blank" or quiet instants. A good point to cut one musical section into another musical section is on the first of a series of drum hits making up a transition. But really, if you're bold, experienced and patient, you will be able to edit any tape anywhere you want to if you do it carefully. There's no better editing training than making recordings of albums at 15 IPS and then purposely cutting them up, seeing how bizarre sounding you can make them. Several weeks of playing around will convince you that you can edit anything. This attitude (if you have paid the dues to deserve having it) is the only attitude to have when it comes to editing. Practice hard to learn your skills and you will never cease to amaze your clients each time you perform the "impossible" edit.

ROCKET PUNCH INS (AND ERASE OUTS) -- Capstan and Pinch Roller Machines Only

Sometimes it is necessary to erase very short pieces of tape. You may wind up with an unwanted noise just barely ahead of a lead vocal recording, for example. Erasing down to the vocal and then "quickly" punching out will not do. You're sure to erase some of the voice. Ducking a noise with the fader can cause you to fumble a mix which otherwise you would be able to do easily, losing you time and prestige. The answer is the rocket erase out -- the opposite of the rocket punch in. To perform the erase out, thread the tape outside the capstan and press play. The tape being on the wrong side of the capstan will not be pulled along, and you can now get the feel of reeling the tape forwards or backwards by gently turning the reels themselves. Now that you can pull the tape in play, play the tape to the problem area, and using a light touch with the grease pencil, mark the very beginning and ending of the unwanted noise by pulling the tape back and forth to "find" the noise positions. Now carefully de-capstan the tape, place a tube of 2 inch paper leader around the record head stack to keep bias, which can erase highs slightly, from getting to the tape, punch in on the proper track with the erase head over the first punch out mark, and manually "walk" the noise off the tape (to the second mark) by carefully turning the supply and take-up reels. Now punch out. The noise is gone with no danger to the valuable vocal.

Erase outs can be taken a step further -- by recording something new as you do a Rocket Erase Out, you perform the Rocket Punch In. As an example let's replace one of a string of 16th notes of a lead guitar solo, erasing the one "off" note and replacing it with a new "on" note at the same time. First have the guitarist attempt to re-cut the whole solo. If the part with the "odd" note is "impossible" to redo, then get the guitar player to redo just the particular phrase with the wrong note. Use a separate track to record the "replacement" notes.

To check the time placement of the important note, solo the old and new tracks to different speakers and slowly reel the tape back and forth to see that the overdubbed note was played at the proper instant to be able to replace the old note.

Now gently mark the tape with a grease pencil at the null sound area just ahead and just behind the note in question. Estimate where, vertically, on the tape the track you are repairing is located (eg., tracks 12 and 13 are in the middle of the 24 track tape). Now using the editing block cut the first mark (considering the vertical point on the tape you're trying to hit) and insert a leader. Do the same to the second mark. Now that you have the note area segregated by a head and tail leader, set up a transfer from the replacement track to the original track (in sync) and record on the original track the one note "overdub." Reconnect the tape to itself and get on with the business at hand. At 30 IPS very, very short rocket punch ins are possible.

You can also walk the note across from one track to the other by reeling the tape by hand watching the punch in and out marks, but the replacement note may sound funny since it will be recorded at your walk speed (3 3/4?) or so, not at the proper 15 or 30 IPS. Be sure that in your efforts to save time, by not leadering out the note overdub area, the erase head does not erase desired notes while you are busy repairing others.

MUSIC PHRASES AND JARGON

Tempo is the term for the number of beats per minute. It is measured as the number of notes that take place in each minute. The composer places the little note symbol with an equal sign and a number at the upper left of the first page of the music. This number is the number of the indicated notes that should occur in one minute. The inscription (half note) = 36 means that there should be 36 half notes per minute.

Staccato means "stabbed". String players are to play with jabs of their bows. It gives music from stringed instruments a sharp, quick, grating sound.

Legato is the opposite of staccato in that the notes are drawn out and played flowing into one another.

Syncopated means putting strong emphasis on places in the rhythm where the beat isn't (not on 1, 2, 3, 4). Many rhythms are hard to play because they are filled with these strong off beats. This is a prevalent feature in some music like Cajun and R&B. Syncopation is sometimes almost completely lacking in music like disco music, where everything is on, on, on the beat.

Acapella means performed without the piano, but now means really without accompaniment from any music or multitrack playback.

"In the pocket" means a part that fits in with the other parts either present or contemplated for the finished groove. (A groove is a togetherness of musical, rhythmic and melodic contributions from the performers, which transmits a strong rhythmic feeling and emotion to the listener). The expression, "What you're doing isn't in the pocket," is somtimes used to get musicians who are overdubbing or tracking to realize that they may be playing too much or too little or not listening to the effect their part has on parts the other musicians are playing. They should then try to get "into the groove."

A score is the master music page holding the music of all the various players. The parts for the players are generally copied off the score by copyists. A composer can look at a score and "hear" what the group or orchestra will sound like in his head.

Pizzicato refers to the sound of stringed instruments being played by plucking the strings with the fingers. This produces very short plucky sounding passages when done properly.

PARTS OF THE POPULAR SONG

A part is the individual music page for only one instrument. It is copied from the score note for note. There must be a part for each musician. The score used by the composer or conductor has all the parts on it.

The top is the beginning of the particular part of the music being worked on at a given time. An artist will ask you to record from the top meaning the whole song or from the top which would mean the first chorus if that is the part of the song that is being worked on.

A verse is the main part of a song where the story is told. It is the song section where the solo voice(s) sing. This part is intended to be listened to, not memorized or sung along with as the chorus or lyrical hook is. Many times one verse is so powerful, that it is repeated again at the end of the song for emphasis.

A bridge carries the song from one musical section to another, perhaps to the chorus or to a re-intro. Many times if there is a key change in the tune the bridge is an instrumental or sung phrase, which makes the key change interesting or at least not jarring.

The chorus is where a group vocal section sings together, and the listener is supposed to sing along mentally or vocally. This area of the tune has a simple melody and words to allow for at least mental participation for the listener.

The release is totally different from the main musical form of the tune. It relieves any monotony that might occur from having the tune be all one kind of groove or music form. A good example is the "Short people are just the same as you and I" section in Randy Neuman's song "Short People." A bridge is often used to return from a release to the song's major theme or groove.

"Double it" means to do the same overdub again but on a new track so that there will be two similar (possibly exactly similar) tracks of the part being recorded.

"Playing eights" means the instrument is playing eighth notes. Eg: "Go to the section where the bass plays eights."

The turnaround brings the song back from its musical wanderings to a recognizable part of the song itself. It is a bridge from a jam or different sounding "interior area" (to use a Zappaism).

A breakdown is where the entire musical layering drops out, leaving only perhaps the drums or the bass and drums. The energy of the beat and force of the music is thereby made to be "carried" by the single instrument.

INSTRUMENTS AND VOICES

The commonly known voice ranges are soprano, alto, tenor, baritone, and bass. Their ranges overlap so that their voices can pass each other in pitch.

Stringed instruments are divided into two main categories, plectrum and bowed. Plectrum string instruments are played with a pick; bowed instruments are played with a bow. Plectrum instruments include the guitar, harp and harpsichord. The piano is not a stringed instrument by definition; it is a percussion instrument (it's played with little hammers).

The strings include the concert bass, the cello, the viola, and the violin. Country music includes the fiddle -- something of a viola.

Wind instruments include brasses, reeds and flutes.

Brasses include the trumpet, trombone, flugel horn, heralding horns, French horn, and tuba.

The reeds are made up of the single and double reed instruments.

Single reeds include the clarinet and saxophones.

Double reeds include the English horn, oboe, bassoon, and contra bass sarrusophone.

Percussion includes drums, blocks, chimes, cymbals, rattles, bells, shakers, piano, xylophone, marimba, and vibrophone (vibes).

Synthesizers include single note, poly tonal, string, keyboard operated tape players (Chamberlins and Mellotrons are capable of sounding like any instrument or sound that can be recorded), Epiphone (wobbly sound pure tone instrument much used by the Beatles), the newer exotic instruments and hybrids like the "Stick," the guitar/synthesizer couplers and programmable computer driven synthesizers like the Oberheim OBX and Synclavier II. These latest computer operated synthesizers can produce sounds of such complexity that human operators would easily tire of resetting them from one sound to another, since it would mean setting up perhaps 25 or more overtone series along with countless other differences in envelope and decay dynamics.

INSTRUMENT DYNAMIC RANGE

Dynamic range is the volume signature of an instrument displayed graph-style over time.

Each classical instrument (non-electronic) has a distinct dynamic range. A complete orchestra has a dynamic range of from 20 to 100 dB above the threshold of human hearing. Each instrument in the orchestra has its own particular dynamic range. Electric instruments are limited in volume range (dynamic range) only by the number and power capability of the amps and speaker systems available.

In recording it is necessary to know the VU characteristics of different instruments so that you stand a chance of keeping the instrument within the recommended recording volume limits. This VU observation is very difficult because the dynamic range of instruments is usually far beyond the 15 or 20 dB dynamic range of the common VU meter.

The art of VU watching is dedicated to trying to get the maximum amount of level on the tape without pinning the needle. Part of the dynamics of any given instrument may "hide" out of view below the -20 db area on a VU meter. The engineer may want to change the instrument's natural dynamic range because:

1. The instrument is too "radical" in its volume excursions (the levels are too wild to record -- usually caused by amateur players) or

2. The instrument's dynamic range is too great to fit properly into the already recorded dynamic ranges of the song's other instruments.

If the instrument's dynamic range is too great and you do not alter it then you have two unhappy choices. You can let the instrument "peek out" of the mix at its loudest passages and let it be "lost in the mix" for the other times it plays too quietly, or you can push up the quiet places so they're audible and let the loudest passages be much too loud and drown out the rest of the other instruments. Having a mismatch of dynamic ranges is the time to apply compression. I will discuss compression and limiting later.

Mismatch of dynamic ranges sometimes occurs when different types of instruments are recorded together like classical instruments and rock instruments. Rock and roll final mixes sometimes have a "natural" dynamic range of only one or two dB. This would not work out with a classical instrument overdub with a dynamic range of perhaps 90 dB.

Classical music dynamics greatly exceed the available dynamic range of standard professional (no noise reduction) recording. To record classical music quietly one would have to use a Dolby noise reduction system, DBX, or record on digital recorders or straight to disc.

Any compression of the dynamic range should occur at the output of the mike preamp stage of the console, since the the signal-to-noise at this point may be as high as 130 dB (down into mike pre amp noise), whereas after a professional tape recorder gets into the act, the new available dynamic range may be down to perhaps 56 dB (down into tape hiss).

INSTRUMENT OVERTONES AND HARMONICS

Only one kind of wave has no overtones -- the sine wave. Overtones that flavor the quality or timbre of the instrument are the musical outputs which are added to the primary note the musician plays. All waveforms besides pure sine waves have overtones. Overtones are frequencies above the tonic or base frequency. Overtones are expressed as musical intervals. (Intervals are the musical expression of relative pitch where a fifth is five notes from the starting note counting the starting note as "one.") Typical overtone intervals are the second, fourth and sixth harmonics, which produce a sweet ringing or singing sound, and the third, fifth and seventh harmonics which produce a variety of not so pleasing sounds, which sound hollow or choked. Higher overtones give the sound brightness; however, if used in excess the note sounds out of tune with the music. Harmonics are special case overtones with frequencies that are whole number multiples of the main fundamental frequency. A frequency 2.4 times above another is not a harmonic. Only frequencies (for example), 2, 3, 4, etc., times the original frequency are harmonics. Some of the overtones of an instrument are harmonics, others are not.

The simplest way to experiment with overtones is with a Hammond B3 organ. The "drawbars" give the various overtones by simply pulling out the ones desired. All the intervals are marked on the organ.

The effect of overtones and timbre (quality of tone) on the listener is made more complex due to the way the ear works. As the sound gets louder, the ear adds overtones to the ones present in the original sound. In addition, listening at different volumes changes the perceived relative amounts of highs and lows as compared with the mid frequencies, which again changes the overtone, timbre and EQ qualities of the sound. This means that as much as possible, one should monitor at the same volumes for all control room festivities -- tracking, overdubbing, mixing, etc.

THE DIGITAL METRONOME

The digital metronome is like the musical metronome except that the little windows read 100, 10, 1 and 1/8 FRAMES PER BEAT. This means that at a window setting of, say 22, 22 frames of a motion picture would pass through a projector for each beat of the metronome. The window indicates, therefore, the number of frames that are projected per click of the metronome.

This is how to determine the dig met settings:

1. The object is to figure the tempo of a written piece of music to the timing of the given piece of film so that the music duration will exactly match the film duration.

2. First, time the film and multiply the number of seconds of film by 24, the number of frames per second at "film speed" using 35 mm film. This will give you the total number of frames.

3. Now count the bars of music including the repeats and multiply

the number of bars by the number of beats per bar (3 per bar in 3/4 time; 4 per bar in common time). This will give you the total number of beats in the music selection or "cue."

4. Now divide the total number of frames by the total number of beats. This gives you the setting for the dig met window -- frames per beat.

5. Remember that the last window is not tenths; it's eighths. So if you have a remainder in your calculator window multiply this remainder by 8 to get the number to put into the eighths window.

An example:

music - 67 bars at 4/4 time, 12 bars at 3/4 time.
film - 5 minutes and 38 seconds

67 bars times 4 = 268 beats
12 bars times 3 = 36 beats + 268 beats = 304 total beats

5 minutes x 60 seconds/minute = 300 seconds
300 seconds + 38 seconds = 338 seconds total film time
24 frames per second x 338 seconds = 8112 total frames

8112 divided by 338 = 24
Set the dig met window to 24 and use this beat for the music.

Note: Be careful using the digital metronome so as not to let the click, click, click get into your program busses. It is best to patch the clicks directly into the cue system at the last point possible, rather than patching the digital metronome's output up to a fader and then bussing the click to a spare buss, thereby chancing getting the clicks into your music recording busses. If you need to record the clicks, perhaps to use later on to overdub drums, plug the metronome directly into the input of the multitrack master recorder; don't run the dig met into the patch bay. Don't record the click much hotter than -7 VU either. It's not that any individual track might get a lethal amount of click leakage, it's that a very little leakage might get onto many of your tracks, and this will sum up in the mix from all the playbacks into a real audible problem. If this should happen to a product that you wind up mixing, juggle the input module phase switches around during mixing so that some combination of in and out of phase nulls out the clicks. Do not phase reverse stereo tracks to one another as this may get rid of the audio as well as the clicks in a mono mix.

P.S. If the answer had come out to 24.7632 on your calculator you would:

multiply .7632 x 8 = 6.1056
Rounding off would give you a 6 to dial into the eighths window.

You can use the digital metronome as a regular musician's type metronome. To do so procede as follows:

1. Find the notation (note symbol) = 120 (or some other number) at the top of the music chart.

2. Remembering that there are 1440 frames per minute (24 frames per second x 60 seconds per minute), we divide the total number of frames per

minute (1440) by the number of beats per minute (given by the composer's little formula at the top of the page):

1440 / 120 = 12

So you set 12 in the window of the digital met. Remember, if you come out with a decimal fraction (eg 12.67), you must multiply the fraction by 8 to get the last digit. Eg 12.67 should be set as 12 and 5/8 in the digital metronome window since 8 x .67 = 5.36 (round the eighths off to 5).

MUSICAL INSTRUMENT SPECTRUMS

All musical instruments have an inherent musical range of notes that they can play. Above these basic notes ride the overtones, which extend sometimes way above the instrument's pitch range. Pitch is the musician's word for frequency. A frequency spectrum is the picture of the volume across the range of audible frequencies (approximately 20 Hz to 20,000 Hz).

*** Please refer to Illus. B3-4 -- Frequency Distribution ***

A cymbal hit with a drum stick produces mostly high frequencies, with some lower level, low frequency content along with the hissing, sizzling sounds. When experimenting with sounds of vastly different frequencies, it is very helpful to use sound effect filters (made by Langevin) or other steep slope filters because they remove vast areas of the audio above and below the frequencies you set them to, letting you listen to what's left.

An electric organ set to a pure tone (sine wave) produces a one line frequency spectrum graph since its only output is one single note.

Horns have very complicated frequency spectrums due to the nature of the complex waveforms making up their sound. What's more, the frequency spectrum can change radically from one note to another. Some advanced thinkers consider instruments as complex filters and equalizers rather than waveform generators. I prefer this realm of thinking when attempting to describe the vastly different sounds that can come from a given instrument.

The wah-wah pedal is a variable frequency boost EQ that raises a limited range of mid frequencies according to the position of the pedal. This device produces, in effect, a variable frequency spectrum instrument.

It is quite important to take the frequency spectrum of the instruments into acount when deciding which mike you are going to commit to what instrument. If you only have three or four condenser mikes, use them on the instruments with the greatest frequency range. Don't use a super hi fi mike on an electric piano that has mostly mids and lows and no real super highs, use a worthwhile dynamic or even a ribbon (keeping in mind the double sided nature of most ribbon mikes).

On the other hand, you might wish to avoid dealing with excess ultra highs or raspiness from a vocal so you might want to use a lo fi mike so that you would get a natural roll off. A superior solidness to the sound is to be had when you use the mike that gives you the sound you want without using EQ or filtering, rather than using the "wrong" mike and making the sound "right" by using EQ. In the case of using more and more console EQ to try to create brightness, using more can give less.

PREPARATION OF DYNAMIC RANGE AND LEVELS FOR SIGNAL TRANSMISSION

Preparing tapes for use has to be done with some forethought. A classical recording intended only for the finest pressing process and disk production would require a better signal to noise product and greater dynamic range than a master tape for an airline in flight eight track cartridge. In the first case the proposed listening environment would be (hopefully) a quiet living room. In the other, the background noise of a jet in flight would easily drown out any program material below around -10 dB on the console VU meter.

The point is that you would want to greatly compress the dynamic range of the classical recording if it were to be used for the jet, otherwise if the passenger "tuned in" during a soft passage and adjusted the volume for a comfortable listening volume, then the very first "hot" passage would likely raise him out of his seat. If the luckless passenger tuned in during a loud passage, the music would fade away almost immediately, requiring him to turn up the volume bringing on the first case again.

Another reason to reduce the original recorded dynamic range (from the original master tape) is to keep the levels up due to poor signal-to-noise conditions in some future medium (like cassettes or optical film tracks). Since cassettes operate at borderline noise conditions at all times, except when used with Dolby or metal cassette tape, huge dynamic ranges are not usable.

AM radio stations are notorious for their vigor in finding new and improved ways to crush a record's original dynamics to almost no dynamic range. The station does this because each time the final output VU meter at the transmitter drops 3 dB, the total watts of the station goes down by 1/2 (or 100,000 watts to 50,000 watts). If the signal were to go down 10 dB then the volume coming out of the little radios all over town would be half as loud. The stations seem to feel that the listeners choose their station solely by its volume tuning into the loudest station, and from some personal observations I feel that they might be right. There is an additional incentive to keep the level up to, but not over 0 VU, the Federal Communication Commission would raise a big fuss if the station's levels rose above 0, since the audio would then spread out all over the AM dial.

Find out all the intended uses of the program material you will be recording at the beginning of the project. If the use is just for high ambient noise environments, Musak, etc., then you should record with less dynamic range from the very start. If the final uses will be varied -- magnetic film, optical film, cassette duplication and record albums -- then you should record in full dynamic range from the start and then compress the program material for the mixes of the assembled masters that are intended for the high noise environments or noisy mediums.

*** Please refer to the discussion of compression and limiting in Advanced 4 ***

If the music sounds at all noisy during the playbacks of the multitrack master, use noise reducton (assuming that you are recording with good levels). Also, for sessions where the music goes up and down drastically from one selection to the next, the use of noise reduction will greatly speed things up since you won't have to worry about low levels producing noisy recordings.

HEADROOM

Headroom is the distance measured in dB from the level an audio component is designed to operate at (its nominal level) to that higher level where the component will distort. Most all modern recording consoles have a headroom of about 24 dB (above the "0" level indication of the VU meter). Some modern equipment has less headroom though. Each 10 dB of increase is approximately twice as loud as before. So if you start at "0" dB then 24 dB of headroom is really a lot of protection from peaks getting up high enough to distort. There are good reasons for this. One is that with percussive instruments there are peaks that may easily go 18 dB or more above the VU meter reading. Another reason is that if a recording system component does distort in the audio chain, it generates square waves amidst the original program material, which can cause distortion in audio devices "down stream" from the original distorting device.

Another reason that designers do not want their devices distorted by peak program material exceeding the device's headroom is that when a short peak event does distort a solid state device, it doesn't recover the very instant the peak goes away. Instead, there is a finite recovery time needed until the device stops distorting. During this recovery time, which may be only slightly longer than the short duration of the offending peak signal, all signals of high or low frequencies and volumes are distorted. Since the distorted area is now somewhat longer in duration than the original peak audio signal that caused it, and since further processing may distort this signal and again make the distortion area longer in duration, the problem grows worse step by step during the processing of distorted audio. With vocals, this process leads to sibilance (unpleasant sizzling, hissing vocal sounds).

GAIN SETTING FOR NOMINAL LEVELS THROUGH THE CONSOLE

*** Please refer to Illus. I2-1 -- The Console Block Diagram ***

There is only one gain adjustable amplifier in the entire console -- the mike pre amp. The EQ may be set at unity gain (meaning a gain of no dB -- straight through -- no level shift) or for a fixed gain from around -10 dB (a loss or negative gain actually) to around +10 or more. The ACA is best set for unity gain. The program amp (if there is one) is set to make up for any desirable loss that the designer deemed proper for the submasters. In the old days when the mixing networks were passive (resistors only -- without amps), the program amp had to make up for the loss of the mixing system as well as any loss in the sub master. Again the trend is for having no program amp, no submasters, and a console with all the gain "up front," with only unity networis past the mike pre-amp. This makes practically all the patch points past the mike pre out patch at nominal or line level.

Deciding on the settines to allow for proper headroom, but preventing noisy operation, used to be somewhat of an art. To some extent it still is, but modern consoles are generally very quiet even with full headroom available. In addition, the submasters and masters are sometimes missing, making their setting loss calculation unnecessary. It is usually possible to determine the settings of the fader, submaster, and master that the original designer had in mind. You do so like this:

1. Put an oscilloscope equipped with a patch cord input wire at the patch bay along with a volt meter, which reads in dB and also has a patch cord input wire.

2. Plug the oscilloscope into a patch bay mult position along with the voltmeter.

3. Plug two other patch cords into the mult and patch them into the following patch points: mike pre out/EQ in.

4. Patch the console's oscillator into the same numbered mike pre in position so that the test setup is: console oscillator to mike pre in; scope and volt meter to both mike pre out and EQ in, manually closing the normal.

5. Turn the console's oscillator up, watching the waveform on the oscilloscope (get some help setting the oscilloscope up if you don't know how to use one) until the waveform frozen on the oscilloscope begins to distort.

6. Measure the mike preamp output level in dB on the volt meter (always read the directions to be sure you know how to test equipment).

7. This reading is the distortion point, also known as the maximum operating level or the maximum undistorted level of the mike pre amp. Measure several of them so that you can get an average and write the average down.

8. Re plug the twin oscilloscope and voltmeter patches into the EQ out/fader in positions, and the oscillator into the EQ in position, and measure the distortion levels of this console item as you did in steps 5, 6, and 7. Write down the EQ's maximum undistorted level.

9. Procede to the patch points·buss assign switches in/ACA out and measure some of the VCAs to see where they distort. Write down the maximum undistorted levels.

10. Go on to the program amp in/program amp out positions and measure the distortion level of the program amp.

11. Now that you've measured the distortion levels of all the active console components and written them down, it's time to draw up a console block diagram and do some subtraction.

To compute proper settings for the fader, submaster, and master, first subtract the number of dB of headroom from each of the distortion level readings you have just made. This will create a set of nominal level readings for the outputs of each active componennt in the console. Eg: suppose the console mic preamp distorts at +30 and the headroom you want to "run at" is 24 dB, so the operating level would be +6 dB.

Plug the oscillator back into the mike pre in and plug the patches feeding the oscilloscope and volt meter into the pre out / EQ in position. Set up the oscillator so that the mike pre out level is at the level written down in your nominal level notes (+6 in the example). Without changing the level of the oscillator, move the scope / voltmeter patch cords to the ACA out bay and bring the fader up to the point where the mixing system is running at the proper headroom (nominal level) that you determined for the ACA's. Move the scope / voltmeter patches to the program amp output / VU tape in bay and set the submaster to the setting so that the program amp is operating at the nominal level. Lastly turn down the master, if necessary, so that the console meters are reading exactly zero. The console is now operating with the proper setting losses. Mark the proper settings for the faders, submasters, and masters down in your little black note book for future reference.

Remember that with less headroom, you have more possibility of distortion; with more headroom you have more possiility of console noise. All you need for these measurements (supposing you already carry an ohm meter with you) is an oscilloscope, and any studio worth its salt has one that you can borrow.

*** This discussion was extremely brief by intent since this is an audio mixing book. Since no real maintenance background may be in the possession of the reader, expert assistance from a recording studio's technical person may be needed to master the foregoing procedure. ***

VU, PLASMA, AND PPM METERS

Most American studios use VU meters. This is not because they are better, its just what people are use to. It distresses most producers to use different equipment while working on a record project, since they then have to think about what the equipment is doing instead of just thinking about the music.

VU meters are very slow. Their indicator needles have so much mass that a very fast (short timebase) signal is not around long enough to get them up to the proper reading. PPM's (peak program meters), on the other hand, have a storage circuit that stores the peak volts to a constant energy, so that the indicator will have time to display the proper reading. With the little storage circuit sending the indicator its volts, the indicator actually rises all the way up to the proper peak volt reading even if the signal is very loud and of very short duration. This allows the engineer to observe true peak levels. Knowing where the true levels are would enable him to record right up to the distortion point of the console amps and circuits without fear of distorting them. Since he could accurately control (or at least observe) the peak level going to tape, it would enable him to record at very high levels. Peak meters are truly the ones to use since you can do your level setting by direct observation rather than by guesswork as with VUs.

Plasma is the term given to the several different types of console readouts that use LED or other light displays instead of VU meters. There is a distinct advantage to the plasma type displays: if they change color when hitting zero level indications and change color again when indicating "in the red," it is much simpler to detect one of thirty-two buss plasma readouts changing colors from green to red than it is to see one of thirty-two VU meter indicators going into the red. In addition, the plasma readouts can be much narrower than VUs, resulting in a more compact level display system. This compactness greatly reduces the eye strain of scanning many VU meters.

Some console plasma displays have variable weighting. This means that you can "weight" the readings to act like a VU meter or unweight it so that it acts as a plasma or peak reading meter. This can be somewhat of a problem though as an engineer usually gets has a feel for the meaning of the console meter readings, and changing the weighting can cause over or under recording until the difference in response time is realized. Almost all plasma metering systems rise instantly to the level indicated and then fall much more slowly. If it weren't for the gradual nature of the descending indication, a peak signal would rise and fall before you got a chance to see it and you would be back where you started with "invisible" audio not showing up on the volume indicator system.

SIGNAL-TO-NOISE RATIO

To measure console noise, you need a VTVM. VTVM stands for vacuum tube volt meter. This kind of meter is needed for measuring very quiet console devices for their noise levels. A technician will tell you that what you really need is a true RMS reading sensitive volt meter, but a common VTVM will do. The noise levels of console devices fall far below the readings available on a VOM -- volt/ohm meter.

Take the maximum undistorted level readings that you obtained during "Gain Setting for Nominal Levels Through the Console" and proceed as follows:

1. To measure the noise of the mike pre amp plug a 10 dB in line mike pad into the mike input jack (XLR) in the studio wall to terminate the mike input circuit.

2. Put the mike pre on any of its gain settings that you normally use for sessions (really you should make a measurement at all the mike pre gain settings for your records to find out which is the quietest one).

3. Plug the console oscillator into the first bay below the mike out bay where you have the 10 dB pad plugged in at the studio's mike input panel, and adjust the fader and oscillator so that the fader is at its proper

setting loss and the buss is reading zero.

4. Measure the level that the oscillator is putting out while it is still connected to the mike input circuit. To do this use the mult connections given in the section entitled, "Gain Settings For Nominal Levels Through the Console."

5. Measure the output level of the mike pre amp.

6. Subtracting the output level of step 5 from the input level of step 4 gives the gain of the pre amp.

7. Now disconnect the oscillator from the input of the mike pre amp circuit, and let the mike pad in the studio be the input to the mike pad and pre, etc. Make sure that no mikes are plugged into the mike circuit that you are using via other parallel mike input panels in the studio.

8. Now crank up the VTVM until the needle reads on the scale -- this is the noise of the mike circuit, which should be a minus (-) dB reading like -85 dB.

9. Add the absolute value (absolute value means to drop the + or - sign) of the noise at the output of the circuit (like -85 dB) to the gain of the circuit (calculated in step number 6 -- let's say a gain of 40 dB was calculated) and you will have the approximate equivalent input noise of the mike pre: 85 + 40 = 125 dB Approx. EIN (Equalent Input Noise).

*** Please refer to Illus. I5-1 -- Equivalent Input Noise ***

Make a note of the reading and continue measuring the rest of the console components. Remember to use the double patch method for connecting any measuring devices up to the console circuits, since this preserves the console's natural loading scheme and will give results for that console as built and operated rather than laboratory results as may be had by measuring isolated console components in the shop under more favorable conditions.

Almost all VTVM's have output jacks that, if patched into a monitor system input (usually line in) during EIN measurements, reveal what the noise being measured sounds like. Console noise should sound like bright hiss with little or no hum, radio stations, clicks, pops, buzzes or whining sounds.

PHASE DISTORTION

Phase distortion is what happens to music (audio) when it is EQ'ed too much. All all equalizers work by changing the phase of the audio but the EQ changes the shape of the audio waveforms as well as the timbre or quality of sound. Waveforms that sound good in a blend in the air like a vocal group may not blend well after being recorded if too much EQ is used on the individual mike input channels. Needing excessive EQ to get a "good sound" is caused directly by the choice of the wrong mike. Mike choice is the key, therefore, to not introducing console equalizer phase distortion.

Phase distortion also results from making recordings on magnetic recorders that do not have coherent phase record circuits. Coherent phase recording is available on only a few machines -- now including some mag machines (35 millimeter movie sound track recorders) and the ATR series of Ampex recorders. Magnetic recorder phase distortion coupled with recording into tape saturation tends to print the highs ahead of the lows so that with each subsequent recording the highs are farther and farther in front of where they should be on the tape. If sibilance (bright sounds recorded too hot

Equivalent Input Noise

Traditionally, the equivalent input noise (EIN) of an amplifier is determined indirectly by measuring the output noise in dBm, and subtracting from this figure the transducer gain in dB. It is essential that a filter of known bandwidth be employed in this measurement since noise is a function of bandwidth, and the amplifier under test may have a passband which considerably exceeds the audio spectrum. A noise bandwidth of 20 kHz may be obtained by the use of a theoretically perfect filter which cuts off sharply at 20kHz. However, it can be demonstrated that the equivalent noise bandwidth is obtained with a 6 dB per octave filter which is down 3 dB at 12.74 kHz.

Since source resistance noise may be an important factor in the determination of EIN, the input resistance and loading effect of the amplifier must be taken into account. For this reason, noise measurements are based upon transducer gain as given by the equation:

$$Gt = 10 \log(Rs/Rl) + 20 \log(V\,out/Vs) + 6dB$$

Where:
- Rs = source resistance
- Rl = load resistance
- Vs = source voltage
- V out = output voltage across the load resistance

In this equation, the first term is concerned with the loading effect of terminations at source and load. This term drops out of the equation when source and load resistances are equal. The second term is the input to output voltage gain, and the third term reflects the 6dB effective gain achieved by bridging amplifiers such as the Automated Processes Model 312.

The equivalent input noise of the Model 312 amplifier is specified as -129dBm with a noise bandwidth of 20kHz. This figure is obtained in the test configuration shown below.

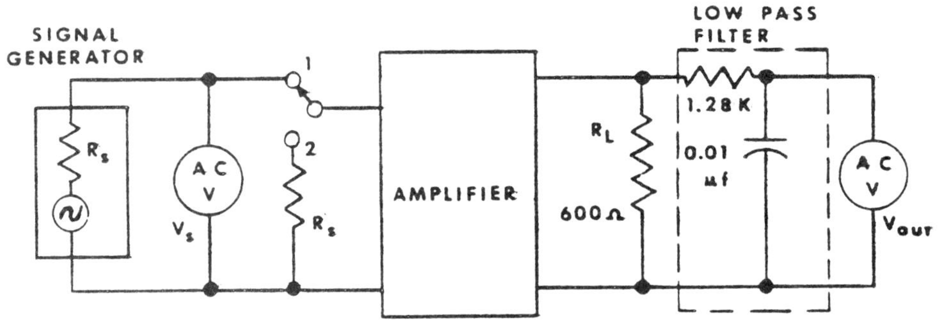

The amplifier is connected for maximum gain in accordance with the instruction sheets furnished. The Model 312 is most frequently used in conjunction with microphones and other low impedance devices, so it is tested in the 150Ω input configuration. Therefore, Rs = Rs' = 150Ω. However, similar results will be obtained if the input is strapped for 600Ω, and it is fed from a 600Ω source.

The voltage gain indicated in the instruction sheets is confirmed with the switch set in position 1, as shown. The switch is then set to position 2 and output noise is measured in dBm.

Although true RMS meters should be used for these measurements, conventional average reading instruments will produce only a 1dB error. The equivalent input noise is obtained by subtracting the transducer gain from the measured output noise.

Thus: EIN = Measured Output Noise - Gt

Typical EIN measurements are as follows:

$$Gt = 10 \log \left(\frac{150}{600}\right) + 66dB + 6dB$$
$$= -6dB + 66dB + 6dB$$
$$= 66dB$$
$$EIN = -63dBm - 66dB$$
$$= -129dBm$$

Amplifier gain may be reduced by as much as 40dB by means of an external gain resistor, as indicated in the instruction sheets. Although this resistor does contribute to the total noise within the amplifier, the gain reduction far exceeds the added resistor noise, so that output noise will always decrease as the gain is reduced.

causing tape saturation) occurs during magnetic recording, the tape phase distortion makes the sibilance even worse from generation to generation.

GENERATION LOSS OF TAPE RECORDINGS

Each time a tape is re-recorded, something is lost. What is lost depends to a great extent on what there was in the first place (so there is something to lose), the quality of the tape machines, the accuracy of their audio alignment, the quality of the tape you re-record the audio on, the physical size of the tracks you are recording on, the condition of the tape recorder heads, and the accuracy of the physical alignment of the heads to the tape on the play and record machine.

As a commonly recorded cassette has very little real highs, almost no brightness will be lost in copying a stereo cassette to a professional two track, but in making a cassette copy of your finished album, for example, from an ATR two track to a common cassette, you will lose a lot.

All tape copies will suffer continued phase distortion except for those few machines that have constant phase (or coherent phase) recording circuits. All tape copies will have more tape noise except for noise reduced tapes that are copied without decoding the audio (tape copy made of an original tape made with the Dolby or other noise reduction system), in which case almost no addition of tape noise is to be expected. All tape copies will usually sound slightly duller.

When not using phase coherent recorders, there is only one way to make a safety copy in which the audio is not further phase distorted or hi frequency displaced and that is to make the copy in the reverse direction. This of course means that the copy will have to be listened through twice, once while it is recording backwards, and again when the copy is played so that you can check for dropouts, wow and flutter, and other tape recording difficulties. Copying in the reverse direction is the only way I know of to actually reduce or eliminate sibilance. EQing or de-essing will do away with sibilance to some extent but will change other sounds you didn't want to change. You will get slightly more tape hiss added to your recording, of course.

In general, you should expect slightly more highs to be lost than usual during the making of a 7 1/2 ips tape copy due to the saturation of the highs, so a small amount of high frequency boost (2 dB @ 10 khz) is a good idea. You should record at a lower level than you do at 15 ips by about 4 dB. Remember that the lower the tape speed, the quicker the high frequencies will enter tape saturation. To find out where a 7 1/2 ips tape copy enters into saturation, make a trial copy while driving the level up and up until you hear the tape copy begin to sound dull. This is where saturation begins. For that matter drive the levels up even more and study the saturation sound carefully so that you will always recognize it when you hear it.

There are other possibilities for lessening the quality when you make copies from one speed to another. Making a 30 ips copy of a 15 ips tape is likely to add less noise (or come up with a copy that doesn't sound that much more noisy) because tape noise at 30 ips is higher in pitch, less noticeable and at a different frequency than the noise of 15 ips recordings -- they don't add together as much as making a 15 ips copy of a 15 ips master would. Making a 15 ips copy of a 30 ips tape is another story. In this case the tape noise being added is much more noticeable than the tape noise you started out with (at 30 ips), plus the hi energy high frequency audio will not print as readily at 15 ips as it did at 30. So the highs will seem to be lessened, as well as the tape hiss sounding greater. This makes for a less satisfactory copy in almost all cases than making a copy at a higher or equal speed. Copying 15 ips to 30 ips you suffer a loss of low frequency content; copying 30 ips to 15 ips adds hiss and loses some highs.

If the heads on the play machine are slightly out of height, when you turn the master tape upside down to make the transfer, the tape hiss of the copy will go up since the master tape tracks are now playing slightly off the play heads, which means they're picking up some unrecorded tape (hiss). To check to see if the heads are recording or playing the tape dead center (also useful to know if you plan to travel to another country to overdub strings or something), record some 1 KHz and dip the tape in Magna See. Use the Ampex head spacing guide and a jewelers loupe (graduated in mils or 1/1000ths of an inch) to see if the tape is recording at the proper height. To see if the play head is centered, play and adjust the play level controls using a 1 KHz tone so that all channels play at exactly 0 on the VU meter, then turn the tape upside down. If the playback levels are within 1/4 dB from the levels you had with the tape right side up, then all is well with the play head height -- otherwise, not.

A problem that is rarely discussed when recording at 30 ips is that of the missing low end. 30 ips tapes are usually down about 3 dB at 50 Hz and about 6 dB at 30 Hz. If you make a copy of the original 30 ips tape again at 30 ips, the new copy will be down at least 10 to 12 dB at 30 hz. Only the use of a special EQ (that you would have to build yourself) would prevent this loss. Those who desire to make low frequency accurate recordings and identical copies should record at 15 ips and, to get rid of noise, use Dolby noise reduction or DBX. Those who want sterling hi frequency response should record at 30 ips and make due with the low frequency loss.

DYNAMIC RANGE

The dynamic range of magnetic recordings is on the order of 56 dB without noise reduction. Using a Dolby system, the dynamic range is about 66 to 71 dB; with DBX, about 76 to 86 dB. Using digital recording methods, 90 dB and more is available. Cassettes have a dynamic range much less than professional recorders even though some selected machine/tape combinations have a dynamic range that is nearly as good (especially when using metal tape).

When noise reduction is included in the recording chain the dynamic range in increased by a factor that is widely published by the various manufacturers. Not so widely published is the increased difficulties caused by different biasing methods, record levels and tape dropouts. For DBX only 1 KHz tone should be used for alignment.

Use noise reduction when:

1. You want to take advantage of the added dynamic range so that you will not have to push up the levels of many recording faders (trying to keep them in a good mix as you do) for recording soft music recordings, you only have to bring them down again for the next recording, which is at loud volumes.

2. You do not want to use compression until the "whole thing" is recorded, leaving as an option the dynamic range of the final product. Noise reduction lets you "record now -- compress later" without bringing up lots of tape hiss.

3. You may have to supply many "dubs" (duplicating masters) to various users throughout the world, who will use your dubs to make yet other dubs, which they will use to mass produce cassettes or cartridges, so you do not want to take the added noise build-up during these other intermediate copies without having noise reduction all along the way except for the final customer copy. Consumer Dolby noise reduction systems are known as B type Dolby and are not nearly so severe in their action as professional (A Dolby) processing.

4. You want to make a large dynamic range recording of an orchestra whose dynamic range exceeds that of recording without noise reduction.

THE NAB AND AES RECORDING EQ (pre and de-emphasis)

*** Please refer to Illus. I6-1 -- Ampex Tape Recording Pre-emphasis Notes ***

If recordings were made with no special recording EQ, the resulting recordings would sound dull and have much tape hiss. Since there is a lot of tape hiss at high frequencies,and very little at low frequencies, the high frequency audio is boosted as it is recorded. This keeps the high frequencies well above the level of the tape hiss on the tape. In order to play back a pre-emphasized recording you have to de-emphasize it first. This is the job of the reproduce EQ. The reproduce EQ reduces the extra amounts of highs on tape due to the record EQ. Thus, the de-emphasis reduces the highs to their proper proportion but as it does this it also reduces all the high frequency tape hiss as well. For each dB that the highs are boosted during recording, one dB of tape hiss is removed during playback due to the

	For 15 IPS and 7 1/2 IPS	For 3 3/4 IPS and 1 7/8 IPS
Frequency (HZ)	Equalization (db)	Equalization (db)
20	-8.6	-8.8
40	-4.1	-4.2
60	-2.3	-2.4
80	-1.4	-1.6
100	-1.0	-1.1
200	-0.2	-0.4
400	±0.0	±0.0
600	+0.1	+0.3
800	+0.2	+0.6
1000	+0.4	+1.0
2K	+1.4	+3.4
4K	+4.1	+7.7
6K	+6.6	+10.8
8K	+8.6	+13.2
10K	+10.4	+15.0
12K	+11.8	+16.6
14K	+13.1	+17.9
16K	+14.2	+19.6
18K	+15.2	+20.0
20K	+16.1	+21.0

action of the reproduce EQ.

The amount of boost is increased for slower tape speeds and decreased for higher tape speeds. The EQ for 15 and 7 1/2 ips is the same however. Please note that in the Ampex pre-emphasis document, that the boost for 20 KHz is 21 dB at 3 3/4 and 1 7/8 which is cassette speed. 21 dB is a lot of boost and will greatly reduce the amounts of high frequency sounds in audio program material the tape can withstand without distorting (saturating) the tape. Notice too that the low frequency information is recorded lower in level. In fact, the lowest lows are printed at -8.6 dB. This means that during pre-emphasis, highs are boosted and lows are cut; and during de-emphasis, lows are boosted and highs are cut. The de-emphasis (reproduce) curve is identical to and the opposite of the pre-emphasis curve -- just change the plus and minus signs of the pre-emphasis curve to get the de-emphasis curve.

TAPE SPEEDS

There is a good tape speed for every job. For storing lots of tape in a small space there's 7 1/2 ips for music and 3 3/4 ips for speech. For convenience (not having to thread a machine or put leaders in the tape for the client) there's cassettes. For general recording at highest quality including all the lows, there's 15 ips. For capturing the high end of the audio spectrum or for especially bright or hot recording (missing a little of the lowest lows) and for real editing power there's 30 ips. What more could an engineer want?

Remember that of the two professional recording speeds, 15 and 30 ips, 30 ips uses twice as much tape, and twice as many reels, which are twice as heavy to carry and take up twice as much shelf space. It is also twice as easy to make "astonishing" hair line edits. 30 ips, of course, doubles the studio's tape sales.

15 ips uses half as much of everything that 30 ips does and in addition allows you to record right down to the lowest lows.

TRANSPORT TYPES

*** Please refer to Illus. I6-2 -- The Mincom M79 Transport Excerpt ***

On page 3-6 of the Mincom M-79 document is Figure 3-3. This type of transport is known as the closed-loop. The closed-loop means that the tape is not subject to the changing tensions of the supply and take-up reels to maintain proper tape speed and tape/head contact. In the closed-loop decks, the reel's motors just keep the tape safely wound on the reels, nothing more.

Starting at the entry point for the tape, we come across the incoming tape guides. These guides should be monitored for signs of tape shedding -- a first clue that you might have a roll of bad tape or that the tape has been cut too wide. The incoming capstan pinch roller presses the tape against the capstan, which pushes it towards the take-up reel. The outgoing pinch roller pinches the tape against the capstan and pulls the tape out of the head area.

The trick to keeping good tension is that the capstan is not the same

SECTION III
OPERATION

3-1. GENERAL

3-2. The 3M Brand Series 79 Recorder may be operated at the console, or up to 30 feet from the console using the remote control assembly provided with the 8, 16, and 24 track recorders. The tape motion controls are conveniently grouped on the tape transport and remote control assembly so as to provide a minimum of movement by the operator. If the optional remote transport control is used with 2 and 4 track recorders, the tape motion controls operate the same as in the 8, 16, and 24 track remote control. All controls and indicators are identified in figures 3-1 and 3-2 with a complete description of each control function tabulated in table 3-1. A study of Section V, Technical Description, is useful for understanding the operation of the controls.

NOTE

For consistently good recordings, the heads, guides, reversing idler, and capstan should be cleaned frequently to remove dust and oxide deposits. The heads should also be degaussed for optimum performance with the power switched off.

3-3 TAPE THREADING

3-4. Threading the recorder is extremely simple, as there are no compliance arms or other mechanical devices in the tape path. Also, there is no tension on the tape until the tape breaks the light path of the photoelectric circuit in the Isoloop assembly and the STOP button is pressed. The use of Scotch Brand Dynarange recording tape, Type 206 or 207, is recommended, although adjustable bias and equalization permits accommodation of a wide range of tape characteristics. To thread the tape, refer to figure 3-3 and proceed as follows:

1. Set the CAPSTAN SPEED switch to HIGH, LOW, VAR, or EXT as desired.

2. Set the TAPE TENSION switch to the HIGH position when using 2 inch and 1 inch tape, or the LOW position when using ½ inch, or ¼ inch tape. LOW is also used if thin tape or small reels are being used.

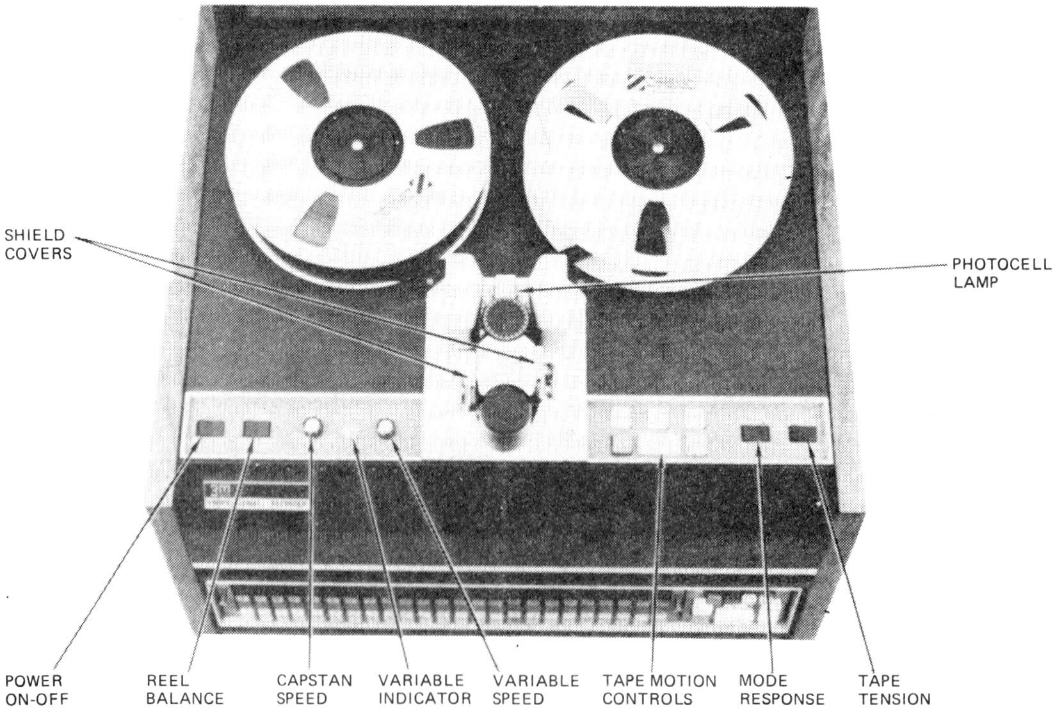

Figure 3-1. Operating Controls

Figure 3-2. Remote Control Assembly

Table 3-1. Operating Controls and Indicators

CONTROL/INDICATOR	FUNCTION
TAPE TRANSPORT	
POWER OFF-ON	Controls ac power to the tape transport and the signal electronics assembly. Press to ON to energize. Power on is indicated by illumination of the VU meter lamps and the end-of-tape sensor lamp; the head shield covers also open.
REEL BALANCE	A three position switch compensates the reel drive motor drivers for difference in reel inertia such as 7 inch reels on one side and 10 inch reels on the other side, and beginning versus end of reel. Normal position is the center position. With a small reel on the right, press the switch RIGHT. With a small reel on the left, press the switch LEFT.
CAPSTAN SPEED	A five position rotary switch selects capstan speed. (3 position for 2 and 4 track recorders.)
VAR LOW VAR HI	The tape speed can be varied between 5 and 45 ips by rotating the VARIABLE SPEED control. A VARIABLE INDICATOR lights when the switch is set to VAR LOW or VAR HI. Low speed equalizer is selected in the LOW position, high speed equalizers in the HI position. A master recording would not normally be made if this indicator is lighted.
LOW	Selects the lower of two fixed tape speeds (7-½ or 15 ips), and the corresponding equalizers.
HIGH	Selects the higher of two fixed tape speeds (15 or 30) and the corresponding equalizers.
EXT	Selects an external input plugged into remote connector J3. The input may be a variable resistor, a voltage source, or a synchronizer input.

Table 3-1. Operating Controls and Indicators (Cont.)

CONTROL/INDICATOR	FUNCTION
TAPE TRANSPORT (Cont.)	
	NOTE The pushbutton switches that backlight when pressed, extinguish when another backlighted pushbutton is pressed.
STOP	Backlights when pressed to place the transport in a standby condition. Stops tape motion from any mode of operation. The switch must be pressed and backlighted to reset after any occurrence of tape runout or power interruption.
PLAY	Backlights when pressed and tape moves at the selected speed. Pressed to reproduce or record.
RECORD	Backlights when pressed simultaneously with the PLAY pushbutton. The record mode is initiated if the remote control master signal electronics RECORD pushbutton is pressed. (Record can also be initiated by the remote control RECORD pushbutton.)
FORWARD	Backlights when pressed, and moves the tape at a rapid speed onto the take-up reel.
REWIND	Backlights when pressed, and rewinds the tape at a rapid speed onto the supply reel.
MUTE DEFEAT *	Allows audible signal in fast forward or rewind when switch is pressed to on and backlights. Press for off.
EDIT *	Backlights when pressed, and disables the take-up reel so that tape can be spilled. Press for on, press for off switch. This mode can only be initiated from STOP or PLAY.
MODE RESPONSE	A three position switch affects the reeling velocity rate of change when changing from one reeling mode to another, or changing from a reeling mode to a stop or play mode.
1	Fast or maximum rate of change permitted by the transport.
2	Fast except for a period of time immediately following a change from FORWARD to REWIND or REWIND to FORWARD.
3	Slow rate of change.

*The EDIT switch is used in the 2 and 4 track recorders instead of MUTE DEFEAT which is used in 8, 16, and 24 track recorders.

Table 3-1. Operating Controls and Indicators (Cont.)

CONTROL/INDICATOR	FUNCTION
TAPE TRANSPORT (Cont.)	
TAPE TENSION	A two position switch affects the reel motor torque, and by making internal adjustments, may be used to accommodate difference in tape width, tape thickness, or reel size.
HIGH	Normally used for widest or thickest tape, and large reel size.
LOW	Normally used for narrowest or thinnest tape, and smaller reel size.
REMOTE CONTROL ASSEMBLY (Mode Control Pushbuttons & Tape Lifter)	
NOTE Remote Control Assembly is used in 8, 16, and 24 track recorders.	
STOP, PLAY, RECORD, FORWARD, REWIND	Same as the transport switches except the STOP switch will not reset the logic to standby after a fail-safe condition.
Tape Lifter (Unlabeled)	The unlabeled switch located above the STOP pushbutton is a three position momentary switch normally set to the center (off) position. In forward or rewind, when pressed to the right, the tape lifter solenoid is defeated which allows the tape to be monitored during the spooling modes. See note under paragraph 5-48.
REMOTE CONTROL (Master Signal Electronics)	
NOTE 2 and 4 track recorder Master Signal Electronics switches are not remote, but are located in the center of the control panel as shown in figure 1-2.	
RECORD (Red)	Backlights when pressed, and allows the record command to be transferred to the individual channel record switch when received from the transport or remote RECORD button.
CUE (Green)	Lights to indicate the command to cue (overdub) playback is being presented to the individual channel function switch. The cue lamp lights when the CUE switch is pressed in play or stop and when the master signal electronics RECORD switch is lighted.
IN (White)	When pressed, will command all channels to monitor the record input signals unless the OUT switch is pressed. A RUNOUT indicator (unassociated with the IN function) is displayed in the switch assembly.
RUNOUT	The RUNOUT indicator lights when tape runs out or breaks, or if tape is not threaded.

Table 3-1. Operating Controls and Indicators (Cont.)

CONTROL/INDICATOR	FUNCTION
REMOTE CONTROL (Master Signal Electronics (Cont.)	
OUT (Amber)	When pressed, overrides the IN button and will command all channels to monitor the playback output signals.
REMOTE CONTROL (Function Switch Assembly)	
	NOTE The function switch assembly is not removable in 2 and four track recorders. Each track contains four switches numbered by channel. The red, green, white, and amber functions correspond to the red, green, white, and amber switches of the master signal electronics switches.
Record, Red (2 through 8, 16, or 24)	When pressed, will route the record command to the appropriate signal electronics channel if the master signal electronics RECORD and transport RECORD switches have been pressed. Record ready is indicated by one half intensity of the indicator (alternating between one half and full intensity of the indicator if wired in some systems). Normal record is indicated by full intensity illumination. A malfunction is indicated by a flashing of the indicator for that selected channel.
Cue, Green (2 through 8, 16, or 24)	When depressed and illuminated at full intensity, a cue (overdub) condition exists on that channel. One half intensity illumination indicates that the switch has been pressed but cue has not been activated.
In, White (2 through 8, 16, or 24)	When pressed and illuminated, but not locked to detent, indicates that that particular channel record input is monitored. When locked in detent position, the channel is locked to record input.
Out, Amber (2 through 8, 16, or 24)	When pressed and illuminated, but not locked to detent, indicates that that particular channel is monitored. When locked, the channel is locked to the playback output.
	NOTE Master IN (white) and OUT (amber) override individual white or amber. If the master IN or OUT, or none of the white or amber are pressed, the channels will automatically transfer to white when recording and to amber when playing back.

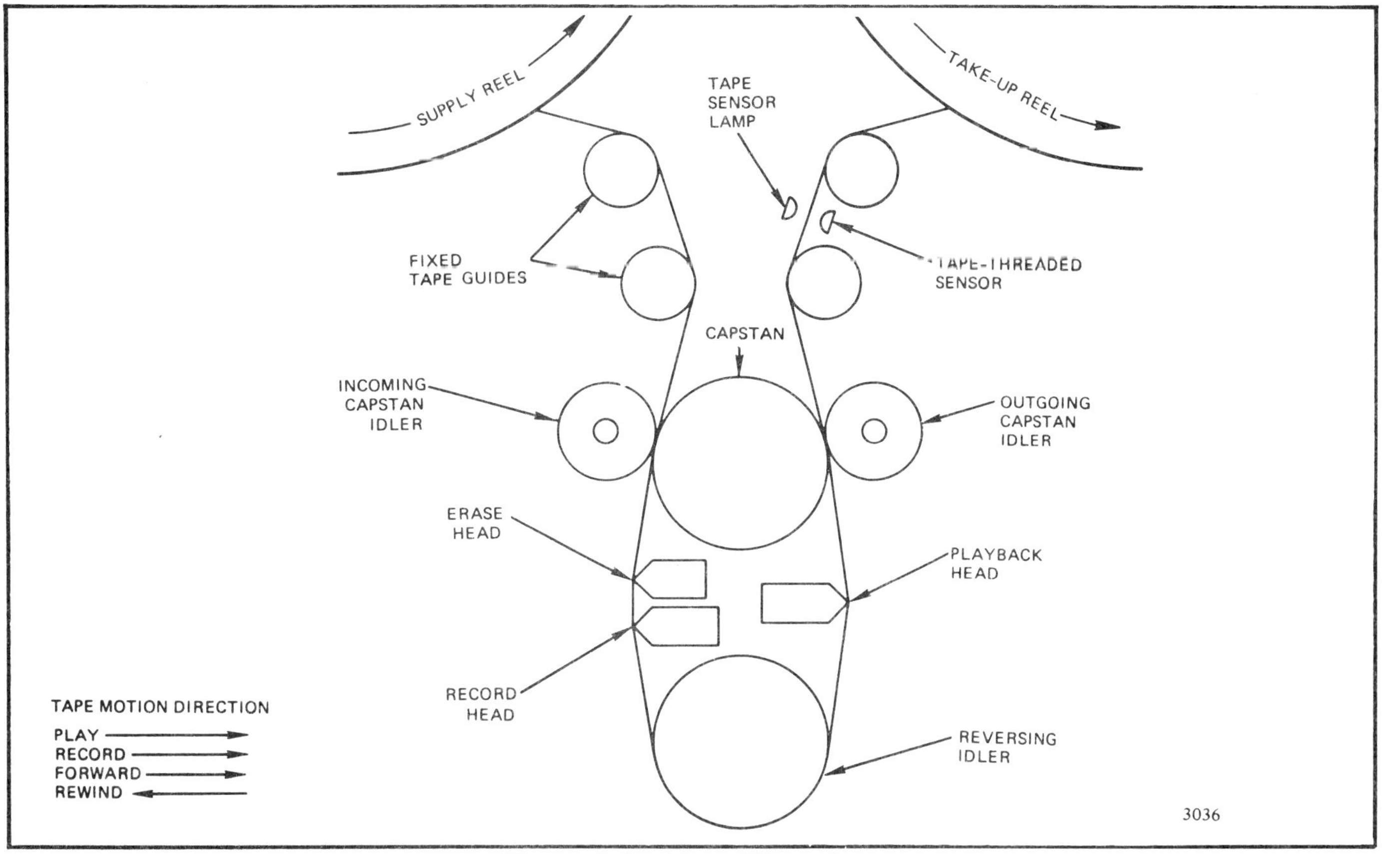

Figure 3-3. Tape Threading

3. Press the POWER to ON. The panel meters lamps light, the tape sensor lamp lights, and the head shield covers open. (Normally, the numbered amber buttons of the remote control will be illuminated. If the white buttons are illuminated, press master amber button and the amber buttons will light.)

4. Place a reel of tape on the left-hand spindle in such a manner that the reel rotates counterclockwise when tape is unwound. Unwind about two feet of tape from the supply reel and drop the tape on the inside of the incoming tape guides, between the capstan idler and the capstan, past the erase and record head, around the reversing idler, past the play head, between the right-hand capstan idler and the capstan, past the outgoing tape guides on to the take-up reel. Thread the tape on the take-up reel hub so that the tape will wind in a counterclockwise direction. As the tape breaks the photoelectric light path at the exiting tape guides, a click can be heard which is the cut out relay deenergizing.

5. Press the STOP button to apply torque to the reel motors which takes up any slack in the tape threaded through the Isoloop, the STOP button should also illuminate at this time indicating that the transport is in standby and ready to operate.

3-5. TRANSPORT TAPE MOTION OPERATION

3-6. Upon initial operation of the transport, observe each mode of operation to gain familiarity with the mechanical actions of the transport as follows:

NOTE

Either the transport or remote control buttons (STOP, PLAY, RECORD, FORWARD, or REWIND) can be used to control tape motion.

1. Press the POWER to ON. The panel meters lamps light, the tape sensor lamp lights, and the head shield covers open.

2. Set the CAPSTAN SPEED, REEL BALANCE, TAPE TENSION, and MODE RESPONSE switches to the desired positions.

3. With tape threaded, press the transport STOP button to light the stop lamp.

NOTE

The tape motion controls may be pressed in any sequence at any time with complete safety to the tape and transport. When the transport is operating in the forward or rewind modes, tape motion has to be stopped before RECORD is initiated at the transport or remote control.

4. Press the PLAY button. Observe that the capstan starts, the head shield covers close, the capstan idlers pull in, and tape starts to move in the forward direction of the selected speed.

5. Press the STOP button. Observe that the capstan idlers release tape tension, the capstan stops, the head shield cover drops back, and the tape comes to a smooth stop.

6. Simultaneously press the PLAY and RECORD buttons. Observe that the PLAY and RECORD buttons both illuminate and that the transport mechanical functions are the same as in step 3. Press the STOP button and allow the tape to stop.

7 Press the FORWARD button. Observe that the tape lifters lift the tape away from the heads, the head shield covers remain open, and the tape rapidly accelerates in the forward direction.

8. Press the STOP button. The tape motion will come to a smooth rapid stop, and the tape lifters will drop back.

9. Press the REWIND button. Observe that the transport mechanical functions are the same as step 7 except the motion will be in the rewind direction.

10. Press the STOP button and allow the tape to stop as in step 8.

11. Observe that in standby (STOP button illuminated), the head shield covers will remain open or closed when operated manually. Note that the head shield covers, if closed in the standby mode, will open when the tape is lifted from the running path between the right-hand tape guides.

3-7. MONITORING

3-8. Signals that are to be recorded or reproduced are monitored on the meter panel above the transport. External monitoring equipment can be connected to the output jacks at the rear of the electronics assembly. Master switches IN and OUT and individual numbered channel white and amber pushbutton switches control the input and output signals, respectively.

3-9. IN/OUT MONITOR SWITCHING. The IN and OUT pushbutton switches are located on the remote control assembly. Pushbuttons IN (white) and OUT (amber) provide simultaneous switching of the output and meter monitoring circuits of all channels to either the input signal being recorded on the reproduced signal from the recorded tape. Monitoring is automatic if none of the switches are pressed. The IN button selects the record input signal, the OUT button the reproduce output signal. Individual switching for each channel is provided by the numbered white and amber monitor select switches. The individual switches allow either the input (white) or output (amber) of any one channel to be selected independently of all other channels.

The individual channel white and amber switches are inoperative if either the master IN (white) or OUT (amber) switch is pressed, but the individual channels will indicate. Master OUT (amber) overrides master IN (white) when pressed.

3-10. RECORDING

3-11. Recording requires the use of a RECORD pushbutton on the transport or a similar RECORD pushbutton on the remote control assembly. In addition, the master RECORD pushbutton on the remote control assembly along with a channel red pushbutton are required to select a channel or channels. To avoid confusion, the following will be indicated to identify the pushbuttons for the recording operation:

1. RECORD (t) (r) refers to the transport or remote RECORD pushbutton pressed simultaneously with the PLAY pushbutton to initiate record.
2. RECORD (m) refers to the master signal electronics RECORD pushbutton on the remote control assembly.
3. Red refers to the individual channel record pushbutton on the remote control assembly.

NOTE

Before preparing to record, the record level adjustment on each board should be set according to the maintenance section.

3-12. The following procedure is for a single recording channel. When multiple channel recording is desired, perform the same procedure for each channel.

1. Press the POWER switch to ON to apply power.
2. Thread a reel of tape through the Isoloop as previously described. For critical recording applications, new or previously degaussed tape should be used.
3. Set the CAPSTAN SPEED, REEL BALANCE, TAPE TENSION, and MODE RESPONSE switches to the desired positions.
4. Press transport STOP button for standby.
5. Press the master RECORD (m) button and the particular channel red button on the remote control assembly. The RECORD (m) is lit, the channel red button lamp is dimly lit.
6. To start recording, simultaneously press the PLAY and RECORD (t) (r) buttons on the transport or remote control. The red channel lamp brightens.

NOTE

When the amber OUT button is pressed, all channels are monitored for output. When the white IN button is pressed, all channels are monitored for input. To monitor individual channels for combinations of IN and OUT, press only the individual channel white and amber buttons. If none of the buttons are pressed, monitoring is automatic for record input.

7. Press the amber OUT button; the playback signal monitored on the VU meter should be approximately the same level as the input signal being recorded. Alternately press the white IN and OUT buttons to ascertain that the input and output signals are at the same level. Listen critically on a good monitor speaker or headphone system to be certain the signals sound identical with the swtich in both the IN and OUT positions.
8. To stop recording, press the STOP button. The tape will stop, the STOP button will light, the individual channel buttons dim.

3-13. REPRODUCE

3-14. When the recorder is used to reproduce prerecorded tape, the following procedure should be followed:

1. Press the STOP switch to light the stop lamp.

2. Check and assure that the master CUE and master RECORD (m) button is not pressed.

3. Press the OUT select button, or individual channel amber button.

4. Press the PLAY button to start the recorder in the reproduce mode of operation. The reproduce output level of the prerecorded tape monitored at the VU meter or output jack will represent the true amplitude level of the signal recorded on the tape.

NOTE

The amplitude level of the prerecorded tape is established by the calibrated adjustment of the reproduce level adjustment located on the reproduce preamplifier in the signal electronics. The adjustment should only be adjusted when performing the playback alignment adjustments covered in the maintenance section.

3-15. CUEING OR OVERDUBBING

3-16. Cueing or overdubbing provides a means of adding one or more audio tracks in synchronism with a first track or set of tracks previously recorded on the tape, and a means of patching tape. Cueing while in the record mode requires that the individual channel green cue button be pressed. The master CUE button can be in either position, but normally out. In play (reproduce) or stop, the cue command is initiated by pressing the master CUE button, and the individual channel green buttons. To disable the cue command, only the master CUE button need be pressed.

3-17. Consider a tape which has had an orchestra previously recorded on track no. 1 and a synchronized vocal is to be added to track no. 2, which has been left unrecorded. To accomplish the synchronized dubbed-in vocal recording on track no. 2, the following procedures should be followed:

1. Thread the prerecorded tape on the transport in the normal manner, and press transport STOP button for standby.

2. Press the master CUE button if the lamp is lighted to extinguish the lamp.

3. Press the channel no. 1 green button.

4. Press the channel no. 1 amber button.

5. Connect an appropriate headphone monitoring set to the output of channel no. 1.

6. Press the master RECORD (m) button and the channel no. 2 red button.

7. With the vocalist listening in the headphones, press the RECORD (t) (r) button. The orchestra previously recorded on track no. 1 will be heard in the phones, and the voice of the vocalist will be recorded on track no. 2 in synchronism with orchestra on track no. 1.

8. After an overdub selection has been recorded, the same tape can be replayed in the normal reproduce mode by pressing the PLAY button without resetting any switches. If a second recording attempt is necessary, starting the recorder in the record mode will again activate the previously selected cue and record tracks.

9. If it is desirable to listen to a prerecorded track for cueing purposes before the recorder is activated in the record mode, the master CUE switch should be pressed. Monitoring of the cue track will now be obtained in the play and stop mode. The master CUE switch should be pressed and released when the overdub recording is played back, otherwise the prerecorded track and the overdub track will not be in sync.

3-18. EDITING

NOTE

The editing procedure described below is used only on 2 and 4 track recorders.

3-19. Editing tape is extremely fast, accurate, and easy utilizing the unique features of the recorder such as simplicity of the Isoloop drive, tape tension and release, and the safety interlocks which permit initiating the edit mode from play or stop modes only. The EDIT switch is also located on the transport for convenience. The take-up reel is inoperative in the edit mode as tape is automatically moved past the playback head. The operators hands are free to handle tape, stop tape motion, and precisely mark the tape for cutting.

3-20. To edit, perform the following:

1. Perform (a) if in PLAY, or (b) if in STOP.

   (a) If in the play mode, press the EDIT button. The EDIT button will light and after a short interval, the take-up reel will come to a complete stop. After stopping, the tape will be moved from the supply reel past the playback head and spill out.

   (b) If in the stop mode, press the EDIT switch to light the EDIT button. Then press the PLAY button. The tape will move from the supply reel past the playback head and spill out.

NOTE

For recorders equipped with variable speed. When tape is close to the cutting point, the CAPSTAN SPEED switch can be set to VAR LOW or HI and the VARIABLE SPEED control can be adjusted for a very slow speed.

2. When the cutting point has been found, press the STOP button.

3. Move the capstan manually to position the tape precisely to the cutting point.

4. After tape is spliced, set the CAPSTAN SPEED switch to the desired position and continue editing.

diameter up and down. There are very precisely machined raised and lowered areas. The incoming pinch roller has raised areas on it, which press the tape only on the smaller diameter areas of the capstan, while the outgoing pinch roller has differently located raised areas, which press only on the larger diameter areas of the capstan. Since the larger diameter part of the capstan is trying to pull the tape out of the closed-loop faster than the incoming idler is trying to permit tape to go into the loop, tension is caused, which keeps the tape against the heads. This small difference in diameter makes the outgoing roller demand tape faster than the incoming roller wants to supply it. The difference of the push ips and the pull ips would eventually tend to rip or stretch the tape, except that the tape is only in the isolated loop for a very short time and then is free again so no damage results.

The reversing idler just turns the tape around. The idler is very delicate in its construction. One good bump and it will "get out of round," which will make the tape hum horribly during rewind or fast forwarding (shuttle modes) due to the wobble introduced by the reversing idler.

The azimuth is adjustable on the record and play head, so it is possible to make recordings that are horribly off azimuth, if you do not double check to see that the azimuth is O.K.

If the deck is set up correctly, the changing tensions due to the diminishing supply pack and the enlarging take-up pack will not affect the tape speed as these tensions cannot invade the closed-loop area.

*** Please refer to Illus. I6-3 -- The Ampex 1200 Excerpt ***

The Ampex 1200 is a good example of the single servo transport type. In the single servo type the supply reel is tension monitored by a servo arm (the second guide from the left in Figure 3-1 page 3-7 of the Ampex 1200 notes). The take-up side is not monitored. As the tape is consumed, the holdback tension on the tape from the supply motor would ordinarily increase, but when the servo arm senses this, it reduces the holdback so that the tension remains almost constant. The adjustments for the servo system are very, very critical and under adjusting will cause the machine to slow down at the end of the reel, while overadjusting will cause tape speed flutter, making recording impossible.

In the standard transport type (with no servos) the tension of the holdback does increase as the reel is used and it is therefore up to the capstan and pinch roller to keep the tape speed normal in spite of the increased holdback tensions.

SECTION 3
OPERATION

3-1. GENERAL

The following paragraphs provide instructions for operating the MM-1200 Recorder/Reproducer in the fixed and variable speed modes. Also included are descriptions of the various operator controls and their uses. Procedures are given for making recordings with and without using the Sel Sync feature, and for reproducing previously recorded tapes.

3-2. CONTROLS AND INDICATORS

The recorder/reproducer may be turned on and off by means of a main power switch located under the cover on the transport control assembly, which is adjacent to the control box assembly. The POWER switch is shown on Figure 1-4 in Section 1 of this manual.

Table 3-1 lists controls and indicators which are located on the control box assembly.

Table 3-2 lists the controls on the search-to-cue and tape timer assembly.

Table 3-3 lists the operator controls on the electronics assemblies.

3-3. SPEED SELECTION

3-4. Fixed Speed Mode

To operate the system at fixed speeds, ensure that the desired speeds have been selected on the capstan servo PWA and the search-to-cue tape timer PWA in the transport control chassis, as described in paragraphs 2-13 and 2-16.

3-5. Variable Speed Mode

To operate the system at variable speeds using Ampex VS-10 Variable Speed Oscillator, remove the dummy plug from the VARIABLE SPEED connector located on the accessory panel and connect the VS-10 electrical connector in its place. The VS-10 can then be used to vary capstan speed as described in the VS-10 Operation and Maintenance manual, Ampex Catalog No. 4890351.

To operate the system with an external signal generator other than the Ampex VS-10, proceed as follows:

1. Connect the external signal generator as described in paragraph 2-15b.

2. Set the speed strapping on the capstan servo PWA to the normal speed range as shown on Table 2-2.

3. Set the TAPE SPEED switch on the control box assembly to HIGH.

4. Adjust the signal generator frequency control for the desired speed. Table 2-3 shows several signal generator frequencies that will produce specific tape speeds.

3-6. LINE/MONITOR OUTPUT SIGNALS

The line/monitor outputs are shown in Table 3-4 for all audio switching combinations in the record, play, and standby (stop, rewind, or fast forward) modes of operation.

Table 3-1. Control Box Assembly — Controls and Indicators

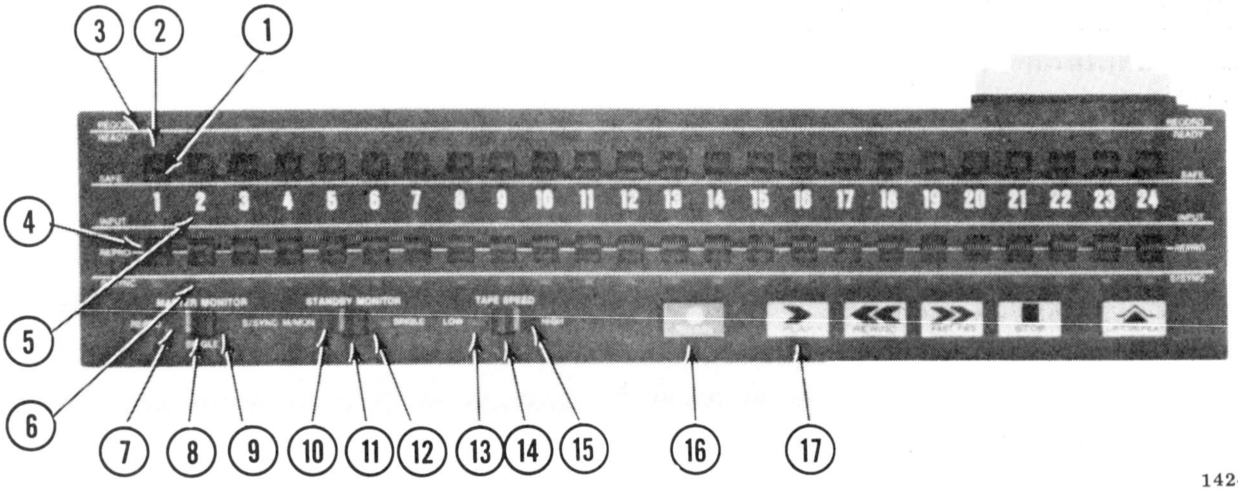

14245-1

ITEM	DESCRIPTION	FUNCTION
1	READY/SAFE toggle switches	Permit channel selection for recording. The READY position enables recording on the corresponding channel when the RECORD and PLAY pushbuttons are depressed simultaneously (see Item 16). The SAFE position inhibits recording.
2	READY indicator LED	When illuminated, shows that control circuits are set to permit recording on that channel.
3	RECORD indicator LED	When illuminated, shows that the recording function is enabled on that channel.
4	INPUT/REPRO/S/SYNC toggle switches	Permit selection of the output signal source for each channel. The INPUT position sets the control circuitry to select the audio input to the machine as the output signal for that channel. The REPRO position sets the control circuitry to select the reproduce head for that channel as the signal source while in the normal reproduce or the record mode. The S/SYNC position sets the control circuitry to select the record head for that channel for use as a reproduce head while the system is in the reproduce or record mode.
5	INPUT indicator LED	When illuminated, shows that the machine input has been selected for that channel as the audio source to be monitored.
6	S/SYNC indicator LED	When illuminated, shows that the record head for that channel has been selected as the audio source to be monitored.
7	REPRO indicator LED	When illuminated shows that the control circuitry has been enabled for the reproduce function.
8	MASTER MONITOR toggle switch	This switch overrides the individual channel switches and permits selection of all record or all reproduce heads simultaneously for the output audio source. In the REPRO

Table 3-1. Control Box Assembly – Controls and Indicators (Continued)

ITEM	DESCRIPTION	FUNCTION
8 (Cont'd)		position, all reproduce heads are selected regardless of the position of the individual INPUT/REPRO/S/SYNC switch. In the S/SYNC position, all record heads are selected regardless of the position of the individual INPUT/REPRO/S/SYNC switch. In the SINGLE position, output selection is determined by the position of the individual INPUT/REPRO/S/SYNC switches.
9	S/SYNC indicator LED	When illuminated, shows that the MASTER MONITOR switch circuitry has been set for the master Sel Sync function.
10	M/MON indicator LED	When illuminated, shows that the STANDBY MONITOR switch has been set to the master monitor mode.
11	STANDBY MONITOR toggle switch	This switch functions in the standby mode only (STOP, FAST FWD, and REWIND). The M/MON position selects whatever configuration the MASTER MONITOR switch has selected; i.e., REPRO, SINGLE, or S/SYNC. With the STANDBY-MONITOR switch in the SINGLE position, the outputs selected will be determined by the position of the individual INPUT/REPRO/S/SYNC switches (same as SINGLE position of MASTER MONITOR switch).
12	SINGLE indicator LED	When illuminated, shows that the control circuitry to select the position of the INPUT/REPRO/S/SYNC switches as a monitor audio source has been selected.
13	LOW indicator LED	When illuminated, shows that the control circuitry for the low tape speed has been selected.
14	TAPE SPEED toggle switch	This switch permits selection of predetermined high or low capstan motor speeds.
15	HIGH indicator LED	When illuminated, shows that the control circuitry for the high tape speed has been selected.
16	RECORD pushbutton switch	This switch, when depressed concurrently with the PLAY switch, enables the selected record control circuitry. This switch remains illuminated while the record function is enabled. To stop the record mode while leaving the tape in motion, press and hold RECORD while momentarily pressing STOP.
17	PLAY pushbutton switch	This switch, when depressed, enables the play mode and enables the selected reproduce circuitry. This switch illuminates when depressed and remains illuminated while the play mode is enabled.

Table 3-1. Control Box Assembly – Controls and Indicators (Continued)

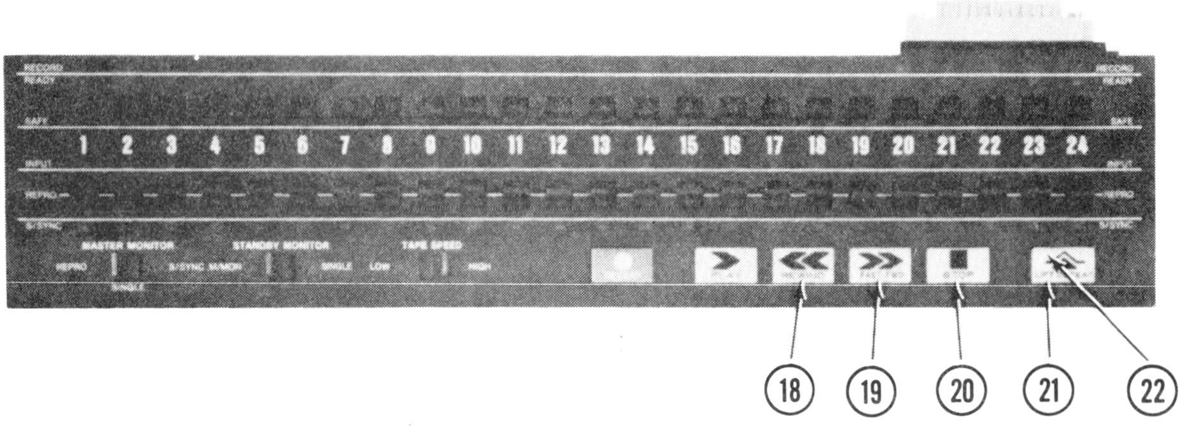

14245-1

ITEM	DESCRIPTION	FUNCTION
18	REWIND pushbutton switch	This switch, when depressed, enables the transport rewind function and transfers the signal electronics to standby mode. This switch illuminates when depressed and remains illuminated until cancelled by entering another mode of operation.
19	FAST FWD pushbutton switch	This switch, when depressed, enables the transport fast forward function and transfers the signal electronics to standby mode. This switch illuminates until cancelled by entering another mode of operation.
20	STOP pushbutton switch	This switch, when depressed, disables any other mode of operation, stops tape movement, and causes the control circuitry and signal electronics to enter the standby mode of operation. This switch illuminates when depressed and remains illuminated during the standby mode. In the edit mode the STOP indicator is extinguished when LIFT DEFEAT is operated to release the reel brakes, but illuminated when STOP is pressed to re-apply the reel brakes.
21	LIFT DEFEAT pushbutton switch	This switch, when depressed, causes the tape lifters to operate in reverse of the normal operation position for any mode of operation. In addition, this switch releases the reel brakes when the transport is stopped in the edit mode. To re-apply the brakes, the STOP button is pressed.
22	SYNC TRANSFER slide switch S58	Slide switch located beneath the LIFT DEFEAT pushbutton. INPUT position: Toward the transport REPRO position: Away from the transport In the record mode only, the sel sync monitor function for those channels selected to record is transferred to either the input (INPUT) or playback head (REPRO) signals.

Table 3-2. Search-to-Cue and Tape Timer Controls and Indicators

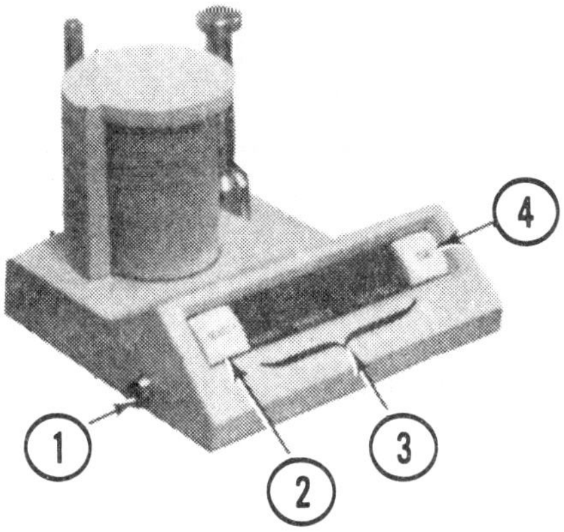

13550-11

ITEM	DESCRIPTION	FUNCTION
1	DISPLAY RESET pushbutton switch	This switch resets the tape timer indication to zero.
2	SEARCH pushbutton switch	This switch initiates the search-to-cue function. The REWIND and FAST FWD indicators are both illuminated during SEARCH.
3	Counter indicator LEDs	When illuminated, these indicators show in + and – hours, minutes, and seconds the distance the tape has moved from timer zero indication.
4	CUE pushbutton switch	This switch selects the cue point for the search-to-cue function.

Table 3-3. Electronics Assemblies – Operator Controls for a Typical Channel

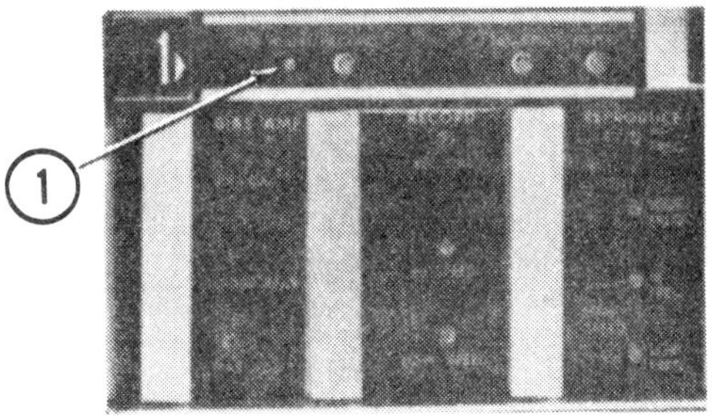

ITEM	DESCRIPTION	FUNCTION
1	BIAS/AUDIO toggle switch	This switch selects the input to the vu meter for that channel. The AUDIO position enables the vu meter to indicate input or output audio levels. The BIAS position enables the vu meter to indicate the record amplifier bias current.

Table 3-3. Electronics Assemblies – Operator Controls for a Typical Channel (Continued)

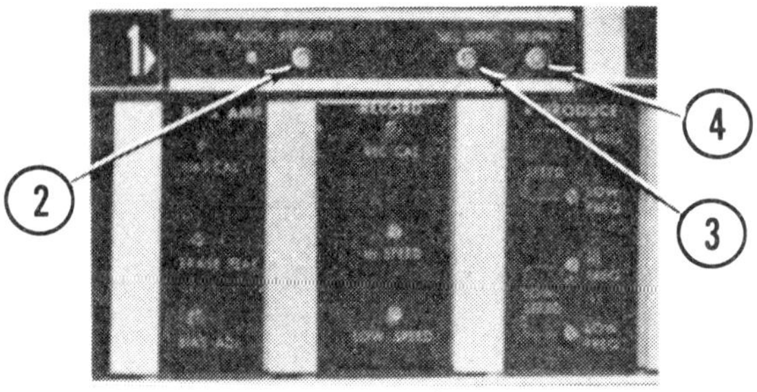

ITEM	DESCRIPTION	FUNCTION
2	RECORD control	This control permits adjustment of the record level audio to the record head for its associated channel.
3	SEL SYNC control	This control permits adjustment of the reproduce audio level from the record head for its associated channel during the Sel Sync mode.
4	REPRO control	This control permits adjustment of the reproduce audio level from the reproduce head for its associated channel during the normal reproduce mode.

Table 3-4. Line Output/Monitor Signals

SWITCH POSITION			MODE		
STANDBY MONITOR	MASTER MONITOR	INDIVIDUAL CHANNEL	RECORD	PLAY	STANDBY STOP, FF, or REW)
M/MON	REPRO	INPUT	REPRO	REPRO	REPRO
		REPRO	REPRO	REPRO	REPRO
		S/SYNC	REPRO	REPRO	REPRO
	SINGLE	INPUT	INPUT	INPUT	INPUT
		REPRO	REPRO	REPRO	REPRO
		S/SYNC	INPUT or REPRO*	S/SYNC	S/SYNC
	S/SYNC	INPUT	INPUT or REPRO*	S/SYNC	S/SYNC
		REPRO	INPUT or REPRO*	S/SYNC	S/SYNC
		S/SYNC	INPUT or REPRO*	S/SYNC	S/SYNC
SINGLE	REPRO	INPUT	REPRO	REPRO	INPUT
		REPRO	REPRO	REPRO	REPRO
		S/SYNC	REPRO	REPRO	S/SYNC

Table 3-4. Line Output/Monitor Signals (Continued)

SWITCH POSITION			MODE		
STANDBY MONITOR	MASTER MONITOR	INDIVIDUAL CHANNEL	RECORD	PLAY	STANDBY STOP, FF, or REW)
SINGLE (Cont'd.)	SINGLE	INPUT	INPUT	INPUT	INPUT
		REPRO	REPRO	REPRO	REPRO
		S/SYNC	INPUT or REPRO*	S/SYNC	S/SYNC
	S/SYNC	INPUT	INPUT or REPRO*	S/SYNC	S/SYNC
		REPRO	INPUT or REPRO*	S/SYNC	REPRO
		S/SYNC	INPUT or REPRO*	S/SYNC	S/SYNC

*Depends on the position of the SYNC TRANSFER switch S58 inside the control box. In the record mode only, the Sel Sync monitor function for all channels selected to record is transferred to either the input (INPUT) or playback head (REPRO) signals.

3-7. OPERATING PROCEDURES

3-8. Making a Recording — Without Sel Sync

To make a recording without using the Sel Sync feature, proceed as follows:

1. Clean and demagnetize the tape path as necessary as described in the Maintenance section of this manual.
2. Ensure that the alignment procedures of paragraph 5-38 in the Maintenance section have been performed.
3. Move the end-of-tape arm fully counterclockwise to the locked position.
4. Thread a bulk-erased reel of tape onto the transport as shown in Figure 3-1, close the head gate, and release the end-of-tape arm.
5. On the control box assembly, set the TAPE SPEED switch to the desired speed.
6. Set the STANDBY MONITOR switch to SINGLE.
7. Set the individual INPUT/REPRO/S/SYNC switches to the INPUT position, the same channel READY/SAFE switches to the READY position, and all other READY/SAFE switches to the SAFE position.
8. Set the BIAS/AUDIO switches on the electronics assemblies to AUDIO.
9. With audio sources connected to the appropriate recording channels, adjust the RECORD control on each switcher module (record channel only) for an indication on the associated vu meter of 0 at audio peaks.
10. Cue the tape at the starting point, press the DISPLAY RESET pushbutton on the tape timer for a zero indication if desired, and initiate the record mode by simultaneously depressing the PLAY and RECORD pushbuttons on the control box assembly.

NOTE

To monitor the tape during the record mode, set the MASTER MONITOR switch to REPRO.

11. The record mode can be ended by any of the following methods:

    a. To stop the recording process and tape motion simultaneously, depress the STOP pushbutton.

    b. To stop the recording process on all channels but not the tape motion, hold the RECORD pushbutton depressed and momentarily depress the STOP pushbutton.

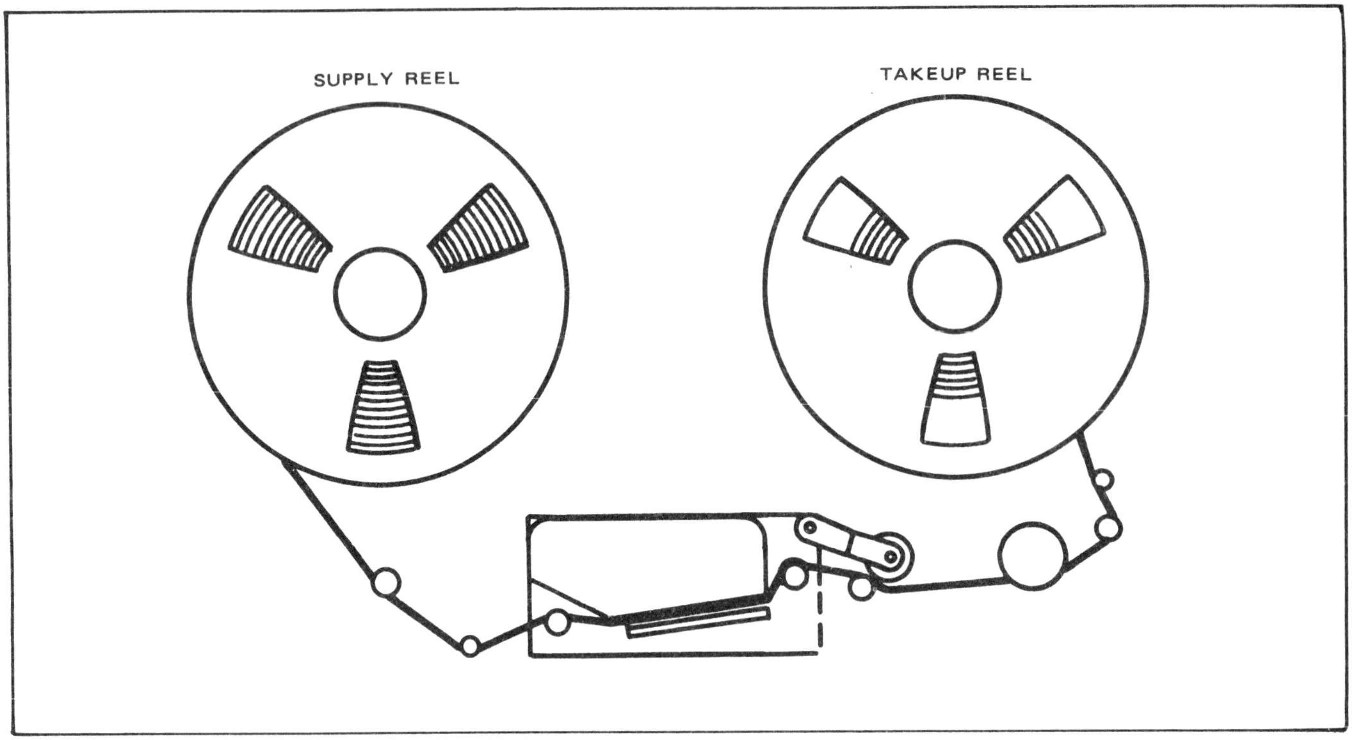

Figure 3-1. Tape Threading Path

c. To stop the recording process on any individual channel without stopping it on the others, set the READY/SAFE switch for that channel to the SAFE position.

3-9. Making a Recording — With Sel Sync

The main function of the Sel Sync feature is to enable the recording of new material precisely in time with previously recorded material. In conventional multi-track recorder/reproducer systems, the different physical locations of the record and reproduce heads along the tape path cause a timing error between material being reproduced on one channel and material being recorded on another. This effect is shown in the upper diagram of Figure 3-2 where material is recorded a fraction of a second before it is reproduced. The result is that the listener who is monitoring one channel while recording on another finishes with the newly recorded track slightly behind the previously recorded track. (See the center diagram of Figure 3-2.) This problem is resolved by using record heads as playback heads such that, while any channel or combination of channels is being monitored by the associated record head(s), the remaining channel(s) can be simultaneously recorded, thereby eliminating the time displacement error. (See the bottom diagram of Figure 3-2.)

To make a recording using the Sel Sync feature, proceed as follows:

1. Clean and demagnetize the tape path as necessary as described in the Maintenance section of this manual.

2. Ensure that the audio alignment procedures in the Maintenance section have been accomplished.

3. Move the end-of-tape arm fully counterclockwise to the locked position.

4. Thread a tape with a previously recorded channel or channels onto the tape transport assembly as shown in Figure 3-1, close the head gate, and release the end-of-tape arm.

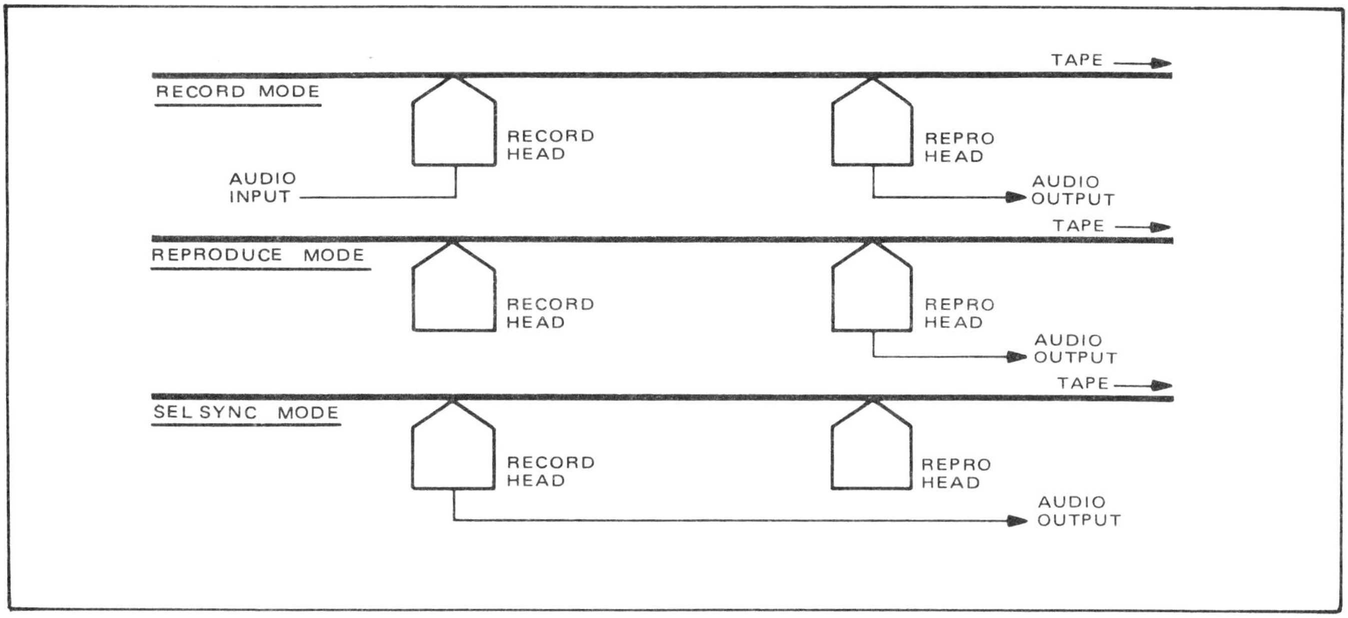

Figure 3-2. Head Usage for Different Operating Modes

5. Set the TAPE SPEED switch on the control box assembly to the position used for recording prerecorded channels.

6. Set the MASTER MONITOR switch to the SINGLE position.

7. Set the READY/SAFE switch(es) of the channel(s) to be recorded to the READY position, and all others to the SAFE position.

8. Set the INPUT/REPRO/S/SYNC switch (es) of the channel(s) to be monitored to the S/SYNC position, and of the channel(s) to be recorded to the INPUT position.

9. Set the BIAS/AUDIO switches on the electronics assemblies to AUDIO.

10. Set the STANDBY MONITOR switch to SINGLE.

11. With audio sources connected to the appropriate recording channels, adjust the RECORD control(s) on the associated audio switcher so that the associated vu meters indicate audio peak of 0.

12. Move the tape to the starting point, depress the tape timer DISPLAY RESET pushbutton for a zero indication if desired, and initiate the record mode by concurrently depressing the PLAY and RECORD pushbuttons on the control box assembly.

13. The record mode can be stopped by any of the following methods:

    a. To stop the recording process and tape motion simultaneously, depress the STOP pushbutton.

    b. To stop the recording process on all channels but not the tape motion, hold the RECORD pushbutton depressed and momentarily depress the STOP pushbutton.

    c. To stop the recording process on any individual channel without stopping it on the others, set the READY/SAFE switch for that channel to the SAFE position.

3-10. Reproducing a Previously Recorded Tape

Reproduce a previously recorded tape as follows:

1. Clean and demagnetize the tape path as described in the Maintenance section of this manual.
2. Ensure that the audio alignment procedures in the Maintenance section have been accomplished.
3. Move the end-of-tape arm fully counterclockwise to the locked position.
4. Thread the tape to be reproduced onto the tape transport assembly as shown in Figure 3-1.
5. Set the TAPE SPEED switch on the control box assembly to the desired position.
6. Set the MASTER MONITOR switch to the REPRO position, and the STANDBY MONITOR switch to the M/MON position.
7. Set all READY/SAFE switches to the SAFE position.
8. Depress the PLAY pushbutton switch on the control box assembly and adjust the appropriate REPRO controls on the switcher modules to obtain the desired audio levels.
9. Depress the STOP pushbutton to stop the reproduce mode. If the tape supply is exhausted before the STOP pushbutton is depressed, the end-of-tape switch will actuate, the tape will stop, and the system will automatically revert to the standby mode.

3-11. Fast Winding

For tape editing or cueing, the tape may be wound in either direction by depressing the FAST FWD or REWIND pushbuttons. These pushbuttons may be depressed at any time without depressing the STOP pushbutton first. When the desired point on the tape is reached, depress the STOP pushbutton to stop tape motion.

3-12. Fast Winding – Search-to-Cue

To use the search-to-cue function, proceed as follows:

1. When it is desired to return to a specific point on a tape, depress the CUE pushbutton on the tape timer assembly when the tape is at the desired return point.

NOTE

The CUE pushbutton may be depressed either when the tape is in motion or when it is stopped.

2. Record or reproduce the tape as described in the preceding paragraphs.
3. To return to the cue point, depress the SEARCH pushbutton on the tape timer assembly.
4. To cancel the search function, depress the STOP pushbutton on the control box assembly.
5. To initiate the play mode as soon as the cue point is reached, depress the PLAY pushbutton on the control box assembly after depressing the SEARCH pushbutton on the tape timer.
6. To enter the fast forward or rewind mode of operation while the search function is engaged, depress the FAST FWD or the REWIND pushbutton on the control box assembly.

NOTE

Depressing the FAST FWD or the REWIND pushbutton on the control box assembly will override the search function as long as the pushbutton is held depressed. When the FAST FWD or REWIND pushbutton is released, the search function will continue until the cue point is reached or the search function is manually cancelled.

3-13. Tape Lifter Defeat

The tape lifter mechanism, which automatically lifts the tape away from the heads during the fast wind modes of operation, may be defeated by depressing the LIFT DEFEAT pushbutton. Holding the LIFT DEFEAT pushbutton depressed during the fast wind modes permits monitoring of the reproduced audio signal at fast wind speeds for cueing purposes. To minimize head wear, it is good practice to avoid unnecessary use of this feature. Alternately, the LIFT DEFEAT pushbutton may be used during the record or play modes to lift the tape from the heads.

The tape first encounters the incoming fixed idler, which should be the roller type. This holds up the tape to the proper height. Next the tape passes over the servo tension arm, where the holdback tension is electronically monitored, then the first head block guide, the erase head, the record head and the play head. Next is the final head block guide and the capstan. Since the capstan is on the backing side of the tape, the pinch roller presses against the recording emulsion side of the tape. Next is the tape timer accessory, the final roller guide, and the tape runout arm, which senses when the tape runs out or breaks and stops the deck.

Changes in the take-up tension due to the changing size of the take-up pack should not really affect the speed of the tape, since the pinch roller and capstan cut off the take up reel from the rest of the tape path. It is always advisable to have a dubbings, or other high quality tape strobe available for the second engineer so that he can occasionally check the actual tape speed. Always record a 60 Hz sync tone on a spare track in case the tape machine does have speed problems during the recording of basic tracks. Remember: the sync tone must be recorded with the tracks and then not be erased to be effective. If the machine does develop speed problems, you can lock the machine back into perfect tape speed using a resolver (for an Ampex 1200 you could use the Ampex Sync-Lock).

*** Please refer to Illus. I6-4 -- The ATR 100 Excerpt ***

The ATR 100 is the first dual servo machine. It uses two sensing arms that continuously monitor the take-up and holdback tensions, keeping them perfectly balanced.

*** Please refer to Illus. I6-4 Figure 3-1 ***

With this system, any changes in reel pack and the resulting tendency towards speed-up or slow down are completely corrected. Even human interference with holdback is sensed and corrected instantly by the servo system. There is no pinch roller at all during normal tape operations. The tape "agrees" to be pushed along by the capstan, since there is neither any tendency for the tape to be pulled forwards or backwards due to the precision balance of the holdback and take-up tensions by the servo systems. All shuttle modes are due to the tape "following" the capstan. The reel motors never do any fast forwarding or rewinding, they keep the tensions applied to the tape exactly equal while the capstan does the shuttle motion.

There are some interesting features to the ATR-100:

1. Constant (coherent) phase record circuit
2. VSO operation from nearly stopped to nearly 100 times normal speed.
3. Automatically drops out of record if the tape speed is in error or if the tape speed "burps"
4. With optional circuit cards, ability to have one machine pre calibrated to record at 30, 15, 7 1/2, and 3 3/4 ips

The only things that are not provided for are a search function and the ability to get back into play mode once the tape is cut during editing (the tape must be in one piece in order to get into play once the deck is in edit mode with the servos relaxed).

SECTION 3
OPERATION

This section of the manual provides locations and functions of the recorder/reproducer operating controls and indicators, a preoperational procedure, and operating instructions for the various modes of operation.

3-1. CONTROLS AND INDICATORS

The location and function of each control and indicator are shown in the following tables. Table 3-1 shows those on the tape transport. Table 3-2 lists the transport controls on the recorder control panel. Table 3-3 lists the signal system controls on the recorder control panel, and Table 3-4 lists those on the input/output assembly (accessory).

3-2. PRE-OPERATING PROCEDURES

See Figure 3-1 and perform the following pre-operational procedure.

1. Clean and demagnetize components in the tape path as described in the Maintenance section of the manual.
2. Set the POWER switch to the ON position. All safe and repro indicators light, the stop and EDIT indicators light and the tape timer indicates zero (0 00 00).
3. Set speed-select rotary switch to desired tape speed; 3.75, 7.5, 15, or 30 in/s. If a speed is selected that the signal system has not been set for, the LOCKOUT indicator lights, play and record modes for that speed are inoperative, and the audio output(s) of the basic recorder/reproducer are muted. If an input/output assembly is connected, the assembly will switch to input signal monitoring.
4. Install reel of tape on supply turntable and empty reel on takeup turntable as shown in Figure 3-1. Reels up to 14 inches (16.54 cm) in diameter and in any size combination may be used. (Note: If reels larger than 11-1/2 inches (29.21 cm) are used, reel motors must be in the "outer" position. Refer to paragraph 2-22).
5. Lower head shield by pressing shield downward until it locks.
6. Thread tape on transport as shown in Figure 3-1. The tape path is from the supply reel, around the inside of the supply constant-tension arm, around the tape timer wheel, across the head assembly, around the capstan, around the inside of the takeup constant-tension arm, and onto the takeup reel.
7. Turn one reel to remove all tape slack.

CAUTION

DO NOT TOUCH THE CONSTANT-TENSION ARMS WHEN RECORDER/REPRODUCER IS IN THE ELECTRICAL THREAD CONDITION. TO DO SO WILL INTRODUCE A REEL SERVO ERROR AND POSSIBLY CAUSE TAPE DAMAGE.

8. While holding the stop pushbutton pressed, turn either reel to slacken tape and then tighten tape with a short quick movement to apply tension to the constant-tension arms. This action causes the reel servos to activate and recorder/reproducer enters electrical thread mode. The electrical thread mode is indicated by the EDIT indicator light going out. (To deactivate the thread mode, press EDIT pushbutton when tape is stopped. Reel servos will disengage and tape may be removed from transport if desired.)
9. Release head shield by pressing shield downward.

Table 3-1. Tape Transport Controls and Indicators

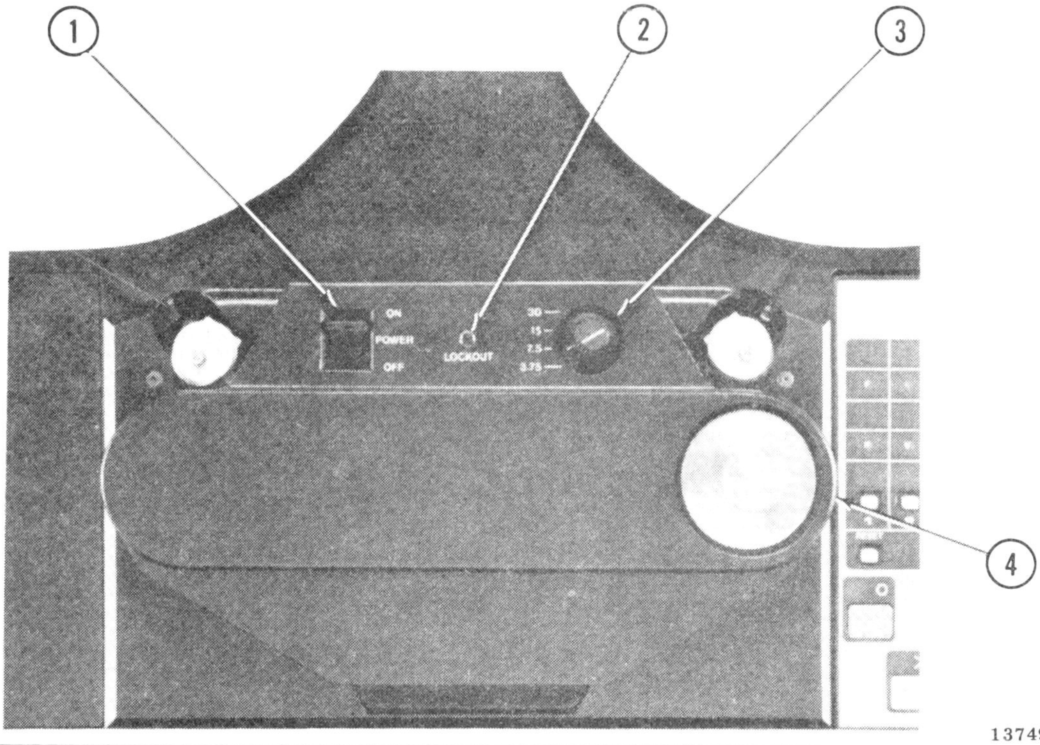

13749-9

INDEX NO.	NAME	FUNCTION
1	POWER swtich	Used to turn ac power to the recorder on and off. When power is first applied, all Safe and Repro indicators light, Stop and EDIT indicators light, and tape timer indicates zero (0 00 00).
2	LOCKOUT INDICATOR	Lights when a speed is selected that the signal system has not been set up for. Play and record modes for the incorrect speeds are locked out of operation. (Note: Indicator also lights for 2 to 3 seconds after power is first applied.)
3.	Speed-Select rotary switch	Used to select tape speed; 3.75, 7.5, 15, or 30 inches-per-second (9.52, 19.05, 38.1, or 76.2 cm/s)
4.	Capstan Edit knob	When tape is stopped in thread mode, the capstan edit knob may be turned by hand in either direction to move tape from reel to reel under reel-servo control.

Table 3-2. Recorder Control Panel, Transport Controls and Indicators

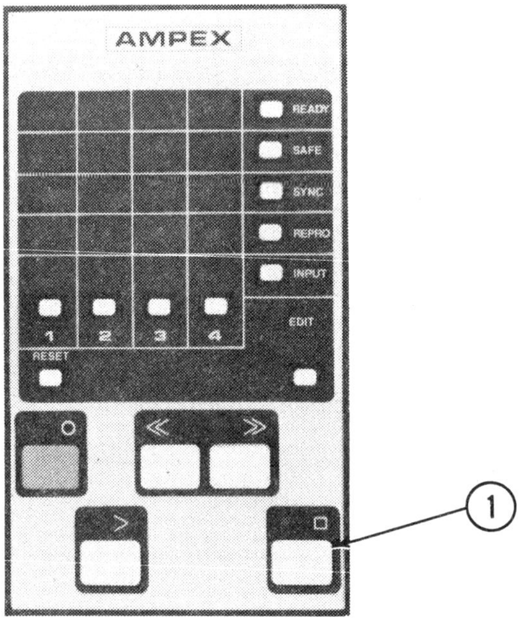

13690-5

INDEX NO.	NAME	FUNCTION
1	□ Stop pushbutton switch	Used to initiate the following functions: *Thread.* After tape is manually threaded on the transport and tape slack removed, either reel is quickly turned by hand to apply tension to the tension arms while the stop pushbutton is simultaneously pressed. This causes reel servos to activate and place transport in a thread condition. Thread condition is indicated by EDIT indicator light (index no. 11, this table) going out. *Stop.* Switch stops tape transport motion and cancels existing mode of operation. Note: When any channel is in record mode, tape motion does not stop until record signal has decayed. *Stop Record.* Also used to stop recording without stopping transport if record pushbutton (index no. 5) is held pressed while stop pushbutton is momentarily pressed. *Momentary Record.* Also used in conjunction with play and record pushbuttons (index no. 3 and 5, this table). If stop and record pushbuttons are held pressed, record mode is entered when play pushbutton is pressed and held. Record mode stops when play pushbutton is released, but transport motion continues.

Table 3-2. Recorder Control Panel, Transport Controls and Indicators (Continued)

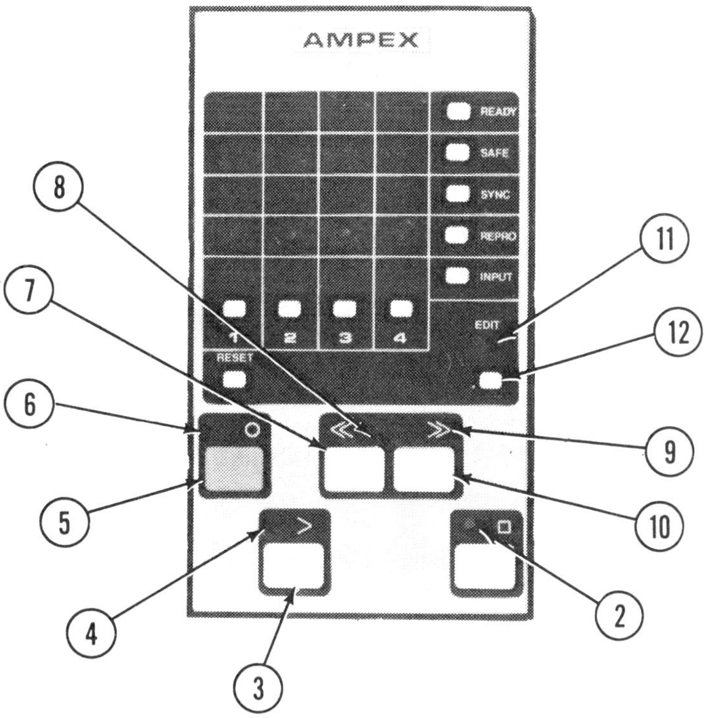

13690-5

INDEX NO.	NAME	FUNCTION
2	Stop indicator (yellow)	Lights when ac power is first applied and when stop mode is active.
3	[>] Play pushbutton switch	Used to select play mode or used simultaneously with record pushbutton (index no. 5) to select record mode. Pressing the play pushbutton during fast forward or rewind mode causes recorder to enter play mode. Also used to initiate one of the following modes: *Momentary record.* Used to select momentary record mode (see index no. 1). *Spooling.* Simultaneously pressing play pushbutton and a fast forward or rewind pushbutton (index no. 7 or 10) causes transport to enter spool mode.
4	Play indicator (green)	Lights when play or spool mode is selected and capstan servo is locked.

Table 3-2. Recorder Control Panel, Transport Controls and Indicators (Continued)

INDEX NO.	NAME	FUNCTION
5	Record pushbutton switch	Used in conjunction with play pushbutton (index no. 4). Simultaneously pressing play pushbutton and record pushbutton initiates record mode of channel(s) in ready mode. Also used to select momentary record mode (see index no. 1, this table).
6	Master Record indicator (red)	Lights when any channel is recording
7	Rewind pushbutton switch	Used to select rewind mode. Also used to select spool mode in the rewind direction when the play pushbutton (index no. 3) and the rewind pushbutton are pressed simultaneously.
8	Rewind indicator (yellow)	Lights when rewind or rewind spool mode is selected.
9	Fast Forward indicator (yellow)	Lights when fast forward or forward spool mode is selected.
10	Fast Forward pushbutton switch	Used to select fast forward mode. Also used to select spool mode in the forward direction when the play pushbutton (index no. 3) and the fast forward pushbutton are pressed simultaneously.
11	EDIT indicator (yellow)	Lights when power is first applied to recorder. Light goes out when stop pushbutton is pressed and thread condition is obtained (refer to index no. 1). Indicator also lights in edit modes of operation.
12	EDIT pushbutton switch	If tape motion is stopped, pressing EDIT causes recorder to go into unthread condition (reel servos disengage) and tape may be removed from transport if desired. Also used to initiate one of the following edit modes: *Stop/Edit:* If tape is stopped, pressing EDIT disengages reel servos for tape threading, editing, or manual tape movement. *Play/Edit:* If play or record mode is active, pressing EDIT pushbotton causes the following action: Tape motion continues in play mode; power is removed from takeup reel; and takeup tension-arm roller engages capstan causing tape to be spilled at takeup side of transport. If in Play/Edit mode and stop pushbutton is pressed, tape will stop, edit mode will be retained, and tape can again be spilled by pressing play pushbutton. To cancel play/edit mode, press stop pushbutton and after tape stops, press EDIT.

Table 3-2. Recorder Control Panel, Transport Controls and Indicators (Continued)

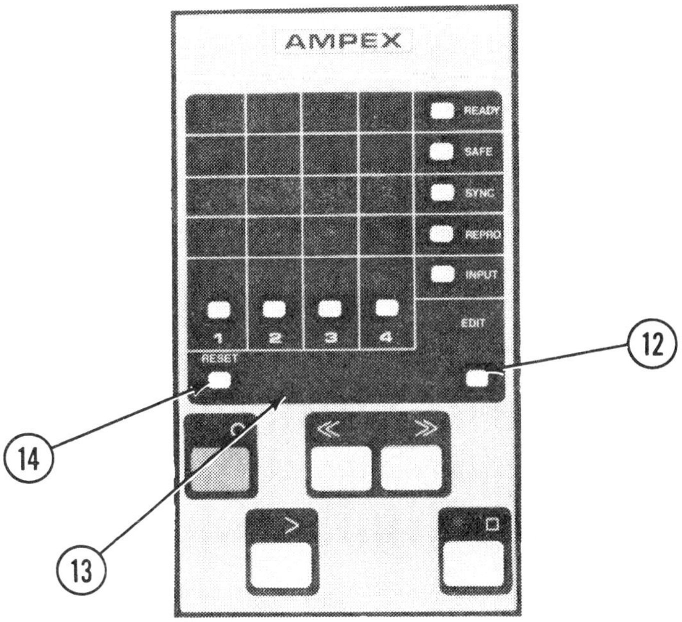

13690-5

INDEX NO.	NAME	FUNCTION
12 (Continued)		*Shuttle/Edit:* If fast forward, rewind, or a spool mode is active, pressing EDIT pushbutton will cause tape lifters to retract while pushbutton is held pressed.
13	Tape Timer display	When power in on, the display shows in hours, minutes, and seconds (or minutes, seconds, and tenths of seconds after changing jumpers – see Installation section) the distance that the tape has moved from zero (tape position when tape timer RESET pushbutton is pressed or recorder/ reproducer power is turned on).
14	Tape Timer RESET pushbutton switch	Used to reset the tape timer display to zero (0 00 00).

Table 3-3. Recorder Control Panel, Signal System Controls and Indicators

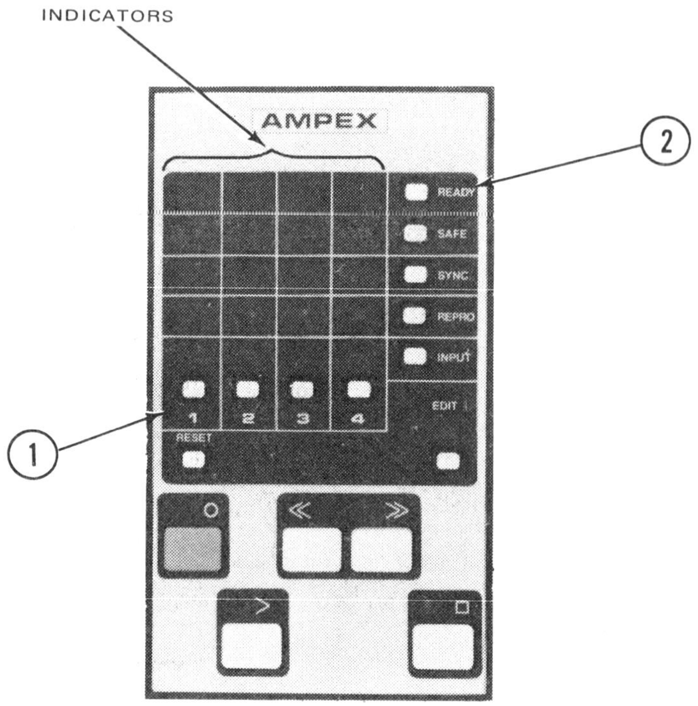

13690-5

INDEX NO.	NAME	FUNCTION
1	Channel Select pushbutton switches (4)	Switches permit channel (1, 2, 3, 4) selection for various modes of operation. Used in conjunction with the READY, SAFE, SYNC, REPRO, and INPUT pushbutton switches (index no. 2 – 6).
2	READY pushbutton switch and Ready (yellow) and Record (red) Channel 1 – 4 indicators	Used in conjunction with channel select pushbutton switch (index no. 1 of this table). In ready, channel is enabled for entering record mode. Ready indicator(s) light when READY pushbutton switch and corresponding channel select pushbutton switch (index no. 1, this table) are simultaneously pressed. Record mode is entered by simultaneously pressing play and record pushbutton switches (index no. 3 and 5, Table 3-2). When record mode is entered, ready (yellow) indicator goes out and record (red) indicator lights, indicating the channel(s) in record mode. If channel is in record mode and READY pushbutton and same channel select pushbutton switch are simultaneously pressed, the channel will go out of record and will return to ready mode. Record mode may be reentered by simultaneously pressing play and record pushbutton switches (index no. 3 and 5, Table 3-2). Tape motion and status of other channels are not affected by this action.

Table 3-3. Recorder Control Panel, Signal System Controls and Indicators (Continued)

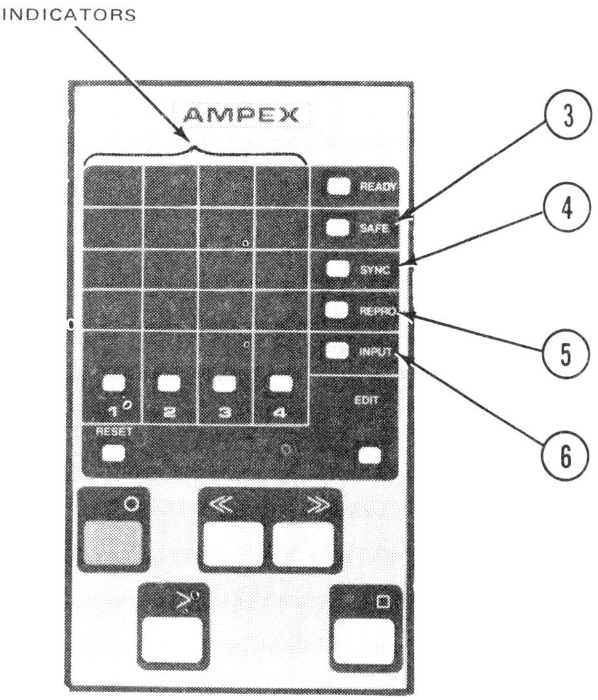

13690-5

INDEX NO.	NAME	FUNCTION
3	SAFE pushbutton switch and Safe Channel 1 – 4 indicators (green)	Used in conjunction with channel select pushbutton switch (index no. 1). In safe, channel is prevented from entering record mode. Safe indicator(s) light when power is initially applied or when SAFE pushbutton and corresponding channel select pushbutton switch (index no. 1) are simultaneously pressed. When lit, recording on channel indicated is inhibited. If channel is in record mode and SAFE pushbutton and same channel select pushbutton are simultaneously pressed, the channel will go out of record mode and enter safe mode. To reenter record mode, channel must first be placed into ready mode. Tape motion and status of other channels are not affected by this action.
4	SYNC pushbutton switch and Sync Monitoring Channel 1 – 4 indicators (yellow)	Used in conjunction with channel select pushbutton switch (index no. 1, this table) to select Sel-Sync mode. Sync indicator(s) light when SYNC pushbutton and corresponding channel select pushbutton are simultaneously pressed. When indicator is lit, signal on tape is reproduced by record head of selected channel(s). This reproduced signal is connected to audio output.

Table 3-3. Recorder Control Panel, Signal System Controls and Indicators (Continued)

INDEX NO.	NAME	FUNCTION
4 (Continued)		In in Sel-Sync mode and ready mode or record mode is entered, input signal being recorded is connected to audio output of that channel. Sync indicator light goes out and input indicator lights. If channel is removed from record mode, system returns to sync mode and sync indicator lights (unless repro or input monitoring were selected while in record mode).
5	REPRO pushbutton switch and Repro Monitoring Channel 1 – 4 indicators (yellow)	Used in conjunction with channel select pushbutton switch (index no. no. 1). In repro, signal being reproduced by reproduce head is connected to audio output. Repro indicator(s) light when power is initially applied or when REPRO pushbutton and corresponding channel select pushbutton switch (index no. 1) are simultaneously pressed. When lit, signal being reproduced by reproduce head is connected to audio output.
6	INPUT pushbutton switch and Input Monitoring Channel 1 – 4 indicators (yellow)	Used in conjunction with channel select pushbutton switch (index no. 1). Input signal to recorder is connected to audio output. Indicator(s) light when INPUT pushbutton and corresponding channel select pushbutton switch are simultaneously pressed. When lit, input signal to recorder is connected to audio output connector for monitoring purposes (E-E mode). Indicator also lights if record mode is entered while in sync mode.

Table 3-4. Input/Output Module Accessory, Controls and Indicators

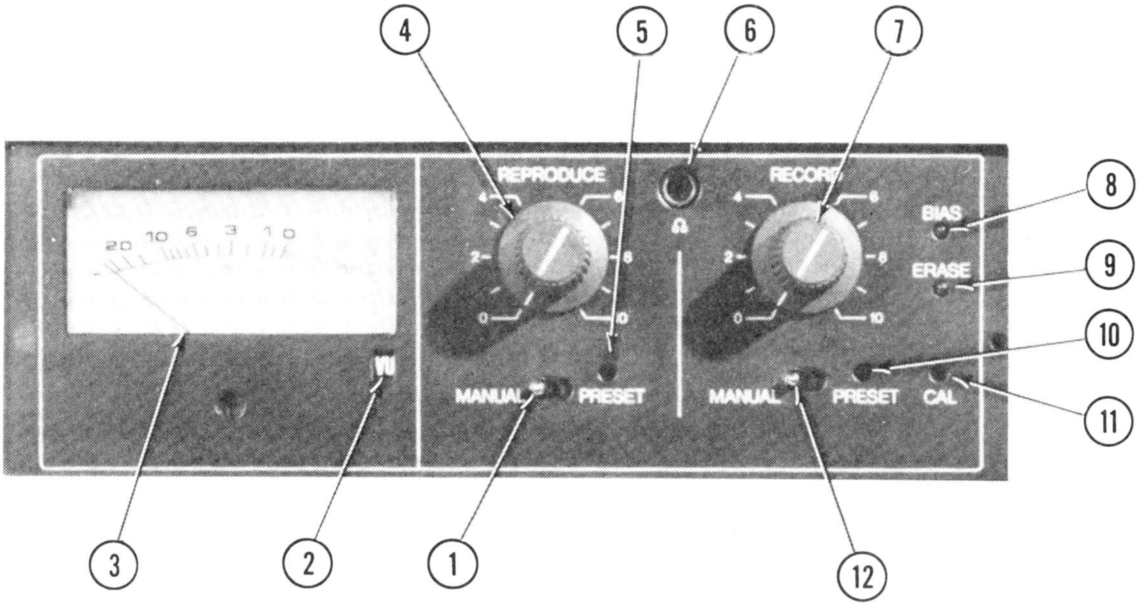

13749-12

INDEX NO.	NAME	FUNCTION
1	Reproduce MANUAL/ PRESET toggle switch	Switch used to select MANUAL or PRESET reproduce level. In MANUAL position, REPRODUCE level control (index no. 4) is used to vary reproduce level of associated channel. In PRESET position, reproduce level is controlled by the setting of the reproduce calibrate control (index no. 5).
2	VU/PK indicator	Indicates type of meter operation selected, volume unit (VU) or peak signal level (PK).
3	VU/Peak meter	Indicates signal level present at audio output of associated channel. Internal switch permits selection of vu or peak signal level reading on meter.
4	REPRODUCE level control	Control used, when MANUAL/PRESET toggle switch (index no. 1) is in MANUAL position, to vary reproduce level of associated channel.
5	Reproduce PRESET adjustment control (R1)	Used to set reproduce line output operating level.

Table 3-4. Input/Output Module Accessory, Controls and Indicators (Continued)

INDEX NO.	NAME	FUNCTION
6	Phone jack	Receptacle for headphone plug. Audio of associated channel can be monitored on headphones (600 ohms or greater impedance).
7	RECORD level control	Control used, when MANUAL/PRESET toggle switch (index no. 12) is in MANUAL position, to vary record level of associated channel.
8	BIAS indicator	Indicator lights in record mode to indicate that bias signal is present at the record head.
9	ERASE indicator	Indicator lights in record mode to indicate that erase signal is present at the erase head.
10	Record PRESET adjustment control (R2)	Used to preset record input level.
11	Record CAL adjustment control (R3)	Used to set input monitoring level and meter indication to match reproduced level from tape.
12	Record MANUAL/ PRESET toggle switch	Switch used to select MANUAL or PRESET record level. In MANUAL position, RECORD level control (index no. 7) is used to vary record level of associated channel. In PRESET position, record level is controlled by the setting of the record PRESET adjustment control (index no. 10).

3-3. OPERATING PROCEDURES

The following general information is applicable to all modes of operation.

1. If the tape runs completely off the supply or takeup reel, the recorder/reproducer automatically stops and enters unthread mode, and the tape timer display "freezes".
2. The following two groups of functions are mutually exclusive in that only one function in a group may be selected at one time: Record, Ready, or Safe; Sync, Repro, or input.

3-4. Recording

Proceed as follows:

1. Perform all steps of the pre-operating procedures.
2. Simultaneously press the channel select pushbutton switch and the READY pushbutton switch for each channel to be recorded. The associated ready indicator (yellow) will light.
3. Simultaneously press the channel select pushbutton switch and SAFE pushbutton switch for each channel that is not to be recorded. (Note: When the recorder is initially turned on, all channels will automatically be in safe mode.) The associated safe indicator (green) will light.
4. Simultaneously press the channel select pushbutton switch and the INPUT pushbutton switch for each channel selected for record mode.
5. Connect signal(s) to be recorded to the appropriate audio input connectors.

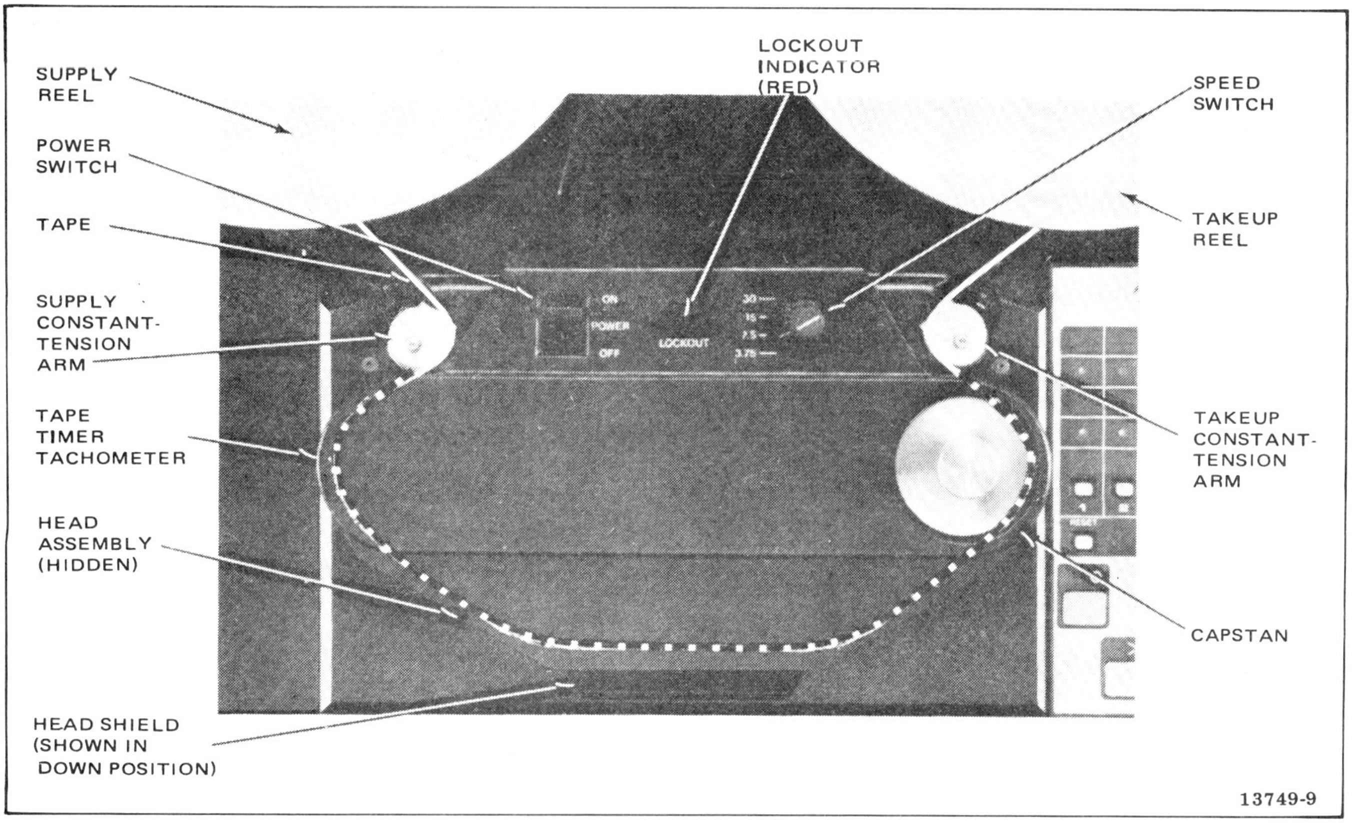

Figure 3-1. Tape Threading Path

6. If system is equipped with an input/output module assembly, perform one of the following steps for each channel to be recorded.

   a. Set RECORD MANUAL/PRESET switch to PRESET and observe level meter for desired indication. If level is not correct, adjust level fed to input/output assembly or perform step b.

   b. Set RECORD MANUAL/PRESET switch to MANUAL and adjust RECORD level control for desired indication on meter.

7. Simultaneously press play and record pushbutton switches to start tape motion and begin recording on selected channel(s). (An alternate method is to depress play pushbutton to start tape motion and then, while holding play pushbutton depressed, press record pushbutton.) The master record indicator (red) and the channel record indicator(s) will light.

8. If desired while recording, the input signal for each channel can be compared with the recorded signal on that channel by holding the channel selector pushbutton depressed and alternately pressing the associated INPUT and REPRO pushbutton switches.

9. When recording is complete, record mode may be terminated by performing any one of the following steps. (Note that internal logic control circuitry will delay any commanded change in tape motion direction or velocity until after all channels have completely terminated record mode.)

   a. Press stop pushbutton to stop tape motion and deactivate record mode.

b. Depress record pushbutton and momentarily press stop pushbutton. Record mode will be terminated but transport will continue to run at the selected tape speed.

c. Press rewind or fast forward pushbutton.

d. Simultaneously press a channel select pushbutton and a SAFE or READY pushbutton to deactivate record on any individual channel as transport continues to run at the selected tape speed.

3-5. Fast Winding

For fast-winding operations (fast forward or rewind modes), press either the rewind or fast forward pushbutton switch. The associated rewind or fast forward indicator (yellow) will illuminate. For editing or cueing operations, these pushbuttons can be pressed alternately without having to press the stop pushbutton between fast-winding selections. Either fast winding mode can be entered from stop, play, or record mode.

Play mode may be entered while in fast forward or rewind mode by pressing the play pushbutton switch. Record mode may only be entered from play mode, provided that capstan servo lock is obtained within 0.8 second.

Two tape lifter arms that move the tape away from the heads are automatically actuated in both fast-winding and spool modes. To monitor audio in either of these modes, press the EDIT pushbutton switch. The tape lifters will retract while the pushbutton is held pressed.

3-6. Reproducing

Proceed as follows:

1. Perform all steps of the pre-operating procedures.

2. Simultaneously press the channel select pushbutton switch and the SAFE pushbutton switch for each channel. The associated safe indicator (green) will illuminate. (Note: When the recorder is initially turned on, all channels will automatically be in safe and repro modes.)

3. Simultaneously press the channel select pushbutton switch and the REPRO pushbutton switch for each channel to be reproduced. The associated indicator (yellow) will illuminate.

4. Press the play pushbutton switch to start tape motion and reproduce recorded material. The play indicator (green) will illuminate.

5. When desired material has been reproduced, perform any one of the following steps to terminate reproduce mode:

   a. Press stop pushbutton switch to stop tape motion.

   b. Press rewind or fast forward pushbutton switch to enter a fast-winding mode.

3-7. Tape Timer Operation

The tape timer displays in hours, minutes, and seconds (or minutes, seconds, and tenth of seconds after changing jumpers, as described in paragraph 2-28) the distance the tape has moved from a zero (0 00 00) reference point established by pressing the RESET pushbutton on the control unit. The timer displays 0 00 00 to 9 59 59 when transport operates in the forward direction, and displays 9 59 59 after passing through 0 00 00 when operating in the reverse direction. If the tape runs off either reel or unthread mode is entered, the tape timer display "freezes".

If a remote control unit is also being used, pressing the RESET pushbutton switch on the remote control unit or on the recorder control unit will reset both tape timer displays to zero (0 00 00).

3-8. Sel-Sync Function

In Sel-Sync mode, the audio signal to be reproduced from the tape is derived from the record head rather than the reproduce head. This mode of operation permits recording of new material precisely in time with previously recorded material and is used normally for Sel-Sync recording and overdubbing as described in the following paragraphs.

In Sel-Sync recording, a performer listens to one or more previously recorded tape tracks using the Sel-Sync/reproduce mode while recording material on another track. For example, assume a four-channel recorder/reproducer is equipped with a tape having two prerecorded tracks and two blank tracks. Typically, the two prerecorded tracks are reproduced (using two of the record heads for pickup), mixed together using studio equipment, and fed to a performer's earphones. The performer then listens to the prerecorded material while recording material on one of the blank-track channels. Thus, the new material is recorded in synchronism with the prerecorded material.

In overdubbing, a performer listens to material that he previously recorded on one or more tape tracks using the Sel-Sync/reproduce mode. The performer can repeat his previous performance (but not record it) and get into proper timing with the original material. At that point where the overdub is desired, the track(s) to be overdubbed is (are) switched to the record mode. The audio that the performer hears is automatically switched from the off-tape audio to the input audio the performer is recording when the record mode is selected.

To place a channel in Sel-Sync mode, proceed as follows:

1. Perform all steps of the pre-operating procedures.
2. Simultaneously press the channel select pushbutton switch and the SYNC pushbutton switch for each channel desired to be in the Sel-Sync mode.
3. Simultaneously press the channel select pushbutton switch and a READY or SAFE pushbutton switch corresponding to the desired Sel-Sync mode of operation.
4. Follow instructions for recording or reproducing depending on desired Sel-Sync mode of operation.

3-9. Editing

The following procedures may be used to perform editing functions.

1. Thread Mode. If tape is stopped in thread mode, the capstan edit knob may be turned in either direction to move tape from reel to reel under reel-servo control.
2. Stop/Edit Mode. If tape is stopped in thread mode, pressing the EDIT pushbutton switch causes recorder to enter unthread mode (reel servos disengage) and reels are free to turn for editing, splicing, or manual tape movement. (Note: This function is disabled when operating recorder from a remote control unit.)
3. Play/Edit Mode. If play or record mode is active, pressing EDIT pushbutton switch causes tape motion to continue in play or record mode, removes power to takeup reel, and causes takeup tension-arm roller to engage capstan which causes tape to be spilled at takeup side of transport. (Note: This function is disabled when operating recorder from a remote control unit.)

   If in play/edit mode and stop pushbutton switch is pressed, tape will stop, edit mode will be retained, and tape can again be spilled by pressing play pushbutton switch. To cancel play/edit mode, press stop pushbutton and after tape stops, press EDIT pushbutton. To enter record mode from play/edit mode, press play and record pushbutton switches simultaneously.
4. Shuttle/Edit or Spool/Edit Mode. If EDIT pushbutton is pressed while in fast forward, rewind, or spool mode, the tape lifters retract allowing recorded material on tape to be reproduced as long as EDIT pushbutton is held pressed.

3-10. Spooling

Spool mode is used to transfer tape from reel to reel at a constant speed (60 in/s or 180 in/s) to obtain a uniform tape pack. (Refer to paragraph 2-24 for procedure to select spool speed.)

1. To select forward spool mode, simultaneously press the play and the fast forward pushbutton switch. The fast forward indicator

(yellow) will illuminate and the play indicator (green) will illuminate when spool speed is obtained.

2. To select rewind spool mode, simultaneously press the play and rewind pushbutton switch. The rewind indicator (yellow) will illuminate and the play indicator (green) will illuminate when spool speed is obtained.

3. If in either forward or rewind spool mode and it is desired to reverse spool mode direction, only the opposite direction pushbutton switch (fast forward or rewind) need be pressed.

4. To deactivate spool mode, press stop pushbutton switch. Another mode of operation, except record, may then be selected before tape comes to a stop.

3-11. Remote Control Operation

All recorder functions may be controlled from the remote control unit with the exception of initiating play/edit or stop/edit (unthread) modes. Except for these two functions, the remote control unit operates in parallel with the local control unit at all times.

3-12. PURC Operation

The recorder/reproducer is capable of operating with or without PURC (pick-up recording capability) operation as desired, by placement of a jumper located on the PADNET PWA. (Refer to paragraph 2-30.) Recorders shipped from the factory are set for normal (non-PURC) operation.

The use of PURC eliminates the problem of overlaps and holes in recordings when inserting (dubbing) new material within previously recorded programs. In a recorder system without PURC, initiating record mode energizes the erase and record heads simultaneously. Since there is a physical distance between the erase and record head, a period of over-recording on unerased tape occurs. When the dub is terminated, a hole is left in the program. The length of time of this hole is related to the distance between the erase and record head and the speed of the tape.

In a system equipped with PURC, separate erase and bias amplifier circuits are provided, and the turnon and turnoff times of the amplifiers are individually controlled. When record mode is initiated, the erase circuit is energized first; then, after a time interval, the bias amplifier is energized. When the recording is terminated, the erase circuit is de-energized first; and then, after a time interval, the bias amplifier is de-energized. Thus, the problem of an overlap and hole is eliminated in the dubbed portion of the recording. These on and off times are illustrated in Figure 3-2.

3-13. Editing Procedures. The approximate on and off time delay to be considered while performing edit functions is calculated by dividing the distance between the erase and record head (1.46 inches) by the tape speed. These approximate on/off delay times are shown in Table 3-5. Because of the long on/off time delay, insert edits at 3.75-in/s operation are not recommended.

Table 3-5. PURC Edit Delay Times

TAPE SPEED (IN/S)	APPROXIMATE ON/OFF DELAY TIME (MS)
30	48.7
15	97.3
7.5	194.7
3.75	389.3

If a single track insert edit is to be made, such as correcting a single word, use a white felt-tip pen or tape-editing pencil to mark the back side of the tape where the correction should start and where the correction should end. While performing the edit function, start the record mode when the first mark passes the erase head and stop the record mode when the second mark passes the erase head.

If an insert edit is to be made on one or two tracks of a four-track recording, monitor one or two tracks of the tracks in Sel-Sync mode. This will aid in synchronizing the new recording with the tracks being monitored.

The on and off delay times shown in Table 3-5 can be used to anticipate when the start and stop recording times are to take place. If one of the tracks is recorded with a timing signal, such as the

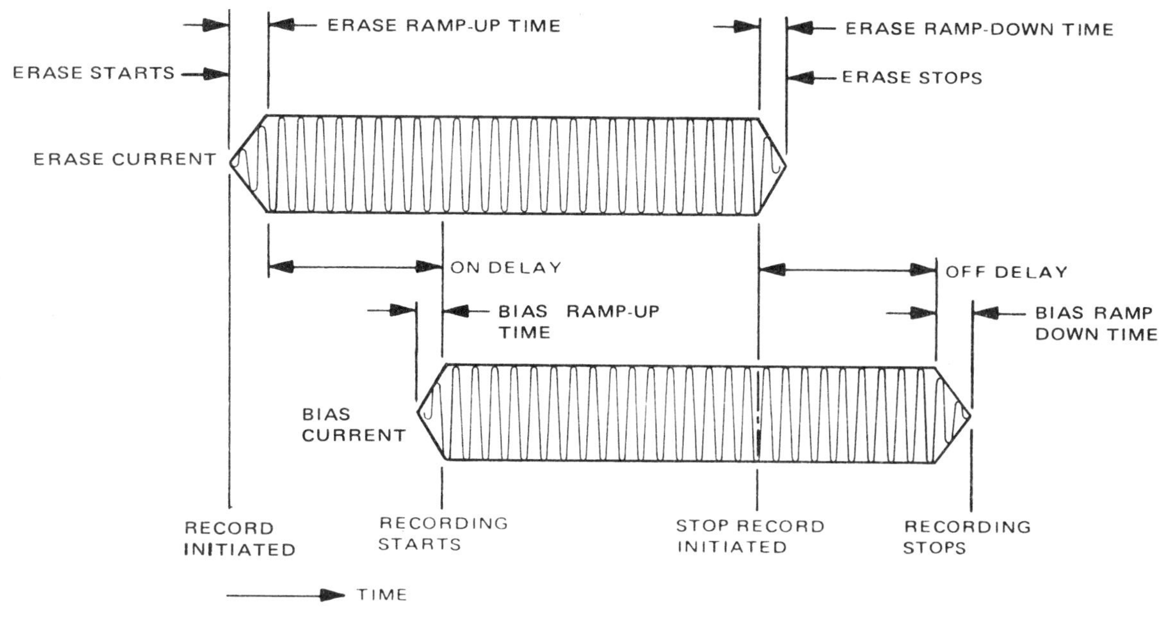

Figure 3-2. PURC On/Off Timing Relationships

SMPTE time-and-control code signal, precise editing can be obtained.

3-14. Insert Edit. A suggested method of performing an insert edit is as follows:

1. For channel(s) to be edited, simultaneously press the channel select pushbutton switch and the READY pushbutton switch.
2. For channel(s) not to be edited, simultaneously press the channel select pushbutton switch and the SAFE pushbutton switch.
3. Rewind the tape to a point before the point to be edited.
4. Press play pushbutton switch to place system in play mode.
5. Press and hold record pushbutton switch, and then press and hold stop pushbutton switch.
6. While still holding the record and stop pushbutton switches depressed, press play pushbutton switch just prior to the edit point (Table 3-5) to start record mode.
7. Continue holding all three pushbutton switches depressed and, just prior to the end of the edit point (Table 3-5), release play pushbutton switch to stop record mode. Transport will continue to run in play mode. Release other pushbuttons and enter any mode desired. (Note that system stays in record mode only as long as play pushbutton switch is held pressed.)

*** Please refer to Illus. I6-5 -- the Ampex 440 Excerpt ***

With these four -- the non servo (Ampex 440 B), the single servo (Ampex 1200), the closed-loop (Mincom M-79) and the double servo (Ampex ATR-100) -- we can consider all the major types of transport covered. Surely there are other makes, and each make has interesting features, but reading the supplied operator manual for each new machine encountered will be the surest and most rewarding means of knowing each of them.

SEARCH TO CUE

Mincom started it all out with the Selectake One. This handy, although simple device, would find a given point on the tape as long as you allowed at least one minutes worth of overshoot tape beyond the selected point. When the search function was activated, the Selectake One would rewind to the cue point and as it passed the point where it had been "cued," it would (going at full rewind speed) change the deck into fast forward. Now at fast forward, it would again scream past the cue point and enter rewind. This process would go on and on until the tape had passed over the cue point slowly enough so that the deck would feel like going into stop mode.

Then came the Selectake Two.

*** Please refer to Illus. I6-6 -- The Selectake Two Instructions ***

This device figures out the stopping torque during rewind and only switches into fast forward at the exact moment when, by doing so, the tape will come to rest motionless at the cue point. (There are also 10 cue points instead of just one.) It is seldom wrong and in fact it is more skillful at locating than a human operator. As the tape plays, the operator can press the down arrow at any time and, by doing so, store the exact moment of the tune into the locate window from whence it can be stored into memories by pressing STORE (1,2, etc.), or just searched to by pressing LOCATE. We will begin our operator instructions with this device.

At the first session put a tiny piece of 1/4 inch splicing tape at about 10 seconds from the beginning of the tape. This is to avoid running off the tape when rewinding to play the tones, or when rewinding to the beginning of the tape manually. From this marker, record one minute of 1 KHz, 10 KHz and 100 Hz tones for setting up the reproduce EQs, azimuths, and levels of all subsequent recorders. Make this marker the tape timer zero point on the tape. The following information for all tunes on each reel will be marked on the box: the song name, the take number, the length of the tune (by using a stopwatch), and the tune's position on the tape (in locator time numbers) timed from the tape's head marker or the end of the tape's head leader. Do not reset the tape time window or the Selectake Two will "forget" how many minutes and seconds it is from the top of the tape. After the reel is nearly used up, place a second marker at nearly the very end of the reel and mark on the box what the locator time of the tail marker (or the tape's tail leader) is. All other reels are head and tail marked, and the tape locator times are written down on their take sheets as well.

When you go to work on some tunes, put the reel on tails out (you're

SECTION III
OPERATION

GENERAL

This section contains the following information: location and function of the operating controls and indicators, operating instructions for the various operating modes, and the tape speed capabilities of servo system equipped tape transports.

CONTROLS AND INDICATORS

Operator controls and indicators are located on the tape transport and record/reproduce unit. Table 3-1 shows the location and function of each control and indicator on the tape transport, and Table 3-2 shows those on the record/reproduce unit.

Table 3-1. Tape Transport, Controls and Indicators

13037-16

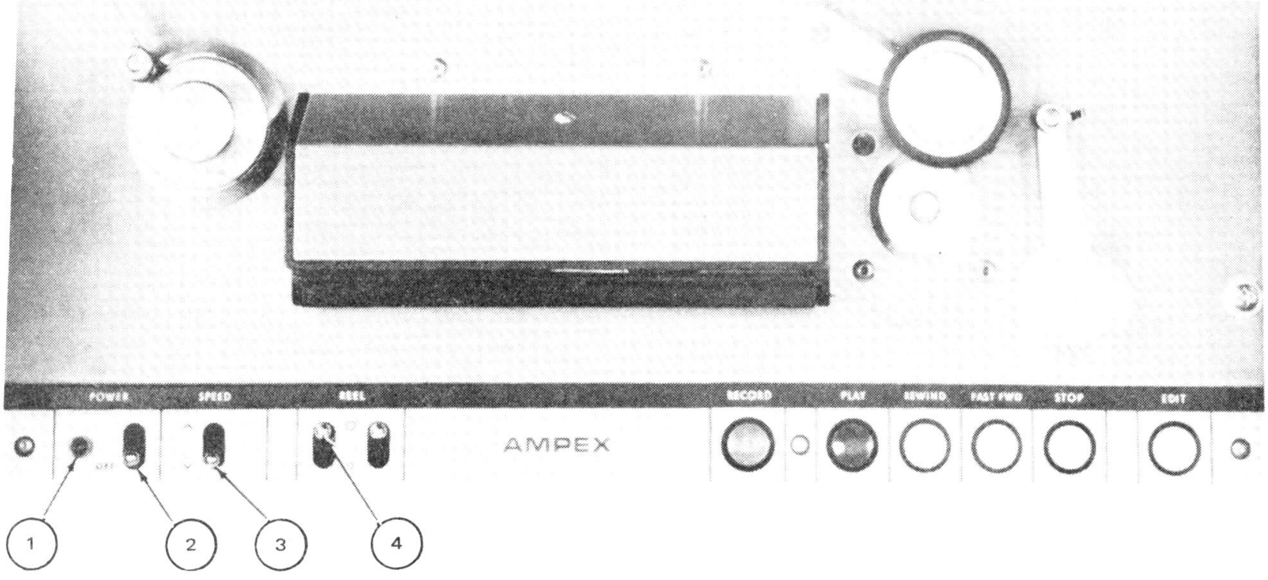

INDEX NO.	NAME	FUNCTION
1	POWER indicator	Lights when AC power is applied.
2	POWER toggle switch	Turns power to the tape recorder on and off.
3	SPEED toggle switch	Selects low (\/) or high (/\) tape speed. (Equalization is automatically switched in accordance with speed selection.)
4	Supply REEL toggle switch	Used to select appropriate tape tension for large (O) or small (o) supply-reel hubs.

Table 3-1. Tape Transport, Controls and Indicators (Continued)

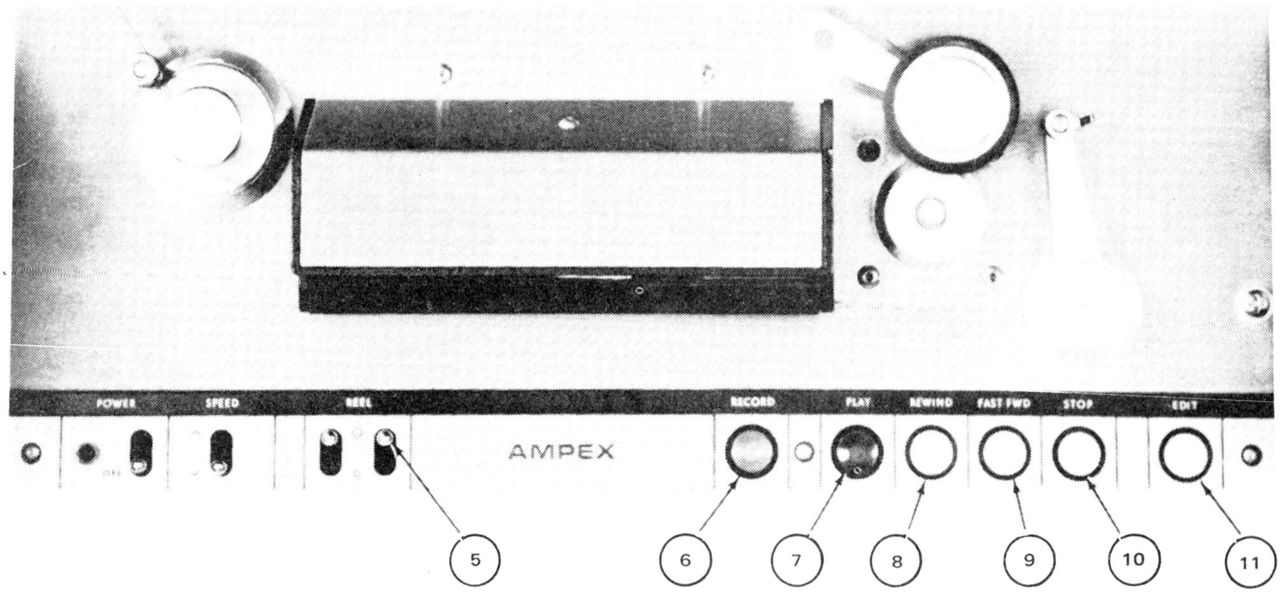

INDEX NO.	NAME	FUNCTION
5	Takeup REEL toggle switch	Used to select appropriate tape tension for large (O) or small (o) takeup reel hubs.
6	RECORD pushbutton	Used in conjunction with PLAY (index No. 7, this table) and READY pushbuttons (index No. 4, Table 3-2). With READY pushbutton depressed, pressing PLAY pushbutton and then RECORD pushbutton initiates record mode.
7	PLAY pushbutton	Used to select play mode or used with RECORD (index No. 6, this table) and READY pushbuttons (index No. 4, Table 3-2) to select record mode. Pressing PLAY pushbutton during a fast-wind mode stops the tape and then automatically starts play mode.
8	REWIND pushbutton	Used to select rewind mode. Rewind can be initiated during any mode except record and play/edit modes.
9	FAST FWD pushbutton	Used to select fast forward mode. Fast forward can be initiated during any mode except record and play/edit modes.
10	STOP pushbutton	Used to stop the tape transport and cancel existing mode of operation.
11	EDIT pushbutton	Used to initiate one of the following edit modes: **Stop/Edit.** If tape is stopped or not threaded, pressing EDIT pushbutton reduces braking force for easier tape threading or manual tape movement. **Play/Edit.** If play or record mode is active, pressing EDIT pushbutton removes power to takeup reel, causing tape to be spilled at takeup side of transport. Play/edit mode can be entered from stop mode by holding down EDIT pushbutton, pressing PLAY pushbutton, and then releasing EDIT pushbutton. **Fast/Edit.** If rewind or fast-forward mode is active, pressing EDIT pushbutton causes tape lifters to lower tape onto head to permit audio monitoring.

Table 3-2. Record/Reproduce Unit, Controls and Indicators

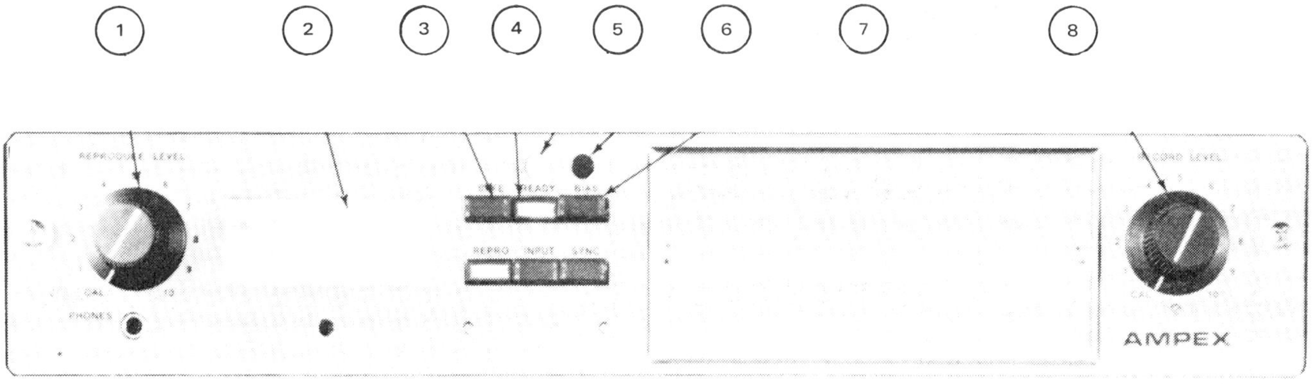

13037-13

INDEX NO.	NAME	FUNCTION
1	REPRODUCE LEVEL rotary control	REPRODUCE-LEVEL range of control used to vary normal or Sel-Sync reproduce level of associated channel. CAL position switches level-adjustment function to reproduce calibrate potentiometer (index No. 11, this table).
2	VU meter	Indicates signal level present at audio output of associated channel except when BIAS pushbutton (index No. 7, this table) is depressed while recording. Meter lights when recorder/reproducer is turned on.
3	SAFE latching pushbutton	In depressed position, prevents selection of record mode.
4	READY latching pushbutton	In depressed position, allows selection of record mode.
5	READY indicator (yellow)	Lights when READY pushbutton is depressed.
6	RECORD indicator (red)	Lights when record mode is selected.
7	BIAS latching pushbutton	In record mode, depressing BIAS pushbutton causes VU meter (index No. 2, this table) to indicate bias-signal level of associated channel. In other modes, depressing BIAS pushbutton disconnects VU meter from audio output.
8	RECORD LEVEL rotary control	RECORD-LEVEL range of control used to adjust gain of record amplifier of associated channel. CAL position switches gain-adjustment function to input calibrate potentiometer (index No. 15, this table).

Table 3-2. Record/Reproduce Unit, Controls and Indicators (Continued)

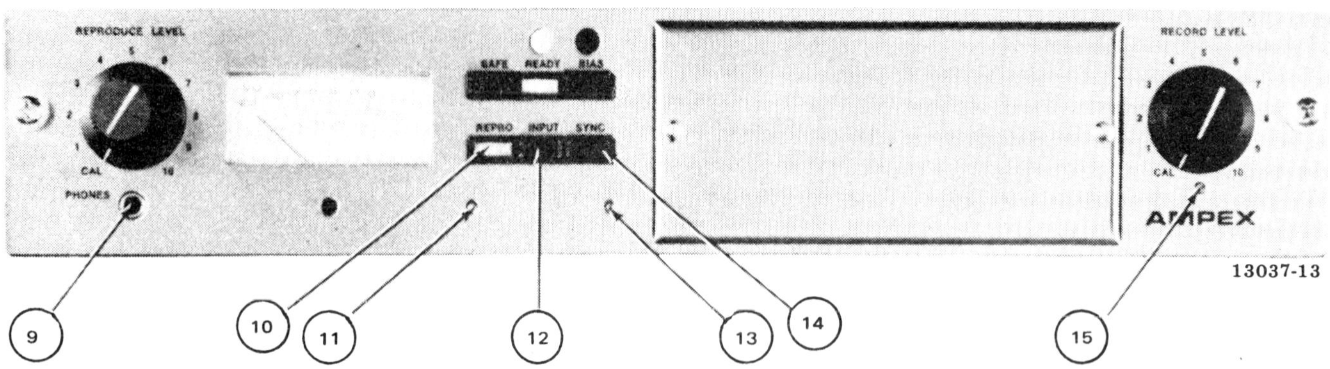

INDEX NO.	NAME	FUNCTION
9	PHONES jack	Receptacle for headphones plug. Reproduced audio of associated channel can be heard on headphones.
10	REPRO pushbutton switch	In depressed position, connects signal reproduced by reproduce head of associated channel to audio output, PHONES jack, and VU metering circuit of that channel.
11	Reproduce calibrate adjustment	Used to set operating level. (See *Operating Level Adjustment* in Maintenance section.)
12	INPUT pushbutton switch	In depressed position, connects signal to be recorded on associated channel to audio output, PHONES jack, and VU metering circuit of that channel.
13	Sync calibrate adjustment	Used to set Sel-Sync signal level.
14	SYNC pushbutton switch	Used to initiate Sel-Sync function. In reproduce mode, depressing SYNC pushbutton connects signal reproduced by record head of associated channel to audio output, PHONES jack, and VU metering circuit of that channel. If SYNC pushbutton is depressed during record mode, input signal being recorded is connected to audio output, PHONES jack, and VU metering circuit of associated channel.
15	Input calibrate adjustment	Used to adjust input level.

OPERATING INFORMATION

NOTE

In order to ensure optimum equipment performance and maximum service life, routine maintenance (refer to *Preventive Maintenance* in Section V) must be faithfully performed.

PRE-OPERATING PROCEDURES

Proceed as follows:

1. Set the transport POWER switch to on position. POWER indicator on tape transport and VU meter on record/reproduce unit will light.

2. Set transport SPEED switch to high-speed (∧) or low-speed (∨) position as required. Tape speeds of 3-3/4 (∨) and 7-1/2 (∧) in/s, 7-1/2 (∨) and 15 (∧) in/s, or 15 (∨) and 30 (∧) in/s are available. Appropriate record and playback equalizations are automatically selected to agree with the setting of the SPEED switch.

3. Set supply and takeup REEL switches to conform with the size of the hubs of the reels being used. For NAB hubs (4-1/2 inch diameter), set appropriate REEL switch to O. For EIA hubs (2-1/4 inch diameter), set appropriate switch to o. If desired, an NAB hub can be used on one turntable and an EIA hub on the other.

NOTE

When using a recorder/reproducer having 3-3/4 in/s and 7-1/2 in/s tape speeds, an AC capstan motor, and an NAB supply reel, set supply REEL switch to o (small reel position) if a fast start is to be made with less than a one-inch tape pack on supply reel. Fast starts may not be possible with an EIA reel on the supply turntable when the supply pack is less than 3/4 inch.

4. Install reel of tape on supply turntable and empty reel on takeup turntable.

5. Thread tape as shown in Figure 3-1. To facilitate tape threading, turn on recorder/reproducer and depress EDIT pushbutton,

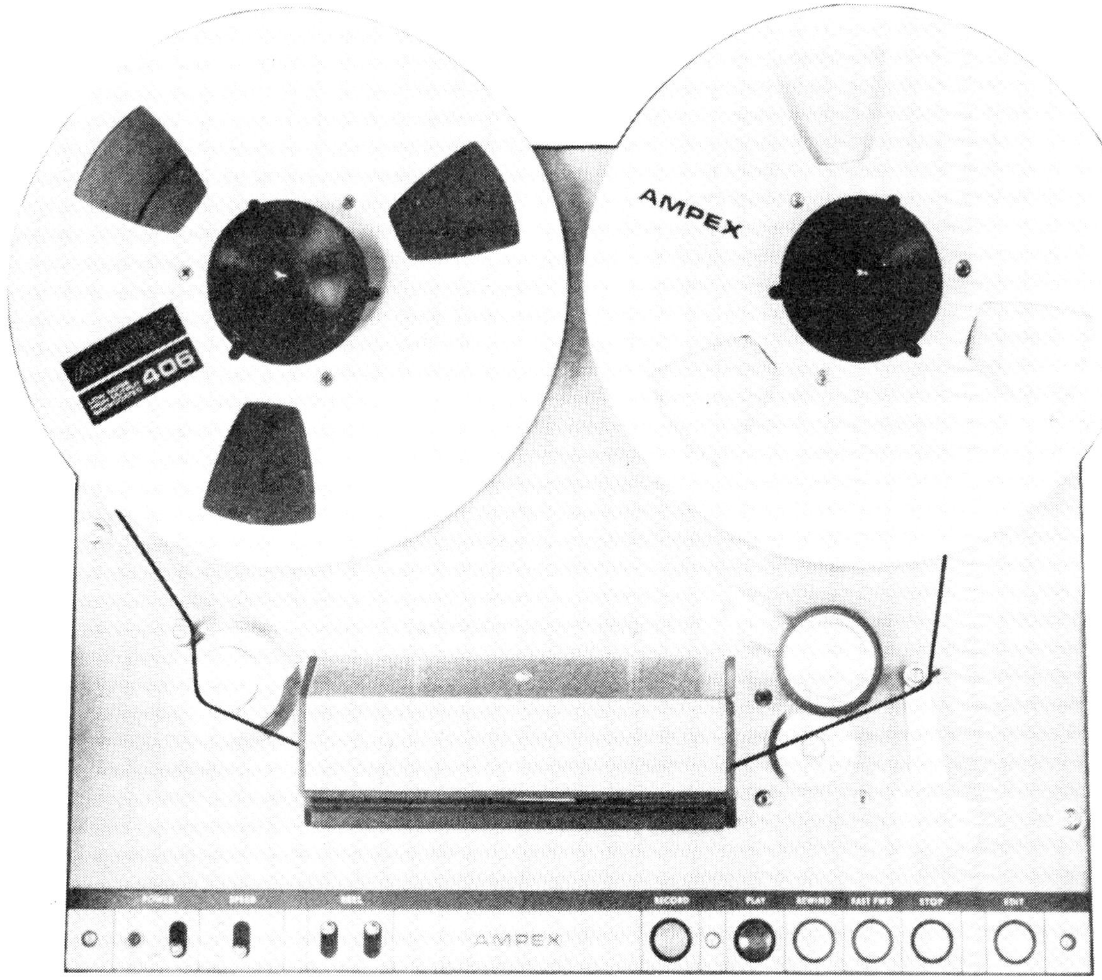

13037-16

Figure 3-1. Tape Path

which reduces turntable braking force so the reels can be easily turned by hand.

6. Anchor tape to hub of takeup reel and turn takeup reel by hand until supply reel moves. This technique removes all tape slack, which causes the tape-tension arm to close the safety switch.

7. If desired, a headset or speaker can be connected to the PHONES jack on a record/reproduce unit of the recorder/reproducer.

RECORDING

Proceed as follows:

1. Perform all steps of pre-operating procedures.

2. Depress READY pushbutton of record/reproduce unit for each channel to be recorded. The associated READY indicator(s) (yellow) will light.

3. Depress SAFE pushbutton of record/reproduce unit for each channel that is not to be recorded.

4. Depress INPUT pushbutton.

5. Connect signal(s) to be recorded to appropriate rear-panel INPUT connector(s).

6. Adjust RECORD LEVEL control so that VU meter indicates 0 for most audio peaks. (Extreme peaks may reach +2 or +3 VU.)

7. Depress PLAY pushbutton to start tape in motion. Then depress RECORD pushbutton to begin recording on selected channel(s). Record indicator (red) will light.

NOTE

While recording, the input signal of each channel can be compared with the recorded signal on that channel by alternately depressing the associated INPUT and REPRO pushbuttons.

8. When recording is complete, depress STOP pushbutton to stop tape motion and deactivate record mode.

NOTE

The recorder/reproducer automatically stops tape motion and deactivates the record mode if the tape runs completely off the supply reel.

REPRODUCING

Proceed as follows:

1. Perform all steps of pre-operating procedures.

2. Depress SAFE pushbutton of record/reproduce units.

3. For two-channel recorder/reproducers having a two-track head and a 2-Ch 4-Tr head, push the knob on head assembly down to select the two-track head or up to select the 2-Ch 4-Tr head.

4. Depress REPRO pushbutton.

5. Depress PLAY pushbutton to begin reproducing recorded material.

6. When desired material has been reproduced, depress STOP pushbutton to stop tape motion and deactivate play mode.

NOTE

If the tape runs completely off the supply reel the recorder/reproducer automatically stops tape motion and deactivates the play mode.

SEL-SYNC FUNCTION

The Sel-Sync function of a given channel can be selected by depressing the SYNC pushbutton on the record/reproduce unit for that channel. Selecting the Sel-Sync function in conjunction with reproduce mode causes the reproduced audio to be derived from the record head rather than the reproduce head. This combination of the Sel-Sync function with the reproduce mode is used in three ways.

1. Sel-Sync recording
2. Over-dubbing
3. Ping ponging

In Sel-Sync recording, a performer listens to one or more previously recorded tape tracks using the Sel-Sync/reproduce mode while recording material on another track. For example, assume a four-channel recorder/reproducer is equipped with a tape having two prerecorded tracks and two blank tracks. Typically, the two prerecorded tracks are reproduced (using two of the record heads for pickup), mixed together using studio equipment, and fed to a performer's earphones. The performer then listens to the prerecorded material while recording material on one of the blank-track channels. Thus, the new material is recorded in synchronism with the prerecorded material.

In overdubbing, a performer listens to material that he previously recorded on one or more tape tracks using the Sel-Sync/reproduce mode. The performer can repeat his previous performance (but not record it) and get into proper timing with the original material. At the point where the overdub is desired, the track(s) to be overdubbed is (are) switched to the record mode. The audio that the performer hears is automatically switched from the off-tape audio to the input audio the performer is recording when the record mode is selected.

Ping ponging is a process of reproducing two or more tracks of previously recorded material using the Sel-Sync/reproduce mode, mixing the reproduced signals together, and simultaneously rerecording the mixed signals on another track.

CAUTION

OFTEN IT IS NOT POSSIBLE TO RERECORD THE MIXED SIGNALS ON A TRACK ADJACENT TO THE HOME TRACK OF EITHER OF THE MIXED SIGNALS. THIS RECORDING PROBLEM IS CAUSED BY THE CROSSTALK BETWEEN THE HIGH-LEVEL SIGNAL BEING RECORDED ON ONE CHANNEL AND THE LOW-LEVEL SIGNAL BEING REPRODUCED FROM THE ADJACENT CHANNEL. THE CROSSTALK CAUSES A FEEDBACK CONDITION THAT CAUSES THE CIRCUITS TO OSCILLATE. THEREFORE, IT MAY NOT BE POSSIBLE TO PING PONG ON A TWO-TRACK RECORDER/REPRODUCER. USING A FOUR-TRACK RECORDER/REPRODUCER, THE POSSIBILITIES FOR PING PONGING ARE RECORDING CHANNELS 1 AND 2 ON CHANNEL 4 OR CHANNELS 3 AND 4 ON CHANNEL 1.

FAST-WINDING

For fast-winding operations press either the REWIND or FAST FWD pushbutton. For editing and cueing operations, these pushbuttons can be pressed alternately without having to press STOP between fast-winding selections. Either fast-winding mode can be entered from the stop or play mode but is locked-out when in record or play/edit mode. To enter the play mode from either fast-winding mode, simply press the PLAY pushbutton.

Two automatic tape-lifter arms, which move the tape away from the heads, are automatically actuated in both fast-winding modes. To monitor audio in either fast-winding mode, press EDIT pushbutton, which overrides tape lifters and allows the tape to move across the heads. When manual override is desired, open head gate and push back either tape lifter.

NOTE

If the tape runs completely off the supply reel the recorder/reproducer automatically stops tape motion and deactivates the play mode.

EDITING

Three edit modes are available; they are: stop/edit, play/edit, and fast/edit. These modes are selected as specified in Table 3-1.

SERVO EQUIPPPED TRANSPORT

Speed Pair Selection

The front panel SPEED switch on the AG-440C permits the selection of the high or low operating speed. The assignment of high or low operating speed is determined by a strapping arrangement on the servo printed wiring assembly. Selected operating speeds may be any two of the following: 30 in/s, 15 in/s, 7-1/2 in/s and 3-3/4 in/s. Strap the speed pair as follows:

1. On the transport control panel, set the POWER switch to OFF.

2. Remove the servo printed wiring assembly (Figure 3-2) from the servo chassis (Figure 3-3).

3. Connect a jumper from the terminal associated with the low position of the SPEED switch, designated E2, to the terminal associated with the desired low tape speed (E1, E3 or E4). Refer to Figure 3-2.

4. Connect a jumper from the terminal associated with the high position of the SPEED switch, designated E5, to the terminal associated with the desired high tape speed (E3, E4 or E6).

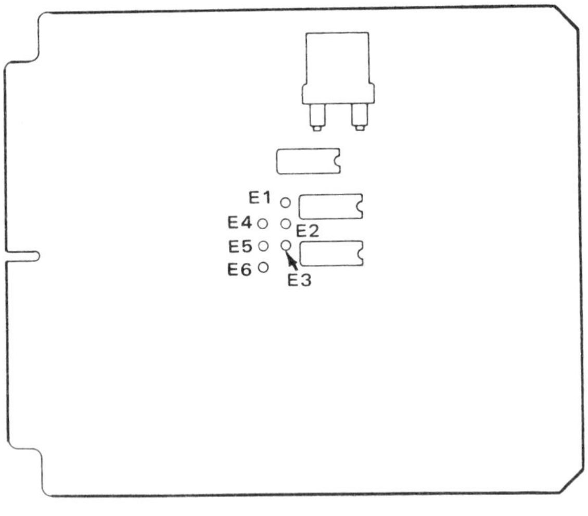

SPEED STRAPPING				
RANGE		30/15	15/7½	7½/3¾
HI	E5 TO	E6	E3	E4
LO	E2 TO	E3	E4	E1

Figure 3-2. Tape Speed Pin Strapping

5. Replace the servo printed wiring assembly in the servo chassis, component side rearward (i.e., facing away from the AG-440C).

Variable Speed Mode

The use of the dummy plug in J4 of the servo chassis (Figure 3-3) causes the capstan servo to operate with a fixed reference frequency of 9.6 kHz. To operate the system at variable speeds, remove the dummy plug from J4 and connect a sine or square wave generator having an output of 3 to 30 Vrms across pins 2 and 3 (ground) of a similar plug. Refer to Figure 3-4.

Insert the new plug into J4 on the servo chassis. If a similar plug is not available, the dummy plug can be modified by removing the existing jumper between pins 1 and 2 and connecting the generator as described above. Once connected, the frequency of the generator can be used to control the speed of the tape in accordance with the values given in Table 3-3.

Table 3-3. Tape Speeds for Various Reference Frequency Inputs

TAPE SPEED (IN/S)	INPUT FREQUENCY
3-3/4	1200
7-1/2	2400
15	4800
30	9600

NOTE

The values given in Table 3-3 are with the SPEED switch in the "high" position and jumper set for 30 in/s.

Capstan Operational Options

Three capstan operational options are available; they are:

1. Capstan rotates at selected speed whenever tape is threaded and recorder/reproducer

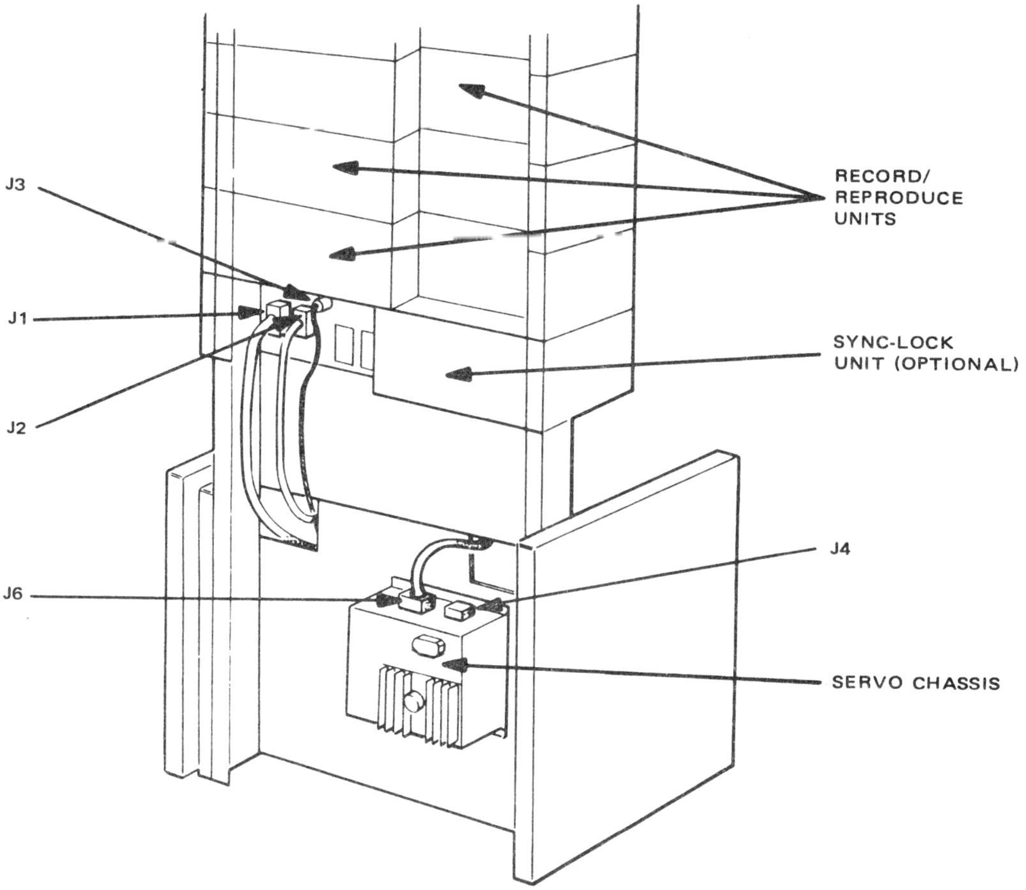

Figure 3-3. Servo Chassis Location

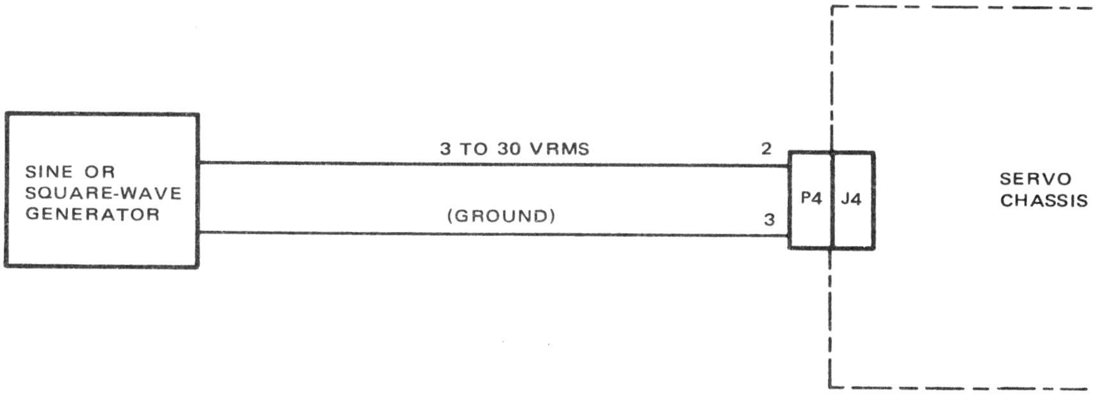

Figure 3-4. Variable Speed Operation

is turned on. To select this option, remove relay K1 from its socket on the servo electronics chassis (located at rear of console). This option may be used at 3-3/4 in/s, 7-1/2 in/s, and 15 in/s tape speeds but not at 30 in/s tape speed because a tape loop is normally thrown that opens the safety switch. Opening the safety switch stops the tape and cancels the current mode of operation.

2. Capstan rotates when tape is threaded, recorder/reproducer is turned on, and low tape speed is selected; however, for high tape speed, capstan only rotates after the play or record mode is initiated. This option is used for 15 in/s and 30 in/s recorder/reproducers when fast starts are to be made at 15 in/s tape speed. To select this option, set the CAPSTAN STOP switch on the servo chassis to HIGH SPEED.

3. Capstan only rotates when in play or record mode at both speeds. To select this option, place the capstan stop switch on the servo chassis in the BOTH SPEED position. If a fast start is desired in this mode, press and hold STOP pushbutton and then PLAY pushbutton. The capstan motor will start and lock up in less than two seconds. When fast start is desired, release STOP and then PLAY pushbuttons. The tape will start moving the instant the STOP pushbutton is released.

REMOTE CONTROL OPERATION

Transport operation from the Ampex Remote Control Unit (Part No. 4010080) duplicates the record, play, rewind, fast forward and stop controls on the tape transport. On the remote control box, the EDIT pushbutton only performs the fast wind/edit function of defeating the tape lifters.

The standby lamp (located between the STOP and EDIT pushbuttons) indicates that the tape is threaded and the remote controls may be operated. If the transport is left in the stop/edit mode, with the tape threaded, the standby lamp will not light. To operate the transport under this condition, press the remote STOP pushbutton to cancel the stop/edit mode; the standby lamp will come on and the remainder of the remote controls will function.

storing tape tails out right?), rewind the few feet to the tail leader or marker, set the reading written down on the box for the tape time window to be showing at this point on the tape into the locate time window with the keyboard, and then transfer these numbers into the tape time window with the up arrow. Now the Selectake knows where it is on the tape, so just read the beginning time of the take you want played and keyboard these numbers into the locate time window -- press LOCATE and press PLAY. The Selectake will zoom to within inches of the proper place on the tape and play back the take.

The set of locator tape times (but not tune times of course) will have to be discarded when you edit the master reel (don't forget to make a safety copy) and a new set of locator tape times will have to be made. There is no need for a head marker now since the head leader on the master tape can serve to re-orient the locator. In fact, any leader can be used to re-orient the tape time readout since you have a timing for each leader written on the box. If the tape time window should "get lost," put the tape on the last of any head leader of any tune, read the name of the tune (you are writing the tune titles on the head leaders right?), and look up the proper locator tape time for this tune; enter this time in the locate time window and transfer it up into the tape time window.

Going all the way with fancy operations would have you making additional logs of locator tape time readings where the verses, choruses, and solos were in each of the tunes, and writing these down on a "locator log sheet" as well. This you can do since the slight errors that build up over the hours can be reset any time you're on a head leader. If you do make fancy logs in locator tape times, the recall zero function can be very useful for finding out the duration of any song section. By setting, say, the lead solo starting point of tune three (18:33) in the tape time window (temporarily losing the proper tape time), and the ending of the lead solo in the locate time window, and pressing RECALL ZERO, you will get the minutes and seconds of the solo. Actually any two points on the tape (or in the tune) set in these two windows, will give an absolute time result on recall zero function.

Another popular, although very simple, search to cue is the Ampex Single Point search to cue. This device simply lets you do two operations: set an arbitrary zero at some point on the tape, and lets you cue any point on the tape (by pressing CUE so that by pressing SEARCH you can return to that one spot9. You cannot set an actual tape location since all you can do with the tape time window is set it to zero (you can't set it to the tape time at the end of the reel or at a leader). The tape time window readout and the search point are not connected in any way so that if you reset the counter to zero the search point you had selected on the tape is not lost. Unlike the Mincom M-79, which makes quiet rewinds, the Ampex machine does not mute, so to make quiet rewinds (or fast forwards) the engineer has to either:

1. Solo an unused fader and turn down all the cue sends to kill the rewind noise in the phones
2. Select console program mode instead of sync (tape) so that rewind noise will not be present
3. Or use the standby monitor function to reduce the highly unmusical rewind noises

SPECIFICATIONS

TAPE TIME DISPLAY

Tape motion driven, digital time display covering times from 00.00 through 99.99 minutes.

Automatic "Hi-Low" speed switching (manual select for 7-1/2 - 15 ips and 15-30 ips ranges).

Associated RESET button allows independent "zeroing" and can be "set" to any desired time by shifting the Locate Time (covered below).

Associated ↓ (shift) button allows shifting of Tape Time into the Locate Time. This allows direct Locate Time entry while playing a tape.

Automatically set to 00.00 upon application of power.

Display "frozen" on tape runout. This allows rethreading without loss of location in the event of tape runout.

LOCATE TIME DISPLAY

Digital display covering times from 00.00 through 99.99 minutes. (The displayed time is the tape time to which the machine will "home" if the locate function is initiated.)

Time entry is via the keyboard, shifting the Tape Time, or from one of nine memory points (covered below).

Associated RESET button allows independent "zeroing" of display without affecting the memory.

Display is automatically set to zero upon application of power.

Associated ↑ (shift) button allows shifting of the displayed Locate Time into the Tape Time display.

KEYBOARD

Numbers 0 through 9 plus STORE and RECALL.

Used to enter locate times and to store times in memory (up to nine points - press STORE, then desired memory position number, 1 through 9) or to recall a previously stored time from memory (press RECALL, then desired memory position number).

MODE CONTROLS

LOCATE – Backlighted pushbutton causes the tape to be driven at high speed to Locate Time and stop within ± 1 count without hunting. (See also PLAY.)

During the LOCATE cycle, a PLAY command (machine or remote) can be stored then entered at the end of the cycle.

Can be entered from any mode.

Is cancelled by any (machine or remote) STOP, REWIND, or FORWARD command or tape runout.

PLAY – Backlighted pushbutton.

Causes tape to be driven by the capstan.

Can be entered from STOP, REWIND, and FORWARD.

If entered in LOCATE the command (machine or remote) is stored until the cycle is complete, then the PLAY mode will be initiated.

RECORD – Backlighted pushbutton.

Energized RECORD buss.

Can only be entered in combination with PLAY when the tape is stopped or capstan driven.

REWIND – Backlighted pushbutton.

Causes tape to wind at high speed on the supply reel.

Can be entered from any mode.

FORWARD – Backlighted pushbutton.

Causes the tape to wind on the take up reel at high speed.

Can be entered from any mode.

STOP – Backlighted pushbutton.

Causes tape motion to cease and places machine in a standby (ready for any mode command) status.

Can be entered from any mode.

GENERAL DESCRIPTION

GENERAL

The Mincom Selectake II, Tape Time Locator uses the latest in microprocessor technology to provide a compact, reliable, tape position locator assembly. The unit is an accessory for use with Mincom Model 79 Professional Audio Recorders. A four digit readout displays tape time from an arbitrarily determined zero time reference. Tape time is derived from sensors in the M79 and is read out in minutes, tenths of minutes, and in hundredths of minutes. (Fractional minutes may be converted to equivalent seconds by multiplying the indicated decimal fraction by 60.) Points of interest on the tape are identified (for relocation) by logging the tape time readout corresponding with each event. The Selectake II has independently addressable memories which allow up to nine such tape time locations to be stored simultaneously. These location times may be entered directly during the recording process or they can be programmed into the Selectake II at a later time. The content of any memory can be recalled or rewritten at any time and in any order.

Locate times are entered on a second four digit display. The time indicated on the locate time display is the time to which the tape will home if the locate function is initiated. Locate times can be entered directly into the display using the data entry keyboard; transferred from the tape time display by pressing a tape time transfer switch, or recalled from a previously loaded memory. The locate time display can also be used to set tape time. The tape time of an upcoming event is entered into the locate display; when the event occurs, the locate time transfer switch is pressed to transfer the locate time to the tape time display.

The Selectake II has a tenth recall address: RECALL "0". In this mode the locate time display reads out the absolute difference between the selected locate time and the tape time. All data entry keys are locked out in this mode. The function control switches grouped in the lower portion of the front panel remain active. The mode is terminated by pressing the RECALL switch.

The Selectake function control switches provide complete mode control of the Model 79 Recorder after power is applied, tape is threaded, and the system placed in Standby operation. (These functions must be accomplished at the console.) The control unit can be located at any site within a radius of approximately 50 feet dependent upon interconnecting cable length. (A 30 foot interconnecting cable is supplied.)

OPERATION

GENERAL

The Selectake II functions as an integral part of the M79 Professional Audio Recorder. It accepts power and tape motion information from the M79 and delivers manual or automatic operating commands to it. Power is applied to the accessory when the M79 power is turned on. The function controls on the Selectake control unit are enabled after the M79 is put into the standby mode. This is accomplished at the M79 console by pressing the STOP switch (after the tape is threaded). When the STOP switch is illuminated, the M79 is in standby. The Selectake II functions as a standard remote control and as a programmable data location device.

A stop mode select switch located on the M79 must be placed in the Selectake II position when the Selectake is used. If the Selectake is disconnected, the switch must be placed in the alternate position for normal M79 operation. The switch is mounted to the right of the cage. Access is gained by raising the top of the M79 console.

CONTROLS AND INDICATORS

All normal operating controls and indicators are located on the sloping front panel of the Selectake II control unit. A two position speed RANGE switch is located on the rear panel. This switch is set to correspond with the speed range of the M79. Figure 1 illustrates all front panel controls and indicators. Locator numbers are assigned to each and a brief description given. The applicable locator numbers are repeated at appropriate points in the individual operating procedures as a convenience and a cross reference.

(1) *TAPE TIME DISPLAY.* Four Digit Display indicates time from 00.00 through 99.99 minutes.

(2) *TAPE TIME DISPLAY SHIFT.* Allows shifting of tape time into locate time display.

(3) *RESET.* Sets time display readout to 00.00 which automatically establishes tape time zero reference.

(4) *LOCATE TIME DISPLAY.* Four Digit Display indicates time from 00.00 to 99.99 minutes. (Time displayed is tape time to which M79 will home if LOCATE is commanded.)

(5) *LOCATE TIME DISPLAY SHIFT.* Allows tape time zero reference to be reset by shifting new time into tape time display.

(6) *RESET.* Sets locate time display readout to 00.00.

(7) *BASIC KEYBOARD.* Numbers 0 through 9 plus STORE and RECALL used to enter locate times and to store locate times in memory. RECALL permits recall of previously stored locate times. RECALL 0 is a special case which causes locate time display to indicate difference between selected locate time and tape time. All keyboard entries are inhibited in RECALL 0 operation. Condition released by pressing RECALL.

MODE CONTROLS

(8) *LOCATE.* Backlighted pushbutton switch causes the tape to be driven at high speed to Locate Time and stop within ± 1 count without hunting. (See also PLAY.) Is cancelled by any (machine or remote) STOP, REWIND, or FORWARD command or tape runout.

(9) *PLAY.* Backlighted pushbutton switch which causes tape to be driven by the capstan. Can be entered from STOP, REWIND, and FORWARD.

If pressed during LOCATE, the command (machine or remote) is stored until the cycle is complete. Then the PLAY mode will be initiated.

(10) *RECORD.* Backlighted pushbutton switch energizes RECORD Buss. Can only be entered in combination with PLAY when the tape is stopped or capstan driven.

(11) *STOP.* Backlighted pushbutton causes motion to cease and returns machine to standby (ready for any mode command) status. Can be entered from any mode.

(12) *REWIND.* Backlighted pushbutton switch causes tape to wind at high speed on the supply reel. Can be entered from any mode.

(13) *FORWARD.* Backlighted pushbutton switch. Causes the tape to wind on the takeup reel at high speed. Can be entered from any mode.

(14) *RANGE SWITCH.* Two position slide switch located on panel. Set to position with nomenclature corresponding to M79 operating speeds.

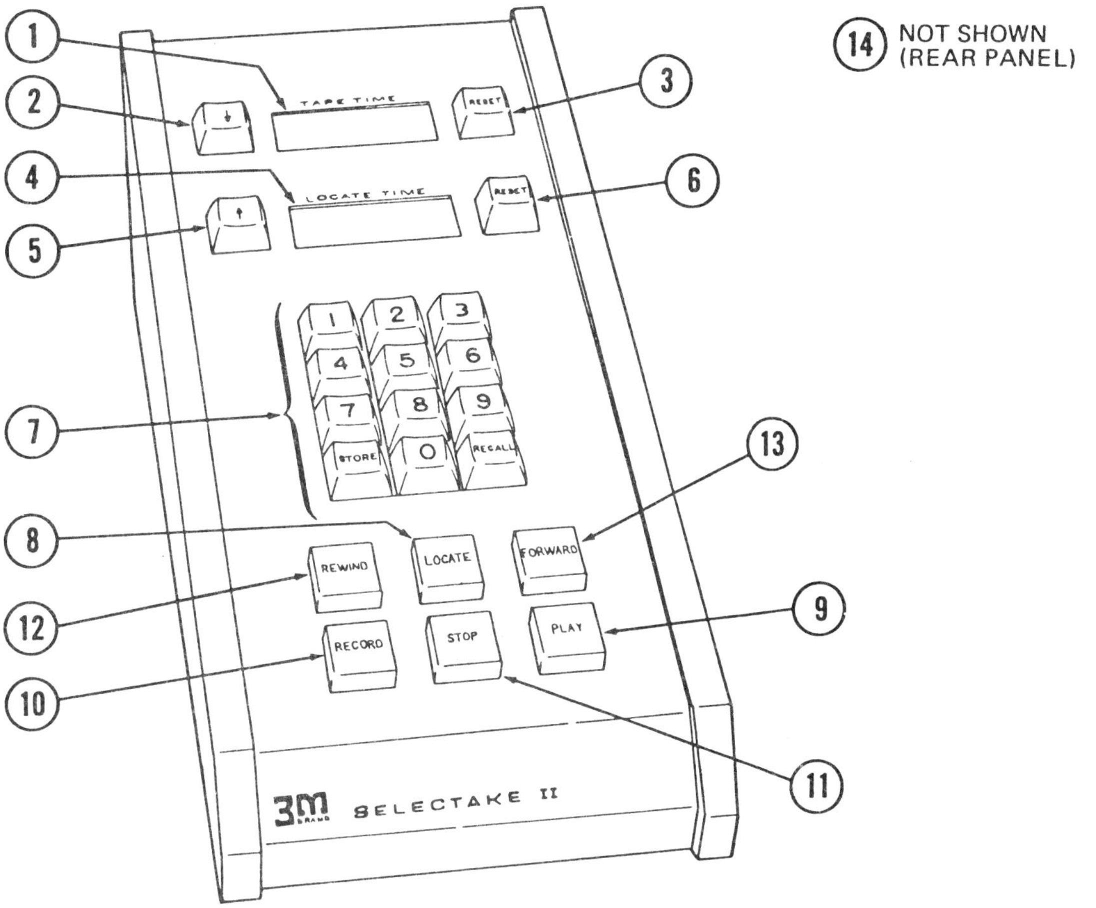

Figure 1. Controls and Indicators

OPERATING INSTRUCTIONS

The Selectake II is a tape time locator. It identifies a specific tape position in terms of tape time elapsed from a zero tape time reference. Data is located by repositioning the tape to the tape time which occurred concurrently with the data. To accomplish this it is necessary to establish an arbitrary tape time zero reference or, in the case of a previously recorded tape, sync to the original reference.

Setting The Tape Time Zero Reference:

1. The Selectake II is automatically set to a zero reference when the M79 is turned on.

2. An alternate zero reference can be established at any tape location by pressing the TAPE TIME display RESET (3) switch. The RESET switch used in conjunction with a slate tone or other electrical or physical reference establishes a particularly convenient zero reference.

3. To recover the zero reference of a previously recorded tape, use the function controls (9) (11) (12) (13) to position the tape on an event for which the tape time is known then:

   a. Use the keyboard (7) to enter the tape time of the event into the LOCATE TIME display (4). Four digits must be entered reading from left to right. The first digit entered appears at the right in the display and advances to the left as additional digits are entered. Multiple digit entries are electronically inhibited, data is only accepted one digit at a time.

   b. Press the locate time transfer ↑ switch (5) to shift the time into the TAPE TIME display (1)

Remote Control Operation:

Control switches (9) through (14) duplicate those on the M79 control console and function identically. The Seletake STOP switch (11), however, cannot be used to initiate the standby mode. There is no priority override. Either set of controls can be used to operate the M79.

Also grouped with the control switches is the LOCATE switch (8). This switch instructs the microprocessor circuitry to perform the locate sequence. The LOCATE command will override any existing machine command and can, in turn, be overridden by the STOP (11), REWIND (12), and FORWARD (13) switches. A PLAY switch (9) command will be accepted during the locate sequence but will not be executed until the sequence is completed. The RECORD switch (10) is disabled during the locate sequence.

Identifying Locate Times:

The tape time readout is presented with data. When a point-of-interest occurs, the concurrent tape time readout must be noted or stored. One event time can be noted and stored directly in the LOCATE TIME display (4) by pressing the tape time transfer ↓ switch (2). Where more than one event is anticipated, each event should be logged to permit the next location to be entered into the display. If immediate use is intended, the locate time can be stored directly into one of the nine memory locations. A log should be made of the significant contents of the memory section to retain the locate times against accidental loss and to identify the memories still available.

Using The Memory Section

The input and output of the memory section is through the LOCATE TIME display (4). To store, the memory section is addressed by pressing the STORE key on the basic keyboard (7). To recall, the memory section is addressed by pressing the RECALL key. The specific memory is selected by pressing the appropriately numbered key on the basic keyboard.

Storing A Locate Time

1. Insert the number to be stored into the LOCATE TIME display (4) by:

   a. Transfer the time from the TAPE TIME display (1) by pressing the tape time transfer switch ↓ (2).

NOTE

There is normally a human response lag between event occurrence and operating the transfer key. Locate accuracy can be improved by changing the LOCATE TIME display reading (subtract normal response time) to allow for the response time error.

   b. Use the basic keyboard (7) to enter the desired locate time into the LOCATE TIME display (4)

2. Press the STORE key.

3. Press the selected memory key.

NOTE

The data entered into the LOCATE TIME display is not effected by the storing operation.

A log should be kept of the significant contents of the memory section to retain the stored locate times against accidental loss and to identify the memories still available.

Recalling A Stored Locate Time:

1. Press the RECALL key on the basic keyboard (7).

2. Press the numbered key (basic keyboard) of the memory containing the locate time to be recalled.

3. The contents of the selected memory will appear in the LOCATE TIME display (4).

Recalling Zero (0):

The RECALL "0" address initiates a difference mode. The LOCATE TIME display (4) reads out the absolute difference between the locate time and the tape time.

1. Press the RECALL key (7).

2. Press the "0" key (7).

3. The LOCATE TIME display (4) will indicate the difference reading.

NOTE

All address keys except RECALL are locked out in this mode.

4. The RECALL "0" mode is terminated by pressing the RECALL key. Operation of the Selectake will return to normal and the LOCATE TIME display (4) will return to the reading which existed prior to initiating RECALL "0".

Locate Operation:

1. Enter the location time in the LOCATE TIME display:

   a. Directly using the keyboard, or

   b. by recalling the locate time from the memory section.

2. Press the LOCATE switch (8). The M79 tape transport will begin a high speed tape transfer in the direction required to bring the TAPE TIME display (1) to the time entered in the LOCATE TIME display (4). The high speed transfer will automatically slow as the locate time is approached and stop when the TAPE TIME display reads within ± 1 count of the locate time.

*** Please refer to the next section on Master Boxes of Tape Machines***

Audio Kinetics makes another fancy locator called the Intelocator. The XT-24 Intelocator is by far the most powerful locator around. With it you can:

1. Search out a given song and play it for a given time (usually the length of the tune) and then automatically rewind to the beginning of the song and stop or repeatedly play the selection over and over
2. Search out a given tune and punch in; record to the spot you specified; rewind; and then play back the punched section once or repeatedy
3. Store four times into memory, whereupon you can then go to these points just by pressing the proper memory button
4. Test all locator control lamps and digit readouts
5. Display tape speed in inches and 10ths of inches per second
6. And keep two separate running tape times, one from the head of the tape and another from the head of the tune, so that you can have the tape time indicator window read either time. That is, you can have the window read so that zero is at the head of the tape (this is master counter time) or so that zero is at the head of, say, the 4th tune on the tape (this is arbitrary zero time).

The XT-24 is used as follows:

Load the tape tails out as before. Rewind slightly to the end of tape marker or the beginning of the tail leader and then set the proper tape end time from the box legend on the Intelocator's "lever wheels" (the number thumb wheels). Now press MASTER SET LEVER. This sets the master counter to the correct actual tape time at the tail of the tape. Now that the Intelocator knows where it is, you can now go to any song by entering the starting master time of the song in the lever wheels and pressing LEVER.

When you get to the song of interest, press SET ZERO and the readout will go to arbitrary zero so that the song beginning can now be located by simply pressing ZERO. The master counter still knows where it is and you can check it by pressing MASTER, which will reveal the master location of the tape for 1 1/2 seconds and then go back to reading arbitrary zero time. Arbitrary zero mode will continue until you press SET MASTER, which will give back the master counter readout in the window and lose the arbitrary zero point.

The only slightly confusing part of this locator is the use of the memory registers. Memory is kept in master time digits even if you have the device in arbitrary zero mode. The way you ordinarily use the memories is to press SET (1), or (2) through (4), as you pass interesting points in the song you want to locate easily. This places the current tape location in the memory you select, but the actual time stored in memory is master time, not arbitrary zero time. If you park the tape at, say, 13:22, then press SET ZERO for arbitrary zero. Now press SET (1) and you will find 13:22, not 00:00, in memory one. To have a look at memory one through four, press STOP and (1) together, or (2), etc. What this means is that when you want to put digits from the thumb wheels into the memories you have to think in terms of master time. To put a time into a memory, say, 27:35 into memory #4, first enter the numbers in the lever wheels and then press MASTER SET (4).

An example:

Suppose you wanted to search out the third song on the reel and have the deck go into play and repeatedly play the song over and over until you pressed stop. You would load the tape, position the tape at the beginning of the tail leader, look up the tail leader time from the legend on the box, set this time into the thumb wheels, press MASTER SET LEVER. The master timer is now set and the locator knows where it is on the tape. Look up the locator time of the third song and set it into the thumb wheels and press MASTER SET (1), memory one now has the locator time of the begining of this tune. Look up the tune length and mentally add the tune length to the locator start time (or use the written time of the next leader) and set the sum in the lever wheels and press MASTER SET (2). Now memory two has the master time at the end of the tune. All you have to do now is set up a cycle between memory one and two. To set up a cycle press SET CYCLE (1) (2). You can set up cycles between any of the locator buttons (lever, zero, mem 1,2,3,4). The only rule seems to be that you do not use a second location that is in front of the first on the tape. Since the tape can not play backwards, the XT-24 will respond with, "Huh?"

If you have the recorder at the beginning of a tune you are working on, with the window set to arbitrary zero, and you want to play the tune over and over, set the lever wheels to the tune length and SET CYCLE ZERO LEVER, which will play the tune. It is important to remember that only the memories relate to master time; the zero and lever wheels relate to whatever's in the window, either master time or arbitrary zero time. So if the master time is in the window SET MASTER and you press ZERO, you will go to the head of the tape; if arbitrary zero time is in the window (set) (zero) and you press ZERO, the tape will go to the head of the tune. Pressing (lever) will cause the lower locator to read the digits set in the lever wheels, so if you lever up 1:00 and press LEVER, in master time this will be 1 minute from the head of the tape; in arbitrary zero time, one minute from the head of the tune.

Once you have set a cycle into the XT-24, you can then command a CYCLE or RE-CYCLE at will as long as you do not set up new cycle start and stop points. There is a PLAY and RECORD CYCLE and RE-CYCLE. The PLAY CYCLE searches, plays,and rewinds to the first point and stops. The RE-CYCLE searches, plays and continues searching and playing forever unless the machine wears out or you break up the command.

The record version is obtained by pressing RECORD and either CYCLE or RE-CYCLE together. These commands are just like the non-record commands except that the first pass is with record function activated. So a record cycle searches, records, searches, plays, searches, stops; (and the record re-cycle continues to search and play forever....).

An additional function, which is not even included in the company's original manual, is the function of simple SEARCH PLAY. To accomplish this function, press the blue search button for the tape location you wish and then SET PLAY.

MASTER BOXES OF MULTITRACK TAPE MACHINES

The first master box to be considered is the Ampex 1200 Master Box since it is more complex than its predecessor (the Ampex 1100).

*** Please refer to Illus. I6-3 -- The Ampex 1200 Master Box Instructions ***

Operationally, there are only four modes of operation for this machine:

1. Alignment mode (for aligning the tape machine): all singles on input (for setting up input); master switch on repro throughout (except for setting input)

2. Mixdown mode: all ready/safe switches on safe; master switch on repro, standby switch on singles, single switches on sel sync (for quieter rewinds)

3. Overdubbing rundown mode (practicing but not yet recording): all ready/safe switches in safe except the track(s) to be overdubbed, which are in ready; all singles in sel sync except the track(s) to be overdubbed which is (are) in input; master monitor switch in the singles position; standby monitor switch in the singles position

4. For true overdubbing while recording, including rolling punch-ins and outs (having the machine switch to input for the console monitor system and cue system): all ready safe switches in safe except the track to be overdubbed, which is in ready; all singles in sel sync except the track to be overdubbed, which is in input; master in sync; standby monitor in single (this just means move the master monitor switch from singles to sync to change from overdub rundown to fully automatic input switching)

These are the basic settings for operation. Understanding and predicting the standby monitor function requires the definition of what the standby modes are.

The standby monitor modes:

Stop
Rewind
Fast forward
Search

The non-standby monitor modes:
Play
Record
Search/play

When the standby monitor switch is to the left (which should really have been called "off"), the machine's monitor does not revert to the settings of the singles during standby modes. If the switch is to the right (which should have been called standby monitor on), then the machine monitor functions do revert to the settings of the single switches, according to the tables above.

There are two additional logical rules that apply above and beyond the standby monitor rules:

1. The master switch in repro, standby "off" rule:
"All tracks are stuck in repro no matter what."

2. The master switch in other than repro position rule:
"Any track put into record will go to input monitor no matter what."

Combining all these rules will enable you to predict which monitor mode to expect at all times given the settings of the 1200 Master Box switches.

*** There is one change in the non-standby monitor table when using search play mode with the XT-24 that is different than Ampex search play mode and that is, the XT-24's search/play command IS a standby mode.

One benefit of knowing the rules is the "quieter rewind trick": put the singles into sync and the master in repro with the standby switch set to singles. This will give you quieter rewinds (being in sync for rewind) except if you press SEARCH PLAY, which is not a standby mode, which will cancel the benefit.

The next master box to be discussed is the Mincom M-79 Master Box.

Basically this box is used in this fashion:

1. For alignment: all individual cue buttons pressed; master buttons: repro, sync, input as needed

2. For mixing: repro master button only (all rewinds are automatically muted unless you press the mute defeat button on the deck)

3. For overdubbing (without actually punching in): all individual cue buttons pressed; master cue pressed; individual input button pressed on track to be recorded; all safe except tracks to be recorded

4. For overdubbing with punching in and out: all in cue. To record press play and record buttons together. Recording requires that you have already pressed the master record ready button and the individual track's record ready button as well (three levels of safety against accidental erasure).

TAPE SPEED PROBLEM PREVENTION (resolving)

*** Please refer to Illus. I7-1 -- the Four Methods of Resolving ***

There are two ways to assure that no tape speed problems will happen if you guard against them starting at the recording of the basic tracks. One is to record SMPTE (Society of Motion Picture and Television Engineers) code and the other is to record a 60 Hz line frequency (reduced in voltage by a suitable line transformer of course) on a track of the multitrack machine as you make your basic tracks.

Four methods are used to deal with tapes that may turn out to have speed problems:

1. Transferring the problem to the next generation by recording to a mix machine that has a free track (like mixing stereo to a four track), and transferring the 60 Hz tone from the 24 track to the mixdown machine. If the four track or 24 track has speed problems, you will resolve them away at the playing of the 4 track mix

2. Self-resolving the multitrack with a resolver and hoping that the two track mastering machine is O.K.

3. Using the 24 track playback of the 60 Hz tone as the audio input to a VSO, which runs the two track capstan motor so that the tape speed (pitch) problems of the multitrack are removed by making the two track run at the proper speed to undo them

4. If you're doing film work, drive the mag recorder (35 MM mag machine) with the sync playback (24 track or 4 track). The mag film (recording tape) has sprockets in the edge making any slippage impossible

The "transferring the problem" method might be used where you are giving out a four track with, say, one channel of mono mix, two channels for recording vocals on, and one channel of sync tone from the playback of the 24 track's sync track. The producer can then go to a smaller studio and overdub his ideas on a 4 track, and re-sync them up to the 24 track machine later. If there were speed problems with the four track at the "little" studio, the artist might sing in the "wrong" key (compared to concert pitch) by a slight amount, but the overdubbed vocals would always be sung "in tune" with the 4 track. When the four track and the 24 track were synchronized back together, the pitch problems would be "resolved." Syncing two machines using only a 60 Hz sync tone is sometimes quite challenging. Both machines are started at a mark on the tape (the position of which is discovered experimentally) and then jolted into temporal alignment during a (hopefully) long count off. If the machines are maneuvered into sync by the time the music hits, the new vocals on the four track (using our example) are then sel-synced over to the master (24 track) machine.

Resolving the 24 track and hoping that the 2 track speed is O.K. occurs most often when tracks are to be cut at a "foreign" studio. Once the tape is returned to the "home" studio, it might turn out that the music tracks were out of proper pitch. If this occurred, then the 24 track could be self-resolved. If not, then the track holding the 60 Hz sync tone could be erased and still nothing is lost. Really there is no percentage in not recording a 60 Hz sync tone during tracking sessions -- no matter what.

FOUR METHODS OF RESOLVING Illus. I7-1
METHOD #1

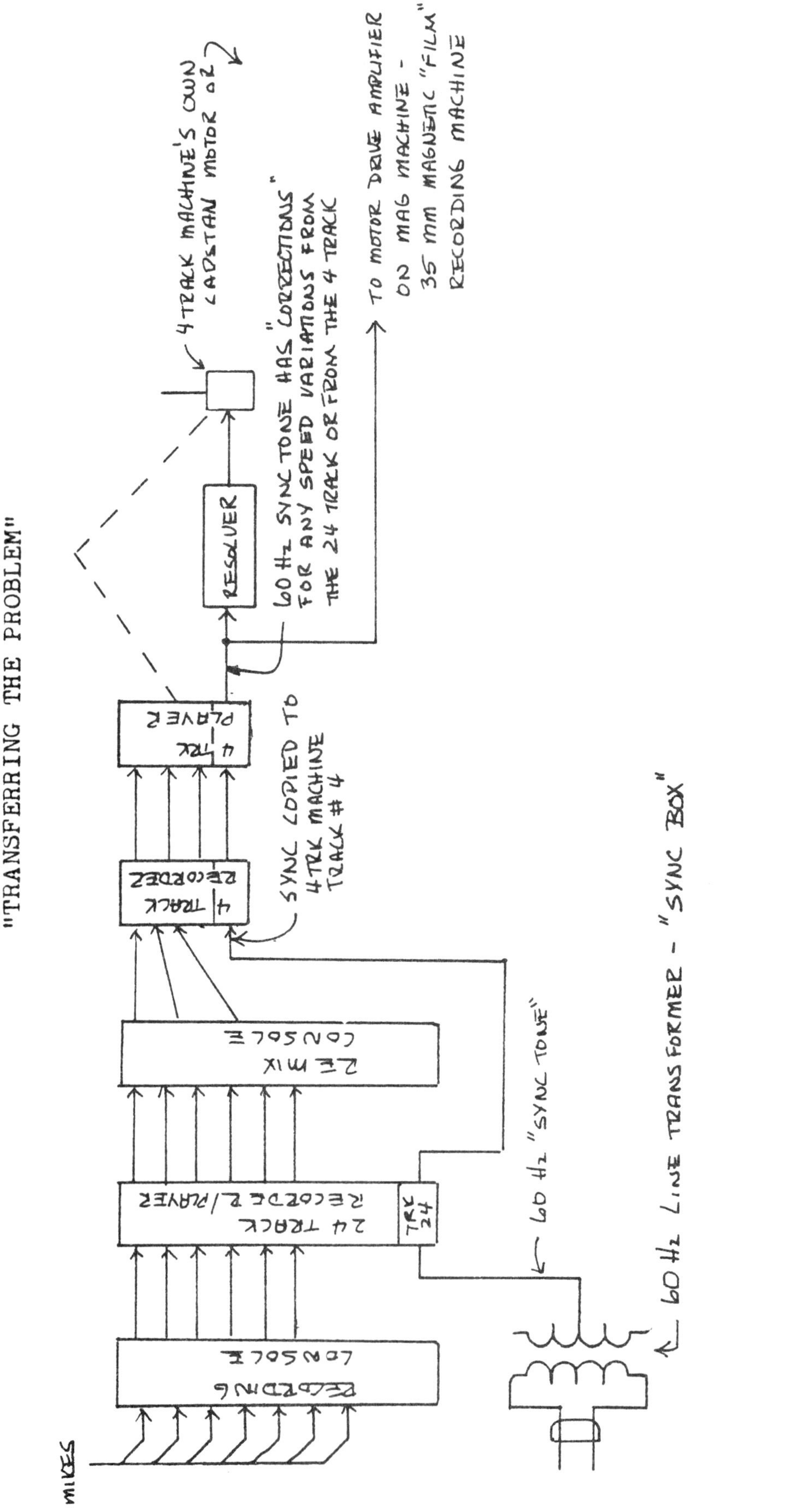

FOUR METHODS OF RESOLVING Illus. I7-1
METHOD #2

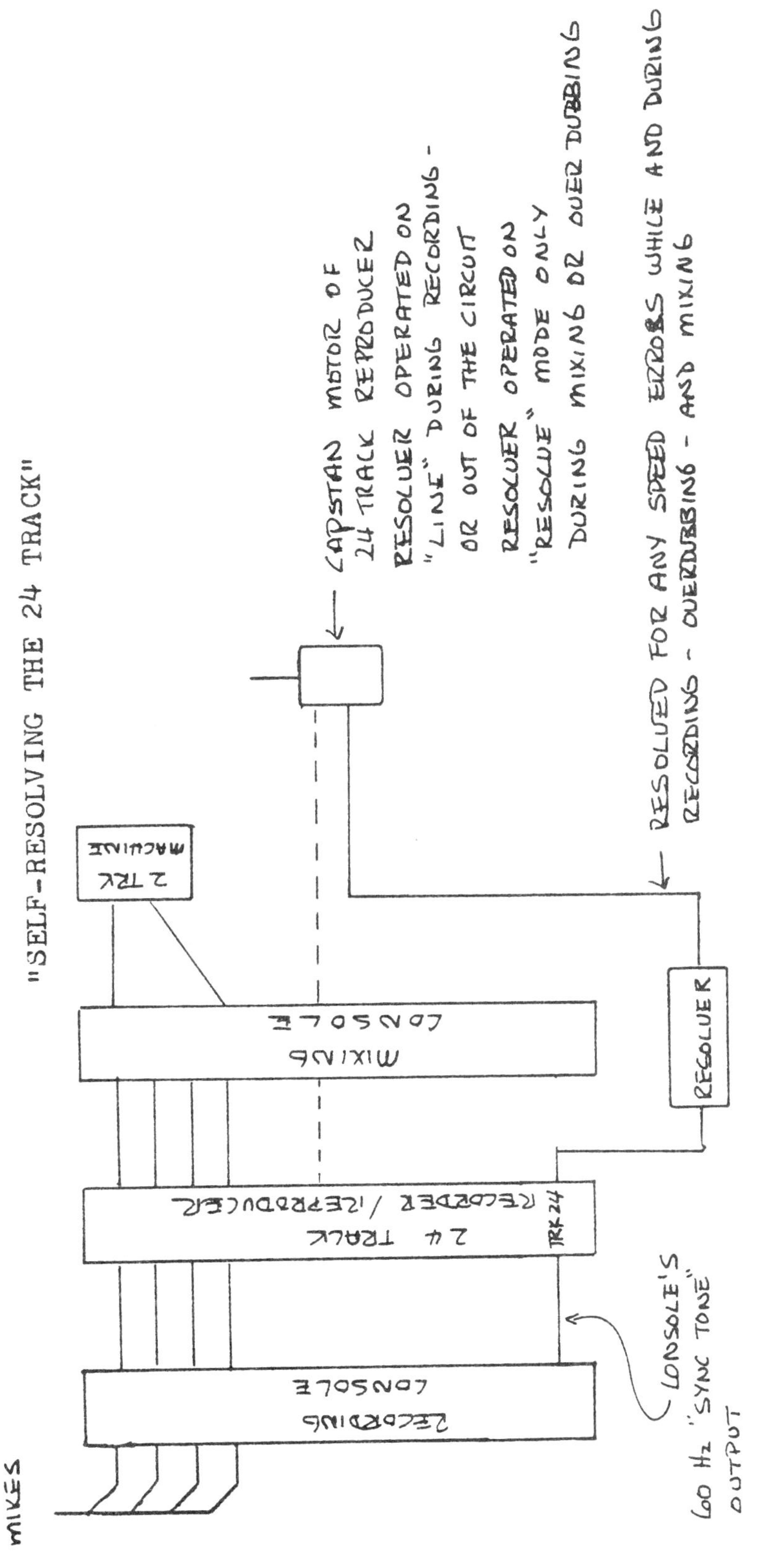

FOUR METHODS OF RESOLVING   Illus. I7-1
METHOD #3

"USING THE 24 TRACK'S PLAYBACK OF SYNC TO OPERATE THE TWO TRACK'S VSO"

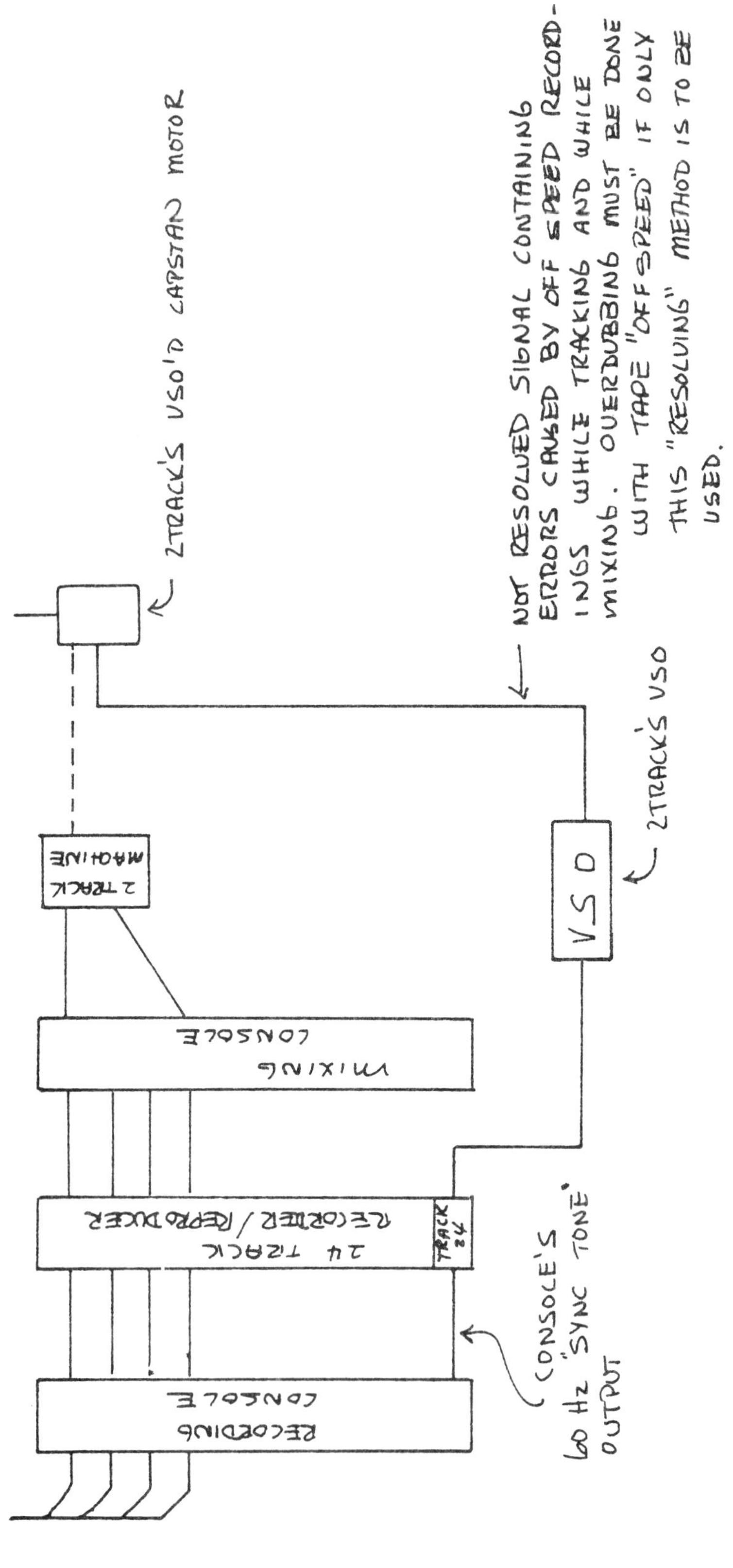

FOUR METHODS OF RESOLVING — Illus. I7-1
METHOD #4

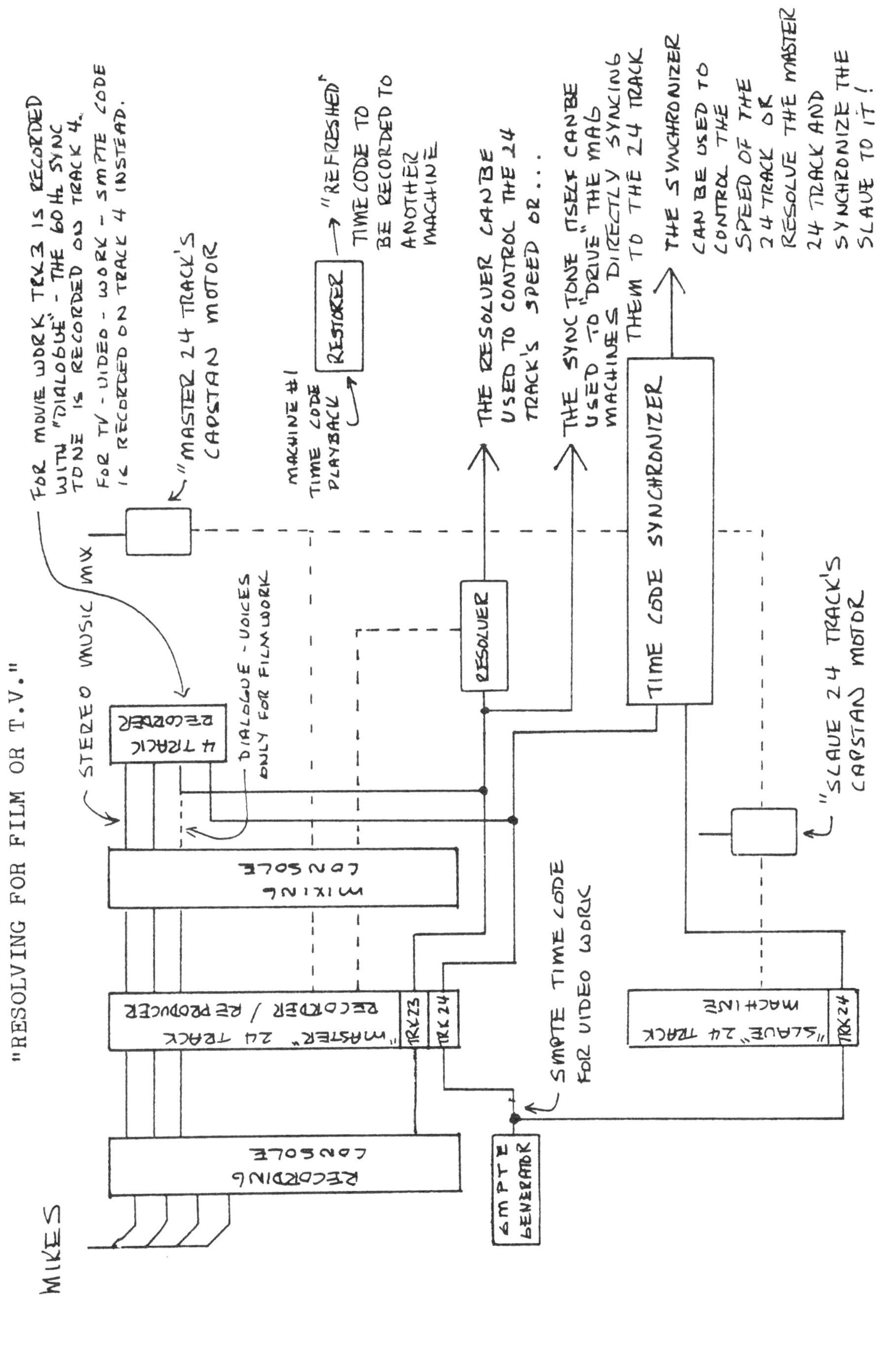

Running the two track's capstan motor with a resolver, using the tone off the mulitrack's sync track playback, works just fine if you are willing to listen to the whole of the project from the 24 track off speed, only to hear the tunes at the proper speed after it's mixed. This is because this method allows the 24 track to run off speed and the 2 track's speed is "corrected" so that the 24 track's speed error is nullified by the speed "error" in the 2 track. The only true pitch playback will then be the 2 track's when it plays its mix back while running off the wall's AC again (plugged directly into the AC rather than the VSO).

In the case of film work, the sync tone is used to resolve the mag recorder.

SMPTE is used for television work since this code can control both speed and the location of all video equipment, including video tape machines, computerized video editors, and all TV sound recording tape machines.

For the film part of the project, the 60 Hz tone is all that's needed to operate the mag recorders resolved to the 24 track or 4 track. You plug the playback of the 60 Hz sync track into the proper jack of the mag machines and press the buss button on the mag machine. From now on, whenever the mag "hears" that the 24 track's 60 Hz sync playback, the mag machine can move magnetic film in exact synchronization.

For the record album mixing part of a large film project, the 60 Hz tone would be used to self-resolve the 24 track to prevent any speed problems.

SMPTE code is more dangerous to use than 60 Hz sync. SMPTE cannot be edited, it cannot be trusted to transfer from machine to machine without "restoring," which means that the code waveform is regenerated fresh electronically exactly in sync with the code that is being read from the playback of the previously recorded code. If you do edit a tape with time code recorded on it, common synchronizers may cause a "burp" in the tape speed as the splice goes through where discontinuous time code exists. In addition, SMPTE code can "leak" from its track to an adjacent track, especially if recorded at too high a level (generally record SMPTE at about -7 VU).

MACHINE SETUP PROCEDURES USING THE DOLBY NOISE REDUCTION SYSTEM

*** Please refer to Illus. I3-1 -- the Machine Alignment Schedule ***

In order to eliminate the confusion for those of you who have had no experience in machine alignment, I have made the setup schedules as simple as I can. The schedules should be memorized. Use the manufacturer's instruction manual for the machine you're going to try these alignments out on to see where the controls are and to get extra information about alignment peculiarities. If you are a beginner (like we all start out being), use a mono or two track machine at first or just align one track of the multitrack machine. Aligning one track successfully means that you could align the whole 24 track if you did everything the same way 23 more times. Be realistic: seek expert help and guidance if you are a beginner.

ALIGNMENT WITH NOISE REDUCTION (Part 1)

*** Please refer to Illus. I7-2 -- Machine Alignment with Dolby ***

Put the noise reduction equipment in BYPASS mode. Play the alignment tones either from the client's master tones or from your studio's alignment tape. Adjust the reproduce levels and sel sync (sometimes called cue) to the 1 KHZ tone so the machine's VU meters read 0.

Play the 10 KHZ tone and adjust the reproduce EQ hi frequency control so that the tone reads exactly zero (there's no satisfaction in aligning a recorder unless you split the zeros for all adjustments).

Play the 100 Hz tone and adjust the reproduce low freq EQ.

ALIGNMENT WITH NOISE REDUCTION (Part 2)

Now that the reproduce level and EQ are adjusted, set up the record levels and EQ:

First set the bias level so that the tracks are recording the audio in the linear region of the tape. There are two distinctly different methods for doing this.

1. Bias method one:

Set the bias with a 10 KHz tone at 15 ips for both 15 and 30 ips work. For most tapes you use 3 dB over bias. This means that you first find the setting of the bias that results in the highest reading on the recorder's playback meters (if the tone is reading too high, don't reset the play level control -- these are already calibrated -- set the console oscillator instead). From the peak level reading point on the bias pot, adjust the individual bias controls so that the tone drops off in level 3 dB as you turn the bias control up. Up is always clockwise in properly designed equipment. This is what is known as 3 dB overbiased using a 10 KHz tone at 15 ips. If you check, you'll find that this is about the same bias setting as using the other method where you overbias 1/2 dB with a 1 KHz tone at 15 ips. If the amount of overbias isn't nearly 1/2 dB, then you should use the 1 KHz bias method instead. The reason for using 10 KHz is the speed and accuracy to be gained by measuring 4 dB instead of 1/2 dB (it's 8 times more accurate).

2. Bias method two:

Set the bias using a 1 KHz tone and over bias 1/2 dB. With some heads, 10 KHz biasing will give no where near the required 1/2 dB overbias condition at 1 KHz, so the 1 KHz method must be used instead. Simply overbias until the level peaks and then drops off by 1/2 dB.

Now that the bias has been set for each recording channel, the record adjustments can be made. The first of these is: use a console tone of 1 KHz and adjust the record level control so that the machine (on record in repro mode) records and then plays back at zero level. Adjust all record level controls so that all the tracks "split the zeros."

Now change the console oscillator to 10 KHz, and set the machine's record EQ to make the recorder indicate zero level. This nearly finishes the record alignment since common professional recorders do not have low frequency record equalizers. Record some 100 Hz tone to see that the machine is printing lows. If you don't have a 100 Hz tone to playback during the

This setup listing is for a first session including the recording of the project's master tone set and the project's master Dolby tone (if M series Dolby is used). This is a simplified schedule; refer to the text for detailed instructions. See also Dolby alignment for a continuing session where the master tone set and Dolby tone are already provided.

Before staring, degauss and clean the heads and clean the tape path

Part One - Machine Playback Alignment

Dolby system: bypassed *
Tone source: reproduce alignment tape
* Dolby bypassed means pressing the "bypass" or "align" Switches on all the Cat 44 cards (the card with the Dolby meter)

[STEP	FREQUENCY	ADJUST
1	1 KHZ	repro level
2	1 KHZ	sync level
3	10KHZ	play head azimuth
4	10 KHZ	hi freq repro EQ
5	1KHZ	re-check repro level

Part Two - Machine Record Alignment

Dolby system: bypassed
Tone source: console oscillator

STEP	FREQUENCY	ADJUST
6	10 KHZ	bias
7	1 KHZ	record level
8	10 KHZ	record head azimuth
9	10 KHZ	record EQ
10	100 hz	lo freq reproduce EQ
11	1 KHZ	re-check record level
12	1 KHZ	input monitor calibrate
13	tone off	erase some 1 KHZ tone to check erase

Alignment without Dolby system is now completed. To complete initial alignment not using Dolbys, skip to step 18 skipping step 19 (do not record Dolby tone) otherwise continue with step 14.

Part Three - Dolby "Line In" to "Rec Cal" Adjustment

Dolby system: non-bypass
Dolby mode: record
Tone source: console oscillator
Recorder: input mode
Individual Dolby Cat 44 n/r swtches: down
Master Dolby n/r switch: in

STEP	FREQUENCY	ADJUST
14	1KHZ	set Dolby unit's line in to read on the dot
15	1 KHZ	set Dolby unit's record cal to regain zero level on each of the recorder's track's VU meters

Part Four - Dolby "Play Cal" to "To Mon" Adjustment

Dolby system: non-bypassed
Dolby mode: Dolby tone mode
Tone source: Dolby tone
Recorder: input mode
Individual Cat 44 n/r switches: up
Master Dolby n/r switch: in

STEP	FREQUENCY	ADJUST
16	Dolby tone	set Dolby unit's play cal so that the tone reads on the Dolby dot
17	Dolby tone	set Dolby unit's to mon so that console VU meters (meters read "tape") indicate zero level

Part Five - Recording Project NAB Tones

Dolby system: bypassed
Tone source: console tones

18 now record one minute of 1 KHZ, 10 KHZ, and 100 hz

Part Six - Recording Project Dolby Tone

Dolby system: non-bypassed
Dolby mode: record, Dolby tone mode
Tone source: Dolby tone
Recorder: recording and reproducing
Individual Dolby Cat 44 n/r switches: up
Master Dolby n/r switch: in

19 record 1 minute of Dolby tone
20 play or wind 1 or 2 minutes of blank tape past last tone for future use by a maintenance person as a record pad

Leave machine in all safe, all repro
Leave Dolby in n/r in, play mode

All alignments completed

This setup listing is for a continuing session of overdubbing or mixing which sets the Dolbys and machine up to the existing project NAB and Dolby tones. Refer to the text for detailed instructions.

Part One - Machine Playback Alignment

Dolby system: bypassed
Tone source: client's set of project nab or aes tones

STEO	FREQUENCY	ADJUST
1	1 KHZ	repro level
2	1 KHZ	sync level
3	10 KHZ	play head azimuth
4	10 KHZ	hi freq repro EQ
5	100 hz	lo freq repro EQ
6	1 KHZ	re-check repro level

Part Two - Dolby "Play Cal" and "To Mon" Adjustment

Dolby system: non-bypassed
Dolby mode: Dolby tone mode
Tone source: client's Dolby tone
Recorder: reproduce mode
Individual Dolby Cat 44 n/r switches: up
Master Dolby n/r switch in

STEP	FREQUENCY	ADJUST
7	Dolby tone	set Dolby play cals for a Dolby dot reading
8	Dolby tone	set Dolbys to mon so that the console VU meters (on tape mode) indicate 0 VU

Part Three - Machine Record Alignment

Dolby system: bypassed again
Tone source: console oscillator

STEP	FREQUENCY	ADJUST
8	10 KHZ	set recorder bias
9	1 KHZ	recorder record level
10	10 KHZ	recorder hi freq record EQ
11	100 hz	check level only - no adjustment
12	1 KHZ	re-check recorder record level
13	1 KHZ	recorder input monitor calibrate

Step Four - Dolby "Rec Cal" Adjustment

Dolby system: non bypassed
Dolby mode: Dolby tone mode
Tone source: Dolby tone
Recorder: input mode
Individual Dolby Cat 44 n/r switches: up
Master Dolby n/r switch: in

STEP	FREQUENCY	ADJUST
14		set Dolby unit rec cal for a Dolby dot reading

Step Five - Dolby "Line In" Adjustment

Dolby sysetm: non-bypassed
Dolby mode: record
Tone source: console oscillator
Recorder: input mode
Individual Dolby Cat 44 n/r switches: down
Master Dolby n/r switch: in

STEP	FREQUENCY	ADJUST
15	1 KHZ	Dolby unit line in for a Dolby dot reading

Leave recorder in all safe
Leave Dolby unit in nr in, play mode

Alignment completed

reproduce alignment, set the lo freq play EQs to read 0 VU.

Lastly check the 1KHz tone again to see that the record EQ doesn't upset the zero indication you had before. If not, then procede to the last step.

Now put the machine in input mode and adjust the input mon cal (or input control) so that the machine meters read zero.

ALIGNMENT WITH NOISE REDUCTION (Part 3)

Now that the tape machine is completely lined up, set up the noise reduction. There are two major noise reduction systems in use -- Dolby and dBX. In addition there are less common types such as Telefunken's Telcon. It is of the utmost importance that the BYPASS switch of any noise reduction system in use be in the bypass position while you are lining up the machine. If you have any doubts, due to being in a strange studio, physically bypass (unplug) the noise reduction and re-plug the recorder directly into the console. The recorder's alignment is now finished; reconnect or non-bypass the noise reduction devices. Put the recorder into input mode and leave it there for the entire alignment of the noise reduction equipment. ** NOTE: If your Dolby package does not automatically switch to the record mode (upper most red LED on the Cat 44 card lights -- "Dolby Reads Line In"), you will have to manually switch the Dolby system to LINE IN mode by pressing the master LINE IN button each time the instructions say to adjust the Dolby LINE IN. Multitrack Dolby packages should be setup so that they automatically "trigger" to LINE IN mode whenever any of the tracks are in input or in record mode. The methods for doing track-by-track Dolby switching are available from Dolby Labs.

DOLBY NOISE REDUCTION SYSTEM ALIGNMENT FOR M-8, M-16, M-24 UNITS

Part one: the session is starting out, and this setup is the initial one and includes recording the project's master tones

*** Please refer to Illus. I7-2 -- Machine Alignment with Dolby Noise Reduction ***

Swab the Dolby unit meter faces off with a damp tissue.

Setting the LINE IN: with the Dolby system non-bypassed, the master noise reduction switch in NR in, the individual NR switches out (on each Cat 44 -- the card with the meter), with a console zero level tone at 1 KHz, adjust the Dolby unit LINE IN control so that the console tones read on the Dolby dot.

Setting the REC CAL: with the same setup as in setting line in, adjust the Dolby unit REC CAL so that the tones read zero on the tape machines VU meters.

Checking the Dolby unit Cat 22 card: for this type of setup it is now the time to see if the actual processor cards have good internal calibration. With the Dolby system in non-bypass, the master noise reduction switch in, with the individual NR switches on the Cat 44 cards up (in), press the DOLBY TONE button. If the readings on the tape machine's VU meters are within + or

- 1/2 dB of 0 VU, then the Dolby Cat 22 card is O.K. (superficially anyway), so we can continue. If it is not, the Dolby Cat 22 card should be thoroughly checked out by a maintenance person before it is used in a session.

Setting PLAY CAL - with the setup as in the previous step, set the PLAY CAL pot so that the Dolby meter reads on the dot.

Setting TO MON -- very simple to do if your particular console has a switch called a VU meter changeover switch, which will make the console VU meters read off the tape playback or the busses; if the console has no such switch, then this step is slightly more difficult.

If there is a VU changeover switch, with the same setup as three steps ago (checking the Cat 22 card), adjust the TO MON calibration control (pot) so that the console VU meters read zero.

If the console has no VU changeover switch, return the noise reduction to bypass, play the 0 level 1 KHz tone and set the console faders (sent to busses) so that the console VU meters are reading 0 back from the tape machine. Now non-bypass the noise reduction (NR in) and go back to Dolby tone mode as in "checking the Cat 22" and adjust the TO MON controls on the Dolby unit so that the console meters once again read zero (now reading from the Dolby tone).

Now the Dolby system alignment is finished, so record a one minute project master Dolby tone.

DOLBY SYSTEM ALIGNMENT FOR A CONTINUING SESSION

*** The session already has master tones including a Dolby tone ***

Swab the Dolby meters off with a damp tissue.

After adjusting the playback and record of the machine with the Dolby system in bypass enter non bypass mode and put the tape machine in repro mode.

Setting PLAY CAL: with the master Dolby NR switch in, with the individual Cat 44 NR switches in (which is up), with the Dolby system in Dolby tone mode, play the the project's Dolby tone and adjust the Dolby unit's PLAY CALs until the Dolby tone reads on the Dolby meter's dots.

Setting TO MON: switch the console meters over to read tape playback and adjust the Dolby unit's TO MON so that the console meters read zero.

Setting LINE IN: with a 1 KHz tone coming from all the console busses, master noise reduction (NR) in, individual NR switches out, Dolby tone off, adjust the Dolby unit's LINE IN control so that the Dolby meter is on the dot.

Setting REC CAL: with the machine in input mode, master NR in, individual NR switches in, enter Dolby tone mode and set the RECORD CAL controls so that the Dolby meter reads on the dot.

The setup is completed.

POSSIBLE TERMINATION PROBLEMS WHEN ALIGNING 361 DOLBY UNITS

*** Please refer to Illus. I7-3 -- 361 Dolby Alignment Schedule ***

Set up the machine with the Dolby units set to NR out, normal, play. Check the terminations. The proper situation is to have the 361's terminations switches in the 10K position (not the 600 ohm position), the tape machine's termination switches on and have the console's monitor system not bridging. If you play a 1 KHz tone and switch the Dolby unit from play/normal to play/check tape and less than 1/2 dB of change occurs at the recorder's VU meters, then it is safe to line up the machine with the Dolby unit in safe/normal (NR out) and have the machine termination switch on and not physically bypass the Dolby unit while lining up the recorder. If more than 1/2 dB change occurs, then your studio has terminations in the wall (this is bad), which means that you should normally leave the machine's termination switch off because the console's monitor system is terminating the machine. When you use Dolby units however, the terminating action on the machine will change if:

1. The 361 is in play (the output of the machine is terminated by the Dolby unit) or
2. If the 361 is in record (the console is terminated by the Dolby unit and the machine output is connected directly to the wall again) or
3. If the 361 is in check tape, where the output of the machine is connected to the playback system, bypassing the Dolby if the Dolby is in play, but the Dolby still terminates the mixing buss when the Dolby is in record

When you use Dolby units, do not let them change the terminations of the machine because, if you do, the Dolby units will not work properly and will change the sound. If this seems confusing, that's because it is, and these Dolby termination shenanigans confound even the best mixers from time to time. In fact, few mixers really understand 361 Dolby termination switching difficulties. If your studio's monitor system does terminate, you have a lot of work cut out for you when using 361 Dolby units. When making recordings using 361 Dolby units, turn off the machine's termination switch and put the Dolby unit's termination switch on 10K ohms; during playbacks you must switch the machine's termination switch on leaving the Dolby unit's termination switch on 10K ohms. For tape transfers from one two track (Dolby encoded tape) to another two track (making a non-encoded copy), set the play machine with its termination switch on and the Dolby on 10K, and the play Dolby units "to mons," going to the copy machine's line inputs. The record machine's terminations should be off, as it would be connected to the console monitor system, which terminates it. When finished, the two track machine (with the Dolby units now removed) must have its termination switches turned back off.

Hopefully you'll find that your studio has no terminations in the console monitor system. I will procede with the 361 Dolby system explanation with that fond hope.

Degausse and clean the heads. Confirm Dolby to machine and Dolby to wall parity. Check termination switches are set correctly.

PLAYBACK ALIGNMENT
==================

Dolby mode: norm/play/NR out
Tone source: client's tones

Step	FrEQuency	Adjust
1	1 KHZ	repro level
2	10 KHZ	play head azimuth
3	10 KHZ	playback EQ
4	1 KHZ	recheck repro level

RECORD ALIGNMENT
================

Dolby mode: norm/play/NR out
Tone source: console oscillator

5	1 KHZ	bias adjust
6	1 KHZ	record level
7	10 KHZ	record head azimuth
8	10 KHZ	record EQ
9	100 HZ	check O.K.
10	1 KHZ	set input calibrate

DOLBY ALIGNMENT
===============

Dolby mode: norm/record/NR out
Tone source: console oscillator

11	1 KHZ	Dolby input control so Dolby reads on the dot
12	1 KHZ	Dolby output control so tape machine reads Zero level while machine is on input

CALIBRATING TO A CLIENT'S DOLBY LEVEL
=====================================

Dolby mode: norm/play/NR out
Tone source: client's Dolby tone

13	Dolby tone	set recorder repro level to read on the Dolby dot

Dolby mode: norm/record/NR out/Dolby tone mode
Tone source: Dolby's own calibration tone

14	Dolby tone	set recorder record level control so that Dolby meter reads on the dot **

** Note: tape machine is now at "Dolby level". Checking the NAB alignment using the 1k, 10k and 100 HZ tones will require re-setting the machine back to NAB levels by using the NAB alignment tones again.

SETTING UP THE 361 DOLBY SYSTEM (INITIAL SETUP INCLUDING RECORDING THE MASTER PROJECT DOLBY TONE)

Dolby unit's termination switch set to 10K, machine termination switches on, machine completely aligned including input (machine alignment done with Dolby units in normal, play, NR out modes)

Setting the Dolby unit input control: send a zero level tone to the two track (machine on input) from the mixing busses of the console. Change the Dolby to normal/record/NR out. Adjust the Dolby unit input control so that the 1KHz tone reads zero on the Dolby dot.

Setting the Dolby unit's output control: with the same setup as directly above, set the Dolby unit output control so that the machine's VU meters again read zero with the machine on input mode. The Dolby unit input and output controls are now set. Press rec, normal, NR in, and the Dolby tone button and record one minute of Dolby tone. If the Dolby tone does not read within + or - 1/2 dB of zero on the tape machine's meters, have a competent maintenance man examine the Dolby unit's Cat 22 card.

SETTING UP THE 361 DOLBY UNIT (CONTINUING SESSION)

To change the recording level to match a Dolby level as shown by the Dolby tone recorded on the client's master tone group:

Play the client's Dolby tone, and with the Dolby units in play, normal, NR out, adjust the two track's play level control so that the Dolby tone now plays at the Dolby dot.

Set the Dolby unit to record/normal/NR out, load some blank tape, and while recording on the two track with it in record and repro mode, press the Dolby tone buttons and adjust the two track's record level control so that the Dolby tone shows up on the Dolby unit's meter on the dot.

A check of the machine's reference tones will show the machine is no longer aligned to the 1K, 10K and 100 Hz tones (which is proper). The two track is now at Dolby level and from now on to check playback or record you will have to start by resetting the machine's play and record controls to the 1 KHz, 10 KHz and 100 Hz reference tones. Be sure that the machine is left adjusted to Dolby repro and record levels again before playing or recording any of the client's music. Leave the recorder in repro, safe, Dolby NR in, normal, play.

Besides the step called "checking the Cat 22 card," there are two tests to determine if the Dolby units are working effectively:

Test #1

1. After all machine and Dolby system alignment steps, using blank tape and with the Dolby unit's noise reduction engaged, record a 1 KHz, 10 KHz, and 100 Hz tone at -20 VU on all the tracks of the machine. Use a tone length of perhaps one minute of 1 KHz, 10 KHz, and 100 HZ.

2. Play back the tones and bring the playbacks up through the console faders reassigned to the busses of the same number (play 1 to buss one, play 24 to buss 24 etc). As the 10 KHz and 100 Hz tone go by, the levels should match the 1 KHz reference level within 1.5 dB. If not, there is trouble somewhere in the terminations, or elsewhere. If the tones come back accurately, this means that your Dolby system alignment has been done properly.

Test #2

Connect the encoding (record mode) Dolby directly into the decoding (play mode) Dolby without any tape machine in the circuit at all. Play various levels of music through the Dolbys (be sure to run some -20 music through them) and A/B the original playback machine to the sound of the 361 Dolbys. Supposedly there should be no difference to the sound, right? Your discoveries may surprise you.

SOLVING CUE SYSTEM SHORTS AND OTHER PROBLEMS

In good studios the cue system includes distribution boxes, which go between the cue system output plugs in the studio wall and the actual phones, which plug into the distribution boxes. In the boxes are arrays of resistors or transformers that feed (or "buffer") the individual headphones so that a problem with one set of headphones will not interfere with the sound of the rest.

*** Please refer to Illus. I7-4 and Illu. I7-5 ***

With these systems, if one pair of phones has a wiring problem then only the phones with the difficulty will go dead and only the associated musician will yell. If there is no distribution system, then all the phones will go dead and all the musicians will begin to yell. It is impossible for a power amp to power up headphones connected to a shorted headphone line.

In studios having no cue distribution system, having all the phones go dead just as you are about to start recording your 35 man union recording session is an electrifying thrill, to say the least. If only one pair of phones were at fault, then by having your assistant unplug one pair of phones at a time, you would eventually turn on the remainder of the phones when he unplugged the guilty headphone. If two or more headphones were at fault, he could go through all the headphones, unplugging them one at a time, and wind up having checked all the phones, and still not have found the guilty headphones. So the proper technique is to unplug all the phones and then plug them in again one at a time until the cue goes dead, and then discard the offending headphone and replace it with a good one.

The preparation of complete cables, with connectors properly installed, is the key to reliable and trouble-free operation of any sound system. For this reason, the following illustrations are included. Experienced audio technicians may wish to review these illustrations, even if they already know how to wire connectors. A few moments of extra care here can save hours of trouble-shooting later on.

As a rule, the amount of insulation removed and the length of exposed cable should be minimized. This reduces the likelihood of short circuits and improves the ability of the clamp to grip the cable firmly. Enough heat should be used to obtain a free flow of solder, but allow leads to cool quickly after solder flows to avoid melting insulation. After each connector has been completely wired, the cable should be tested with an ohmmeter or a cable tester. Continuity between the various conductors and their associated connector pins must be established, and there should be infinite resistance (an open circuit) between all connector pins. In most cases, especially in portable installations, XLR connectors should not conduct at all between the shell and pin 1. This avoids grounding problems from inadvertent touching of the shell to other devices.

Cables to be connected to terminal strips should be prepared by stripping the ends and installing crimp-on or preferably, solder type lugs. If there is any chance the cable will be strained, use a cable that is constructed with internal strain relief cord, such as Belden No. 8412. Crimp a lug onto the cord, and secure the lug to an unused terminal. (The cord should be drawn slightly tighter than the wire leads in order to take the strain first.)

WIRING AN RCA-TYPE PIN PLUG*

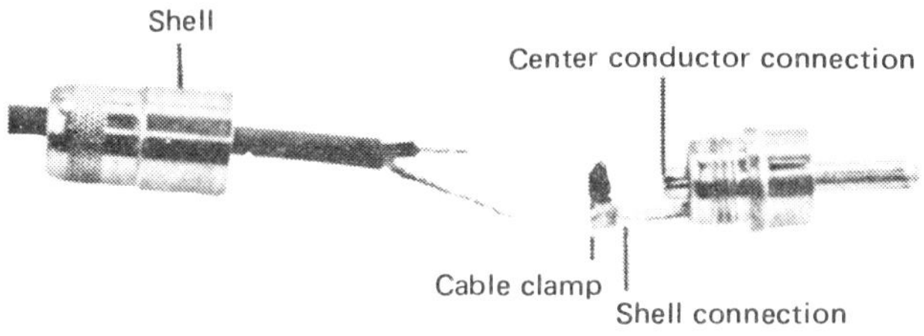

Parts identification and cable preparation.

Strip approximately 1/2" of outer insulation. Unwrap or unbraid the shield and form a lead. Strip approximately 5/16" of insulation from the center conductor. Tin both leads.

Solder the shield to the outer surface of the shell connection, allowing enough free shield to wrap the cable around to the center of the connector. Cool the connection immediately with pliers.

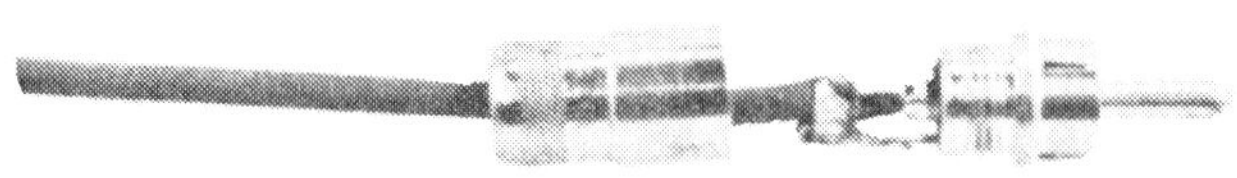

Insert the center conductor in the hollow pin, and fill that end with solder. Cool the connection immediately with pliers. Clean any solder splashes and inspect for burned insulation. Pinch the clamp around the outer insulation with pliers, firmly, but not so tight as to cut the insulation.

Slide the shell forward and screw it tightly to the threaded plug.

Switchcraft No. 3502 connector illustrated. Many large diameter cables are more easily wired to "simple" RCA type pin plugs without a shell (Switchcraft No. 3501M, or equivalent). The braid can then be soldered directly to the shell of the plug.

WIRING A MALE XLR CONNECTOR

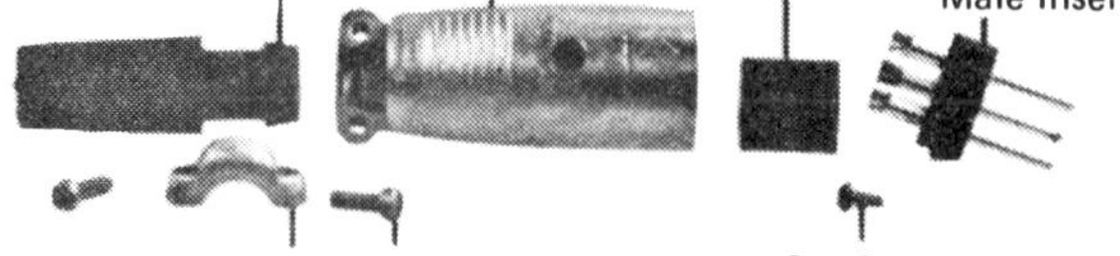

Parts identification (as the connector is usually packaged).

Insert strain relief in rear of shell. Then slip shell onto cable end, followed by insulating collar. Strip outer insulation 1/2". (No. 8412 cable illustrated here.)

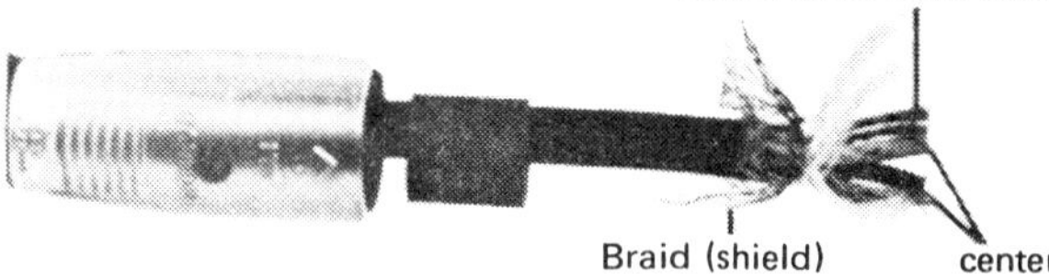

Cut tracer cord, unbraid shield, cut cotton strain relief cords.

Strip approximately 1/4" of insulation from center conductors, tin, and trim to approximately 1/8" exposed wire. Then twist shield, positioning it in the correct orientation to mate with the insert. After tinning the shield, cut it to the same length as the center conductors.

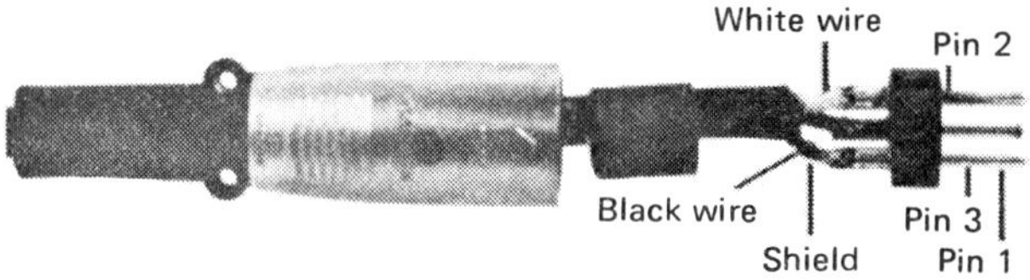

Solder the center conductors to their respective pins, using just enough solder to fill the end of the pins. Yamaha's wiring standard dictates that the black lead mates with pin 3 and the white (or red) with pin 2 (see footnote on page 10 of this section). Then solder the shield to pin 1. Clean any solder splashes and inspect for burned insulation.

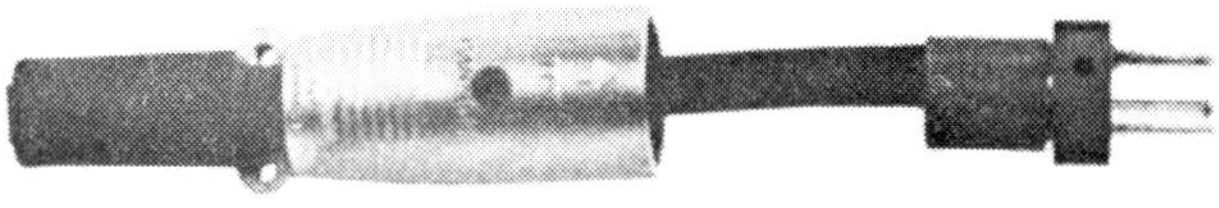

Slide the insulating collar forward, up to the flange of the male insert. The outer cable insulation must be flush with, or covered by the end of the insert. If any of the center conductors are visible, the cable clamp may not be able to firmly grip the cable. Then slide the collar back into the shell.

Slide the shell forward, orienting its internal keying channel with the raised lip (key) on the insert. Secure the insert in the shell with the set screw. Place the cable clamp over the rear of the shell, with careful attention to the clamp's orientation; a raised lip inside the clamp should be aligned immediately over a lip in the shell for thinner cable (No. 8451). The clamp should be turned around for heavier cable (No. 8412) to provide clearance. Insert the clamp screws and tighten fully.

WIRING A FEMALE XLR CONNECTOR

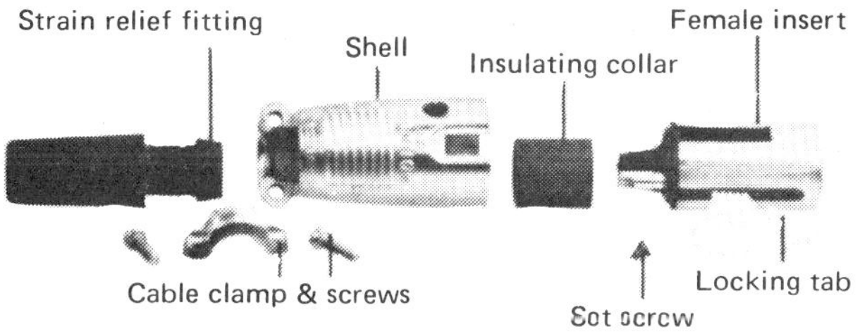

Parts identification (as the connector is usually packaged).

Insert strain relief in rear of shell. Then slip shell onto cable end, followed by insulating collar. Strip outer insulation approximately 9/16". (No. 8451 cable illustrated here)

Pull off foil wrap. Strip approximately 5/16" of insulation from the center conductors, leaving approximately 1/4" of insulation between the bare wire and the outer insulation. Tin the center conductors, and trim so that about 1/8" bare wire remains. Then tin the shield conductor, orienting it with the center conductors so they are aligned with the proper pins of the insert. Cut the end of the shield so that it extends 1/16" beyond the center conductors.

Solder the center conductors to their respective pins, using just enough solder to fill the end of the pin. Yamaha's wiring standard dictates that the black lead mates with pin 3, the white (or red) lead with 2 (see footnote on page 10 of this section). Then solder the shield to pin 1. Clean off any solder splashes, and inspect for burned insulation. Insert the locking tab in the female insert, as illustrated, with small nib facing front of connector.

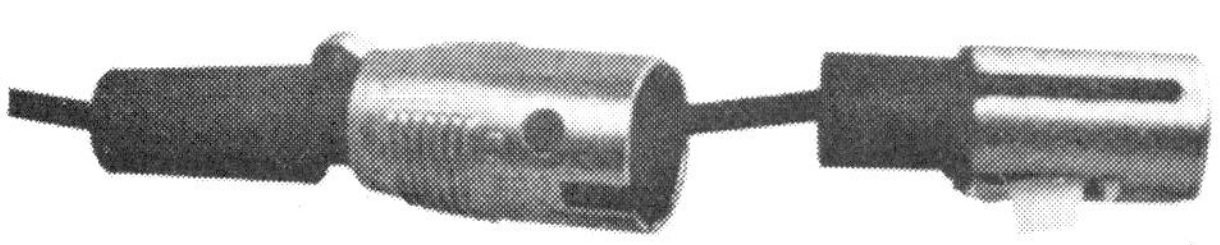

Slide insulating collar foward, up to rear edge of female insert. The outer insulation of the cable must be flush with, or covered by the end of the insert. If any of the center conductors are visible, the cable clamp may not be able to grip the cable firmly, and the connector leads will soon fatigue. Then slide the collar back into the shell.

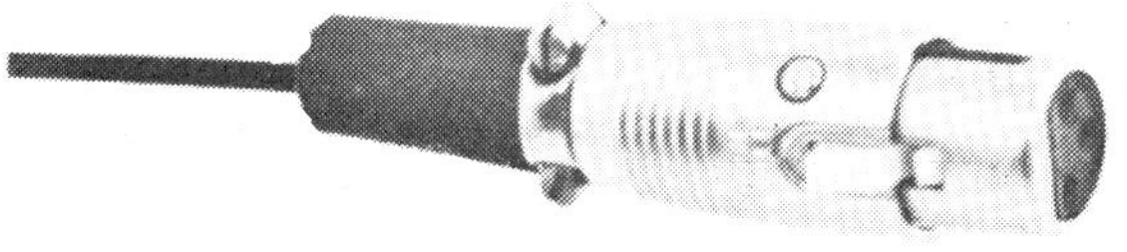

Slide the shell forward, orienting the notch in the shell with the locking tab in the insert. Secure the insert in the shell with the set screw. Place the cable clamp over the rear of the shell, with careful attention to the clamp's orientation; a raised lip inside the clamp should be aligned immediately over a lip in the shell for thinner cables (No. 8451). For heavier cables (No. 8412), the clamp should be turned around to offset the lips and provide more clearance for the cable. Insert the clamp screws and tighten fully.

WIRING A STANDARD PHONE PLUG (2-conductor)

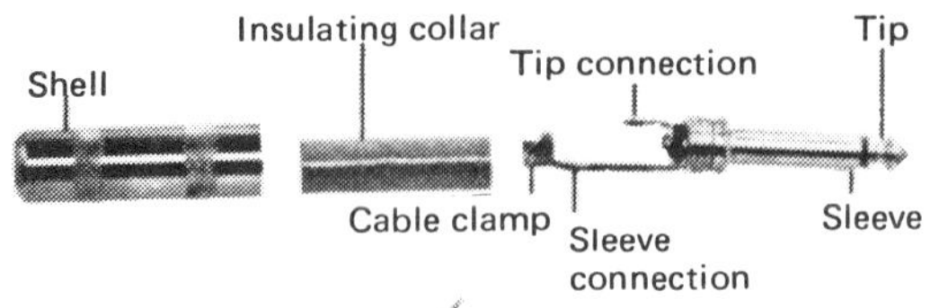

Parts identification.

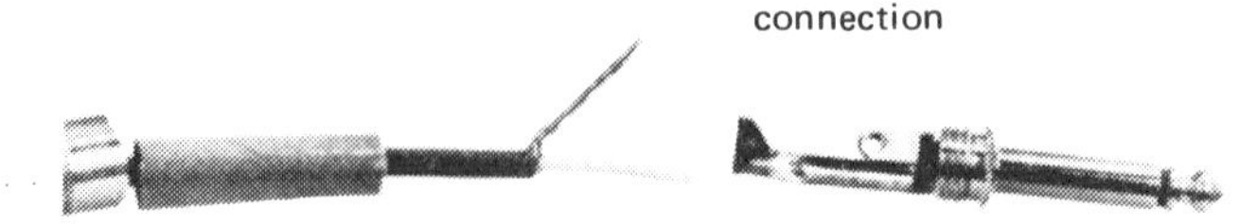

Slide shell, then insulating collar over cable end. Strip outer insulation for length equal to length of sleeve connection. Unwrap or unbraid shield, twist to form lead.

Position outer insulation just ahead of cable clamp, strip center conductor from point just behind tip connection. Tin center conductor and shield. Bend shield as illustrated, solder to outer surface of sleeve connection. (Cool immediately with pliers.) Insert center conductor in tip connection, solder, cut end flush. Bend the end of the tip connector (slightly) toward the sleeve connection to help prevent the burr (from the cut wire) from cutting through the insulating collar.

Using pliers, bend cable clamp around outer insulation. Clamp should be firm, but not so tight as to cut insulation.

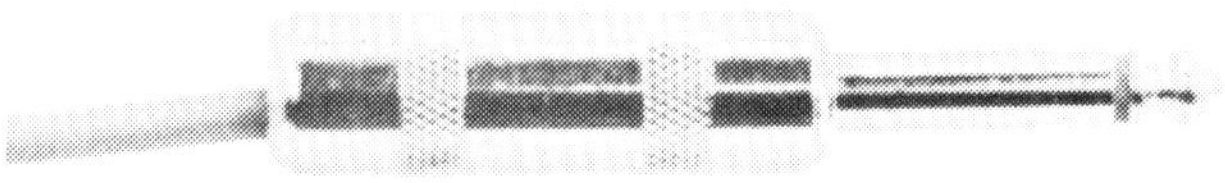

Slide insulating collar forward, until flush with rear of threads. Slide shell forward, screw tight to plug assembly.

WIRING A TIP, RING & SLEEVE PHONE PLUG (3-conductor)

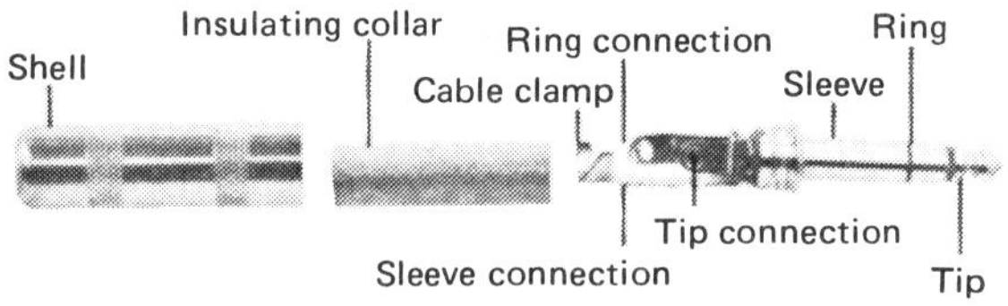

Parts identification.

Slide shell and insulating collar over cable end. Strip outer insulation for length equal to length of sleeve connection. Remove any tracer cords and strain relief cords. Form lead from shield. Hold cable with outer insulation just ahead of cable clamp, and strip the red (or white) conductor just behind the tip connection. Then strip the black conductor just behind the ring connection. Tin all leads, and cut the center conductors so approximately 1/8" of bare wire remains.

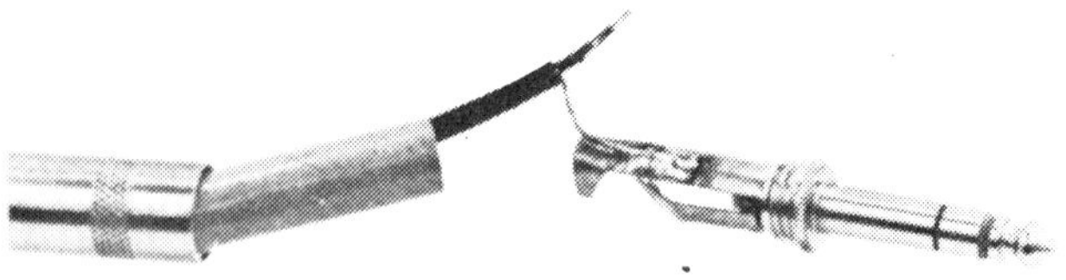

Solder the shield to the outer surface of the sleeve connection, allowing enough free shield to bend around to the other side of the cable clamp. Cool the connection immediately with pliers.

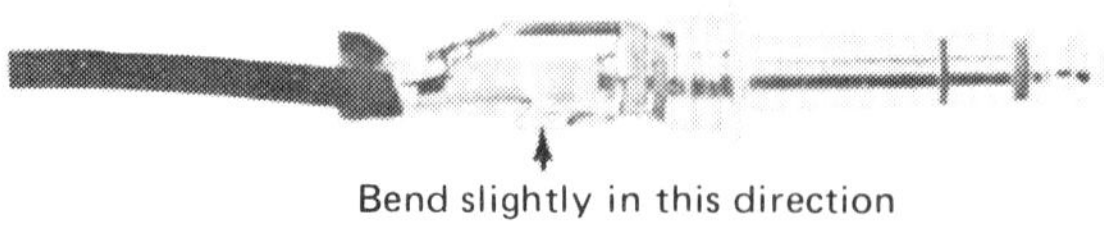

Insert the center conductor leads in their respective connection points, and solder in place. Trim the leads flush. Bend the end of the tip connection (slightly) toward the ring connection to help prevent the burr (from the cut wire) from cutting through the insulating collar.

Using pliers, bend the cable clamp around the outer insulation. The clamp should be firm, but not so tight as to cut the insulation.

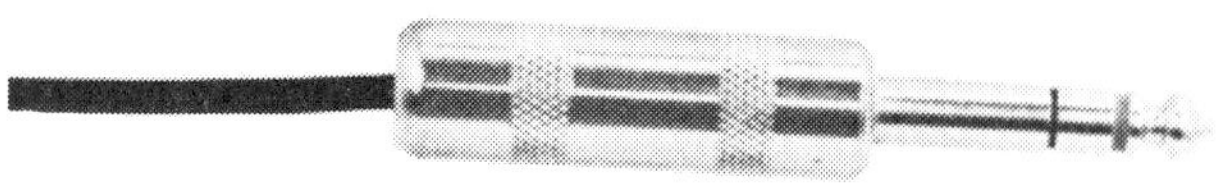

Slide the insulating collar forward, until flush with rear of threads. Slide the shell forward, and screw tightly onto plug.

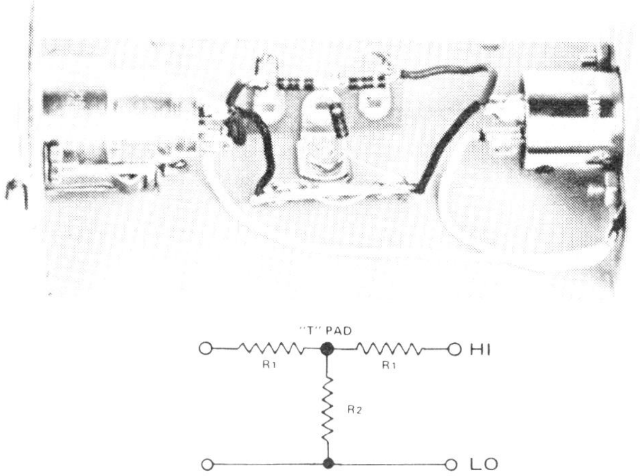

Fig. 39A — Pads Constructed in Mini-Boxes.

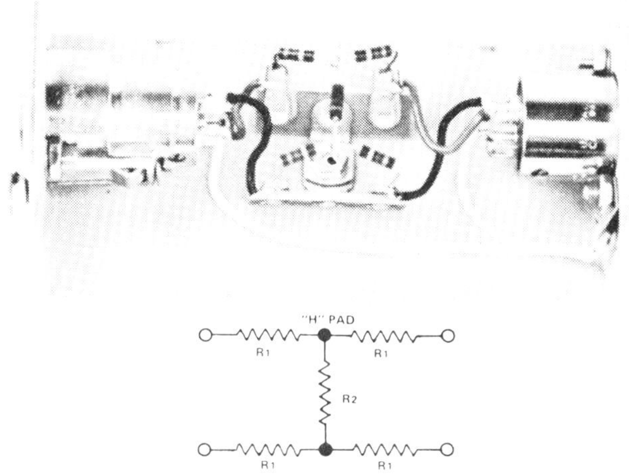

above +30dBm, use 1 watt, low inductance resistors. 10% tolerance is acceptable for most pads.

Fig. 39B — Pad Constructed in Switchcraft Model S3FM

It is possible to construct a pad within an XLR connector, but the extremely tight fit can adversely affect reliability. The Switchcraft model S3FM is a tube with a male A3M (XLR) at one end, and a female A3F (XLR) at the other end. Pads using 1/4 watt resistors can be constructed inside this device. Cover the entire pad with insulation tubing before final assembly into the S3FM.

A "mini-box" fitted with male and female XLR connectors is an easy to build, rugged housing for a pad. Use stranded wire for best results.

Illustrated are three typical pad construction techniques. For most applications, it will be sufficient to construct only a few types of pads: 20dB, 24dB, and 40dB pads cover almost any requirement. Consult Figures 37, 38 and 39 for schematic, construction and resistor value information.

Transformers

Audio transformers (as distinguished from power supply transformers, RF transformers or other transformers) are primarily used for ground isolation, impedance matching and level matching. The following paragraphs detail several applications of audio transformers at low signal levels. Speaker-level transformers are discussed on Page SEVEN 6; the Appendix gives further details on transformer operation.

Matching Transformer Box:

Impedance matching transformers can be used to connect a high impedance source to a low impedance load, or vice-versa (see Page SIX 5 for a discussion of matching versus bridging inputs). The box shown below may be used to run a 600-ohm balanced or floating line to the P-2200 input, or it may be used between any 600-ohm source and high impedance input. Use a transformer capable of handling nominal +4dB (1.23V) inputs with at least +24dB (12.3V) peak capability.

The transformer should be mounted in a mini-box, wired to the XLR connectors with stranded wire, and connected to the auxiliary equipment with one of the cables previously illustrated. In line transformers, such as those manufactured by Shure Brothers, Sescom, and others may be used, with suitable adapters.

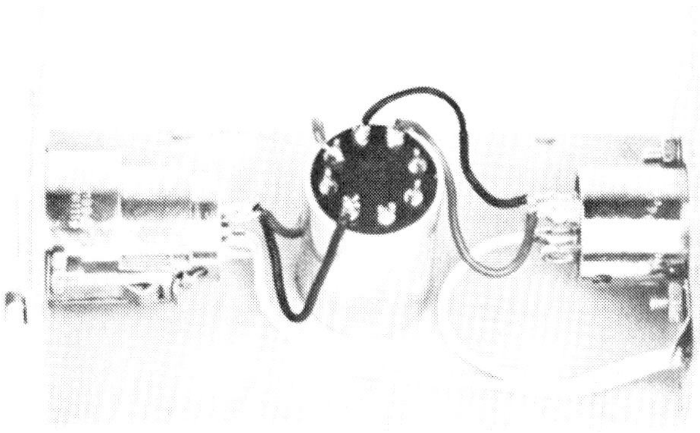

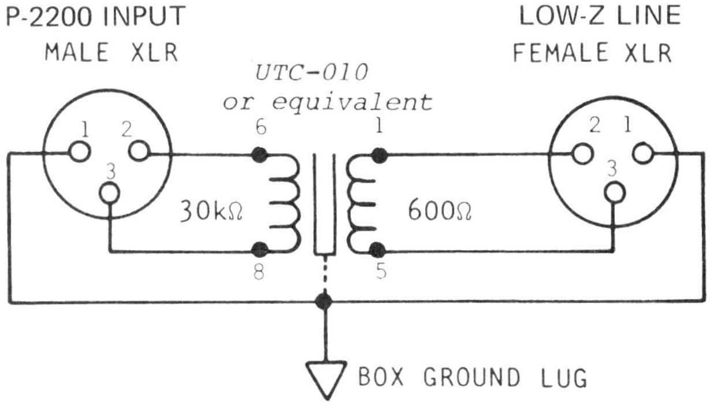

Fig. 40 — Matching Transformer Box

Step Up Transformer Box

The step up transformer box illustrated here is similar to a pair of matching transformer boxes. This configuration provides voltage step-up for optimum drive levels when connecting the output of a low impedance, low level source, such as the headphone output of a mixer, to the two inputs of the P-2200. It has a stereo phone jack input, but if the input source is monaural, the transformer lead to the ring of the T.R.S. input jack may be moved to the jack's tip so that a standard T.S. phone plug input will feed both transformers. Alternately, the box may be built with separate T.S. phone jack inputs, or with XLR inputs. Two standard (2-wire) phone jacks outputs are provided for connection to the "left" and "right" inputs of the P-2200. Construct two cables from dual conductor, shielded cable and T.S. phone plugs to connect the transformer box output to the P-2200's input. Locate the step up transformer box at least 5 feet from the P-2200 to avoid hum pickup from the amplifier's power transformer. However, the cables from the transformer box to the amplifier should be no longer than 10 feet, since this is a high impedance circuit. Use low capacitance, coaxial, hi-fi type cable between the box and the amplifier. Since the inputs of the P-2200 are unbalanced, connecting two cables to its input forms a short ground loop as shown in Figure 60 (see discussion of grounding on Page SIX 13). To keep hum pickup at a minimum, run the two cables close together; this minimizes the area (and therefore the hum) enclosed by the loop.

The two diagrams show circuits using a Triad A-65J transformer, and a UTC A-24 transformer. Similar 600 ohm to 15K-ohm transformers are acceptable. The 1/4 watt, 10%, 15K-ohm resistors are used to terminate the transformers, for lower distortion, and improved frequency response.

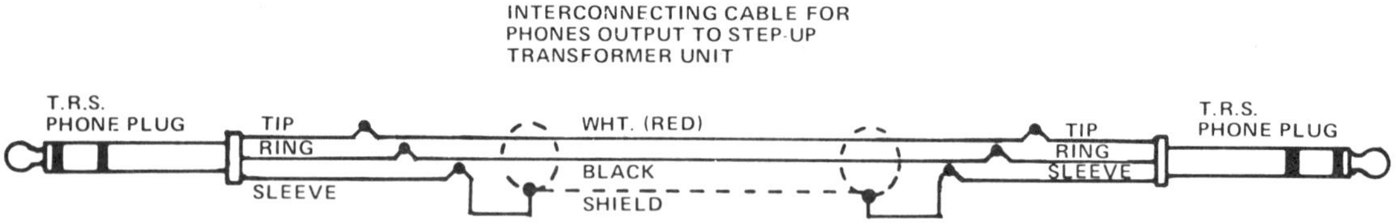

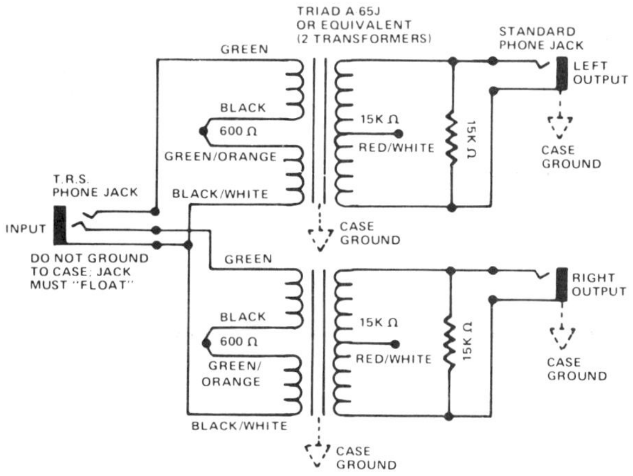

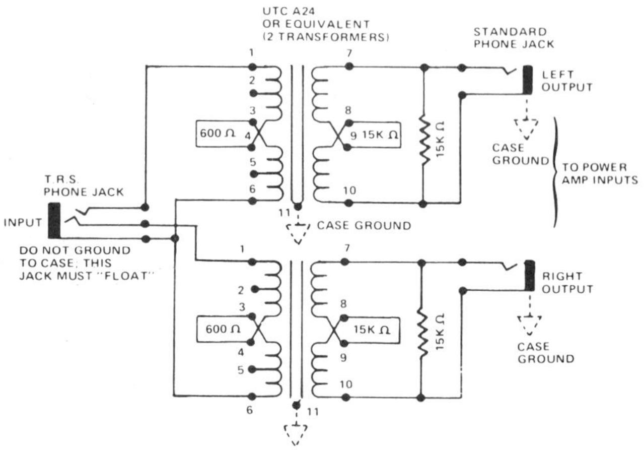

TRANSFORMER AVAILABILITY

The matching and step-up transformers mentioned in the preceding subsections are available from many electronic parts dealers. Yamaha does not endorse specific products by citing them herein; rather, these transformers are mentioned for convenience only. If you are unable to locate the transformers from your local electronic parts dealer, contact the manufacturer at the address shown below.

Sescom, Inc.
P. O. Box 590, Gardena, CA 90247
Phone (800) 421-1828 (213) 770-3510

Shure Brothers, Inc.
222 Hartrey Ave., Evanston, Illinois 60204
Phone (312) 328-9000 Cable: SHUREMICRO

Triad
305 N. Briant St., Huntington, Indiana 46750
Phone (219) 356-6500 TWX: 816-333-1532

UTC
150 Varick St., New York, NY 10013
Phone (212) 255-3500 TWX: 710-581-2722

A line of very high quality transformers, suitable for the most critical applications, is available directly from:

Jensen Transformer Company
1617 N. Fuller Ave., Hollywood, CA 90046
Phone (213) 876-0059

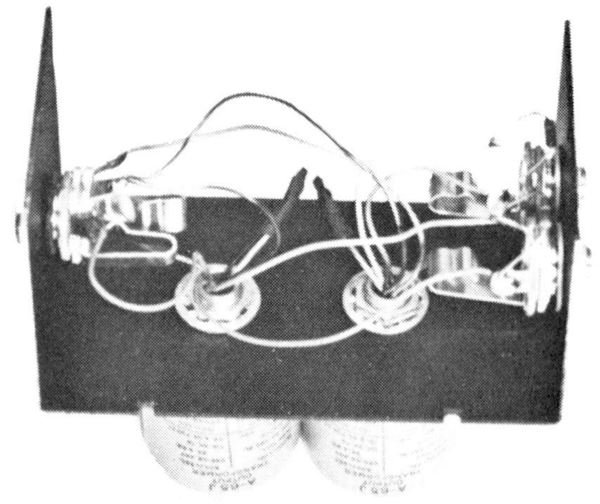

Fig. 41 – Step-Up Transformer Box

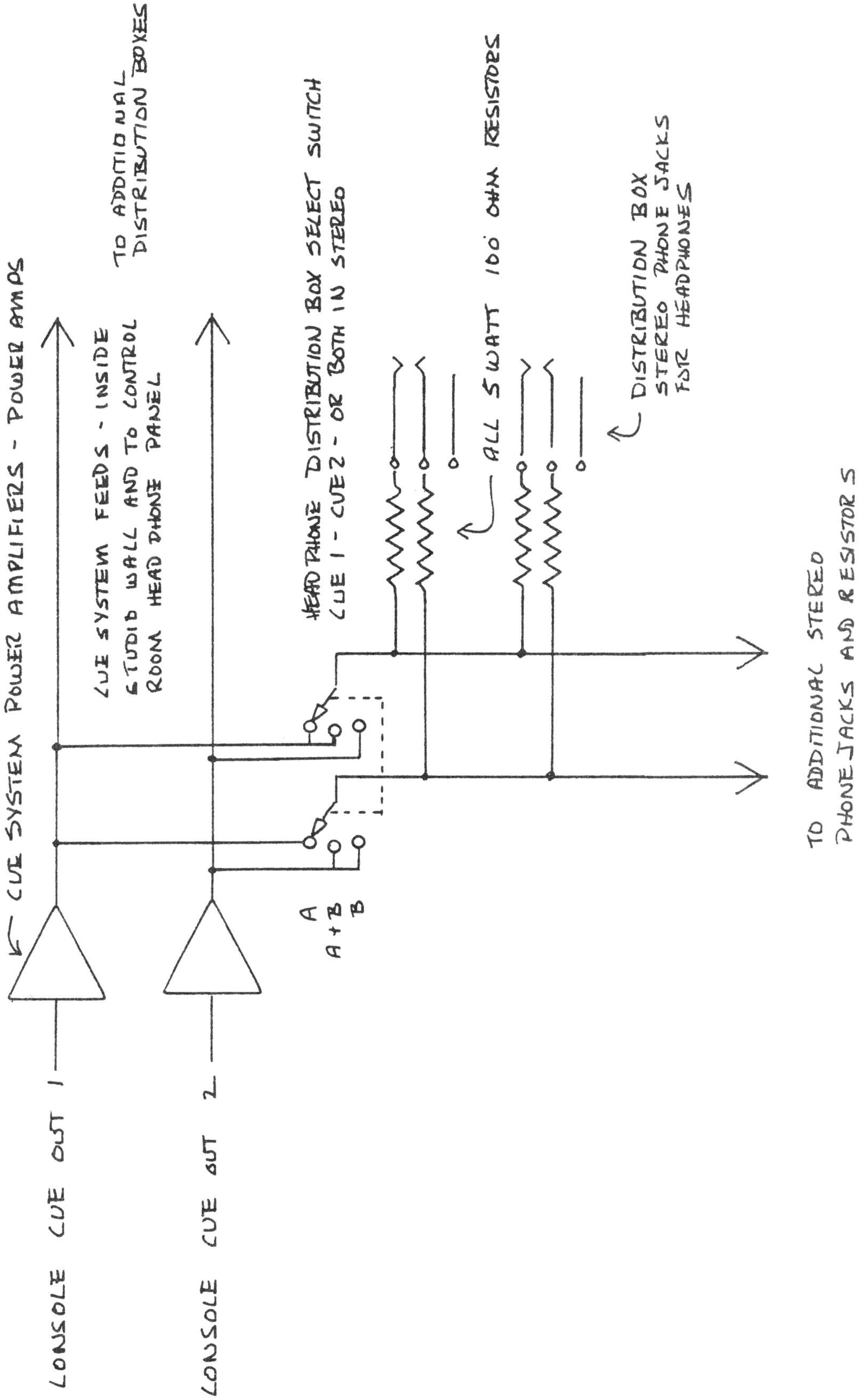
CONSOLE CUE OUT 1
CONSOLE CUE OUT 2
CUE SYSTEM POWER AMPLIFIERS - POWER AMPS
CUE SYSTEM FEEDS - INSIDE STUDIO WALL AND TO CONTROL ROOM HEAD PHONE PANEL
TO ADDITIONAL DISTRIBUTION BOXES
HEADPHONE DISTRIBUTION BOX SELECT SWITCH CUE 1 - CUE 2 - OR BOTH IN STEREO
ALL 5 WATT 100 OHM RESISTORS
DISTRIBUTION BOX STEREO PHONE JACKS FOR HEADPHONES
A
A + B
B
TO ADDITIONAL STEREO PHONE JACKS AND RESISTORS

Very important to the recovery of any recording studio equipment failure during a session with a client present is: 1) doing something immediately in such a way that you reassure the producer that the problem will go away quickly, and 2) protecting still functioning equipment, which might be damaged due to possibly panicky attempts to fix the original problem. Before beginning cue repair operations, it is a good idea to put a 400 hertz tone into one of the established recording busses that you are sure is included in the cue mix (like the bass). This way, the cue system will have a dependable, constant sound source to hear when it starts working again.

Before the Monday morning of a 9 AM, 45 musician session, check the phones by running some loud music through them. Wiggle and stress the headphones and their cables, listening for crackles and other potentially hazardous signs of mistreatment or old age.

It is really to put on a bad "show" due to faulty headphones. Be sure that you are aware of and responsible for the condition of all the studio's session accessories including the phones.

SOLUTIONS FOR PHANTOM MICROPHONE PROBLEMS

*** Please refer to Illus. I7-6 -- The Phantom Power Diagram ***

The phantom supply design is used in the studio to replace battery operation of mikes such as the Neumann U-87, and to turn on other mikes requiring 48 volts such as the Neumann KM-84, KM-86, AKG 414 and AKG 451E. By studying the circuit diagram provided, it is seen that the phantom volts battery or battery substitute device (power supply) powers each mike by sending the battery minus to the mikes via the sheild of the mike cable, and sending the battery positive to the mikes via the hi and lo signal wires of the mike cables. In case of mike shorts or other problems, there are two 6,800 ohm resistors between the battery plus pole and each hi and low mike cable wire, where the battery connects to the console end of the mike cable. There should be a spare heavy duty 48 volt power supply where it can be plugged or switched into service if the main phantom supply goes down in the middle of a session. Not having spare power supplies ready to go in case of an emergency for all the console needs and the phantom supply is a dangerous way of cutting costs. Power supplies are cheap insurance compared to the costs of having to make good to a producer for lost session time (called down time), possibly including the cost of many union musicians.

Since the very first patch bay in almost all consoles is the mike out bay, and since this bay is physically the console end of the mike cable, then it is clear that the mike out bay has 48 volts on the hi and low contacts of the jacks (remember that the female patch bay connector is the "jack," the patch cord has "plugs" on each end). If you patch a mike out to pre in, and you plug the mike out position first then the tip and ring of the other end of the patch cord has 48 volts on it. As you try to patch the pre in part of the patch, you will assuredly touch either the tip or the ring to the jack field and momentarily short out the particular mike's phantom volts. In some cases, this act is enough to make a tremendous popping noise in all the other mike input circuits and may damage your monitor system's speakers and/or drivers. In addition to damaging the speakers the loud "crack" may well interfere with your hearing for the rest of the day. The "hot switch" may

PHANTOM POWER DIAGRAM Illus. I7-6

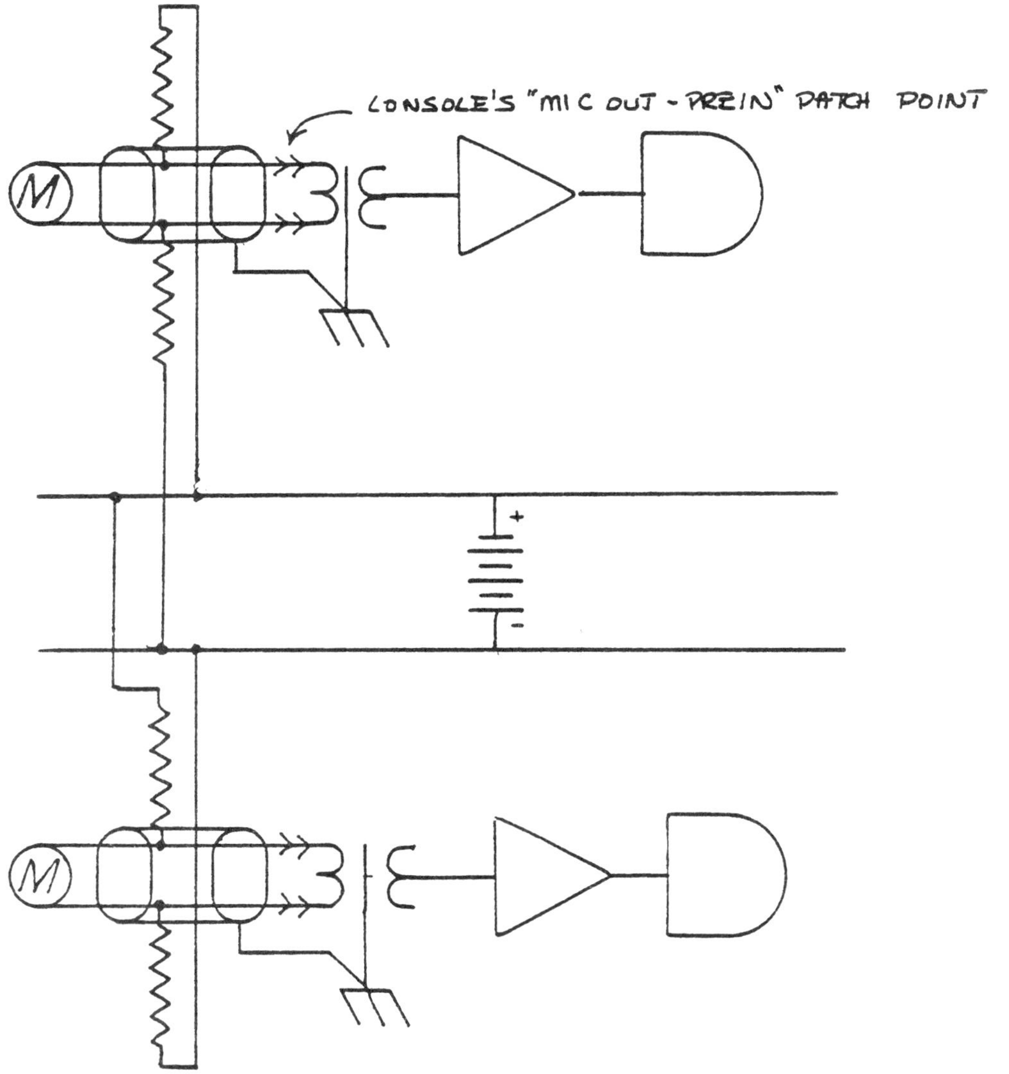

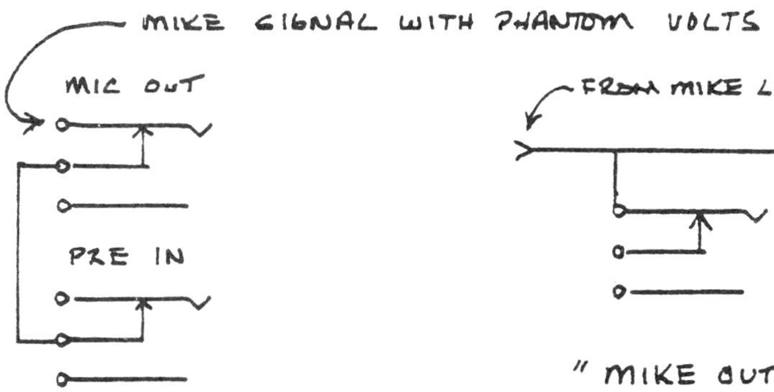

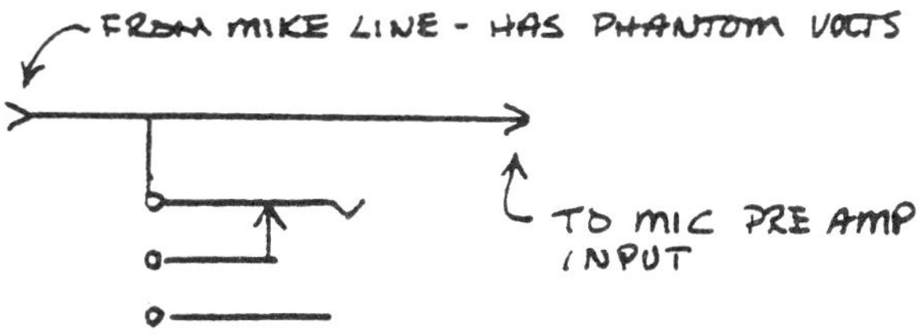

"FULL BREAKING" NORMALS

IF EITHER PATCH IS MADE - PHANTOM VOLTS will NOT APPEAR AT "PRE IN" PATCH POINT

"MIKE OUT MULT"

ON PATCHBAY MEANS NO NORMALS AT ALL - MIC MULT PATCH POINT IS A "PICKOFF POINT" OF ACTUAL MIC INPUT WIRE. DO NOT USE FOR DIRECT BOXES OR MICS IN CONTROL ROOM WITH MIKE CONNECTED IN STUDIO!

also damage the reverb chamber speakers and damage all the headphones in the studio. The thing to do, therefore, is to mute the originating and destination modules of the patch that is getting made or broken and make the "pre in" patch first, and then make the "mike out" patch. When pulling a mike out patch, pull the mike out patch first and then pull the pre in patch.

Since phantom volts cannot go into the console, due to the common mode rejection activity in the mike input transformers, they will probably not be present at the pre in patch bay. If you were to plug a phantom type mike directly into the pre input jack, the mike wouldn't turn on. The same holds true with direct boxes that feed on the phantom volts: they will not work if you try to operate them into the pre in patch bay. There are two ways around this little problem:

1. The "mike mult out" form of manufacturer's patch bay design makes the mike out patch point a pick-off point, which doesn't disturb the mike's signal from getting through to the mike pre. If you plug a phantom direct box in a mike out mult patch bay point, like some Automated Processes consoles have, the direct box will turn on, but if you also have a mike plugged into that same numbered mike in jack at the studio wall, both the direct box's and the mike's outputs will be connected together, probably ruining the sound of both.

2. The old "using a patchbay mult to get the phantom volts through the open normal trick" -- in other kinds of patch bays, notably in older consoles, the mike out/pre in patch point is double breaking and the phantom volts can't make it to the pre in bay once you plug the direct box in there since the normal opens. The solution is to patch the mike out to a mult (a mult is four or more patch field points labeled mult 1,2,3, etc). These mults attach the patch cord audio signals together. So when you take another patch cord from the mult to the pre in position, the phantom volts are then brought to the mult (from the mike out patch cord), and since the mult is going to the pre in, all you have to do is plug the direct box into the mult. This sends the phantom power to the direct box and sends the direct box's signal to the mike pre input.

SOLVING CONSOLE AUDIO PROBLEMS

A sure sign of a professional engineer is one who doesn't let anyone know when he's having problems with the console systems. It is a real challenge sometimes to "patch around" problems in such a way that you don't get confused, and the client is not distracted from the music. Of course, complaining about terrible equipment problems is a favorite ploy of the visiting engineer who thinks it makes him look good, and he wants his personal client to think of him as a marvelous engineer for being able to get such good results considering the "shabby condition" of the studio. This is bunk except in cases where the equipment is really so poorly maintained that even creative techniques can't make it work. Complaining about anything during an artistic musical session will be a bummer to anyone present and is not a suitable or professional contribution, especially from a highly trained and talented technical/musical person -- the engineer.

In the old days there were many more patch points available than in today's consoles. More patch points mean more patching power. It is quite practical to patch EQ #32 down to console channel #10 if the EQ in module 10 stops working. Simply note "EQ from 32" on the fader strip at fader 10. In fact, you can steal any patchable console device from anywhere it isn't needed. Inventive patching will keep a console operating long after it should have been put out to pasture. Console problems are a headache for the engineer, but the revelation of any difficulties should be saved until after the session is over. If, after revealing to the producer that the studio is a turkey and on its last leg, so to speak, he says that it's impossible to change studios, and he wants you to continue to work around the difficulties, then that's that; at least the producer knows the score. The above discussion is assuming that you are the visiting engineer with your private client. Obviously, if the client is visiting you at your studio, you should say your piece to the studio owner or maintenance man and not to the visiting producer.

Other patching techniques have to do with a console that is working properly but isn't really big enough to accomodate your session:

1. Not enough busses -- use the reverb sends. Patch the reverb send out to the VU tape in patch point. Use the machine on input monitor mode (the monitor system on "tape" or "play"). If you still run out of busses, use the fader out and send this directly to the tape machine (run the fader nearer the top than usual) and use the VU meter on the machine. These "behind your back" VU situations are very dangerous however, so it might be best to use a limiter to guard against level excursions, and then convert the limiter VU system to "output" monitor mode and use the limiter VU meter as your VU.

*** We will discuss limiters and other outboard equipment in Advanced 4 and 5 ***

2. No monitor reverb -- this problem can be corrected if there is a spare monitor buss, eg. a two monitor room with a four buss monitor system. Take a spare monitor buss output to the chamber in patch and return the chamber out or reverb return to an unused track of the 24 track, where you can record it or just leave the track on input and monitor the reverb at the monitor system at the position used. All you have to do is keep track 24 on input all night to have your reverb.

Another way to do it is to use the reverb sends on the input modules and actually record the reverb on one track of the 24 track. Then it will always be available for use even if you don't set anything up, by just playing the reverb track back along with the rest of the recording. Using this method however, it becomes difficult to add new sounds to the reverb mix, if overdubs are done on a track before going on to record new tracks.

3. No monitor system to cue -- this cue mix technique should be rarely used, but when it's needed, it makes things go very easily. It is a system where the cue mix is derived from the monitor mix heard in the control room. It is most often used when the producer or artist is one of the musicians in the studio. He is then concerned with the sounds and the mix to a critical degree. If he can hear exactly what you're hearing (the same mix), then he has a better chance of getting what he wants from the session. Patch the left and right monitor busses to any two faders or outboard devices with gain pots and patch the now volume controlled monitor signals to cue 1 and 2.

4. No program/tape VU changeover switch -- it is very handy to

changeover the console VU meters to read the output of the tape machines. This is especially true if the tape you're working on was poorly recorded and there are many recorded level changes on, for example, the vocal tracks. Your control of levels will be improved if you can see which tracks have the meteoric recorded levels without having to twist your head around to see the tape machine meters. There is a way to get a console VU function that will serve this aim, and you get it by assigning (in addition to the stereo buss) each desired fader to its same numbered buss. When the playback occurs, the console buss VU meters will be energized to a reading determined by: 1) the mixdown fader settings and movements, 2) the volume (the recorded levels) coming from the tape, and 3) the relative level of the mix buss to the main busses. If you need more VU indicator deflection, simply reset the sub masters higher (since you're not using the busses for recording you won't care if extra noise is present). This use of "dummy busses" is very handy when doing multi-mike overdubs or many track combining operations. Use the dummy buss technique often. Never substitute mechanical techniques like VU watching for careful listening, however.

MAkNG SAFETY COPIES

*** Please refer to Illus. I7-7 -- Tape Copies ***

Safety copies are for the protection of audio projects especially those facing a strict deadline like T.V. shows, where the loss of a generation of audio quality is not as important as the ability to recover the audio portion of the damaged portion of the finished product. Safety copies protect time and money. Occasionally a splice breaks during rewind or fast forward and the resulting mess of wrinkled tape might include a master take. Wrinkled tape is unusable due to the speed fluctuations and dropout problems that will occur should you attempt to run the wrinkled portion through the deck. Recorder deck logic can fail once in a while (perhaps a .60 cent ic chip blows up) and the pinch roller might clamp down on the tape during a high speed shuttle. Clients will not be happy about having to use a copy of the master, but happier than they would be if the master tape was damaged without a backup copy.

Safety copies should be made:

1. When planning to fly tapes in the baggage compartment on airliners
2. When finished recording basic tracks
3. When finished with overdubbing and preparing to mix (use the same safety copy tape over if you wish)
4. When the project is for T.V. with an imminent air time when even a generation loss would be better than a garbled or missing musical number

When making the safety copy, proceed as follows:

1. Using the master's tones, adjust the reproduce of the play machine to perfection
2. Adjust the input monitor calibrate on the play machine even though the record functions won't be setup
3. Adjust the playback of the record machine to perfection using the same master tone set

Written by Sherman Keene

TAPE COPIES INTRODUCTION

This is an essay on how to make stereo tape and safety copies using professional recorders and 361 Dolbys. The techniques covered would, of course, work with four, eight, sixteen and 24 track machines too. The only difference between the techniques for making a safety or tape copy are in the choice of tones used to align the record machine (either client's tones or a standard reproduce alignment tape). Please refer to the sections called "Record Machine - Reproduce Alignment".

There two types of copies: safety copies and tape copies.

Safety copies are made for the protection of the master tape and must be so identical to the master that pieces of the safety copy can be substituted into the master without any remarkable change in level, azimuth, or recorded EQ or quality. Only a slight increase in tape noise and a slight dulling of the sound might be heard but only if using the finest monitor systems. For safety copies, the record machine's reproduce system is aligned using the client's master tone set (the 1 KHz, 10 KHz and 100 Hz tone on the client's master reel - or project tone reel).

Tape copies are made for the product's users such as for duplicating purposes. No matter how "off base" the original tape's levels, EQ or azimuth are, the tape copy "starts fresh" and "comes out" with standard levels and perfect recorded EQ and azimuth. We do not "pass the buck" with poorly recorded tapes when we make tape copies; we correct the recording to industry standard levels and correct azimuth and EQ. For tape copies, the recorder's reproduce system is aligned using a standard reference alignment tape such as those produced by ampex and standard tape laboratories. Naturally, a mixing project started out fresh by you would have a master 2 track tone set which was a near a duplicate of the standard alignment tape as you could get. Examples of user's tape copies might be:

1. A copy for the producer to take home & elsewhere to listen to his current project for judging the correctness of his mixes, EQ etc.
2. For duplicating (dub) purposes eg. "We need three A-Type Dolby copies of the master for our East Coast tape cartdridge duplicating plant's use".
3. For distribution of the product to the users eg. "We need 30 copies of our 15 IPS radio station spot called 'The time is right - 60 second version'".

Examples of a safety copy might be:

1. A Dolby to Dolby copy of the entire album for storage in a different vault than the original tape
2. A safety copy to be used to edit in pieces of a song to make it longer or change the arrangement of the song

3. A safety (24 track) copy before the master tape is flown to a new location for mastering, overdubbing or other use

INITIAL SETUP

Note: This essay is for recording studios where the monitor system of the console does not terminate the outputs of the 2 track machines.

1. Degause and clean both the play and record machine
2. Check the machine's terminations - should be on for both machines
3. Check that the machine to machine, machine to Dolby, and machine to wall (the machine's monitor and input cables) are all track one (left) to connector one and so forth (parity check)
4. Exercise (switch back and forth) safe/ready/sel and repro/input/bias switches
5. Check the Dolby's terminations - should be set to 10K for both (all) Dolbys. Refer to "Taking Care of Possible Termination Problems When Aligning 361 Dolbys" in Intermediate 7

PLAYBACK ALIGNMENT - PLAY MACHINE

Tone source = client's tones played back

Tone	Adjustment
1KHz	PB level
10KHz	PB head azimuth **
10KHz	PB hi frEQ EQ
100Hz	PB lo frEQ EQ
1KHz	re-check PB level: if off, repeat from the top skipping azimuth

Tone source = console 1KHz 0 level tones

Tone	Adjustment
1KHz	input monitor calibrate

Leave machine on input so that the console's tones can be "passed on" to the record machine. This saves you the inconvenience of having to continually rewind the play machine because you need "Just a little bit more of the 1 KHz tone". The console's tones are very steady and will last as long as you like.

Place calibration sticker on deck - speed, PB only, (client's name) tones, date, your first name; see placard format later in this instruction.

** When adjusting azimuth, play the 10 KHz tone and bring the left track up to buss 1 and 3 and track 2 up to buss 2 and 3. Bring up track 1 to read -6

VU on buss 1, and then bring up track 2 to read -6 VU on buss 2; this should make for a 0 reading on buss 3 when the azimuth is adjusted perfectly. Now adjust the azimuth for a peaking of buss 3 which should peak at around 0 VU (depending what the two tracks are reading on the two track). If you have an oscilloscope, patch it to the console busses 1 and 2. Once the console summing buss has been peaked, look to the oscilloscope for a precision indication of azimuth adjustment. Keep in mind that the hi freq tones at low tape speeds will be recorded at -5 or -10 dB to avoid saturation. When recording tones at low speeds, use -5 or -10 for the same reason.

PLAYBACK ALIGNMENT - RECORD MACHINE

Tone source = Ampex (or other) reproduce alignment tape (for a tape copy) OR client's master project tones (for a safety copy)

Tone	Adjustment
1KHz	PB level
10KHz	PB head azimuth
10KHz	PB hi freq EQ
100Hz	(for a tape copy) skip this step
100Hz	(for a safety copy) set lo freq PB EQ
1KHz	re-check level and PB EQ - re-start if necessary skipping azimuth

RECORD ALIGNMENT - RECORD MACHINE

Tone source = console tones relayed through the play machine (which is still on input)

Tone	Adjustment
1KHz	bias
1KHz	rec level
10KHz	record head azimuth
10KHz	hi freq rec EQ
100Hz	(for tape copy) repro lo freq EQ
100Hz	(for a safety copy) just check that the lo freq are within plus or minus 2 DB (+/- 2 db)
1KHz	re-check rec level - continue checking level, and hi freq rec EQ and lo freq PB EQ until all read 0 VU
1KHz	set input mon cal - continue recording the 1 KHz tone during the setting of input so that you will have some uninterupted 1 KHz tone to erase during the next step

Place tape at beginning of 1KHz tone passage (the tones you have been recording on blank tape) and after shutting off the console oscillator, begin recording (erasing) the previously recorded 1KHz tone. Turn up the control room monitor to full volume cautiously and check that the machine is erasing properly and is recording quietly without an excess of "rocks" which indicate a possible bad batch of tape. Aside from the very beginning or ending of a tape reel, there should be very few rocks heard. There will always be an occasional rock, but do not use tape which has the sound of an avalanche of rocks for making tape copies.

Place sticker on deck - speed, tape type, the date, your initials so that others will know who aligned the machine and to what levels

FINAL ALIGNMENT CHECK

Now instead of using the console's tones relayed through the play machine (on input), change the play machine to repro mode and play the client's tones and record them on the copy machine. If the alignment is correct, they will transfer O.K. (all tones reading zero on the play and record machine) and this means that you may begin to copy the program material as well. If the tones do not transfer perfectly (which is usual due to differences in VU meters), continue making the necessary fine adjustments until they do transfer perfectly.

MONITOR SYSTEM USE DURING COPYING

Except when initially establishing that the copy A/B's with the original in a suitable fashion, listen only to the playback of the recording machine(s) while copying!

Leader all tape copies except when asked not to do so. Do not leader media tape copies (for radio station spots - commercials etc.)

Remove all tones from the record alignment setup (don't use too much tape for aligning record!!!!) before beginning the copies.

Reject a copy with any (!!!) dropouts. If you have dropout problems more than once on a reel of tape, reject the tape back to the manufacturer and get another reel.

Turn off console oscillator before beginning copies.

CUE TONE USE FOR MULTIPLE COPIES OF THE SAME SELECTION

If you have to make, say, 30 copies of the same 60 second tape, it will save you a lot of time if you use the cue tone method of knowing where to cut the copies apart when you finish recording them and are now ready to cut and spool them onto their little shipping reels. What you do is prepare a 40 hertz tone to come up on the two busses feeding the two track which is the reproduce machine. After ecah 60 sec. spot finishes, you count to 10 while you rewind the master. At the 10 second count, you switch the machine to input and then back to output (leave on input for one second). As soon as you have switched back to output on the play machine, start counting up to 10 again. At the count of 10 press play on the play machine. Now when you are finished with all the copies, you just rewind the tape copy back onto the shipping (7 inch or 5 inch) reels and stop the tape and cut it each time you hear the cue tone's "beep" at rewind speed. No one will ever hear your cue tone since it should only be a few inches long at the head and tail of each copy.

SPECIAL DOLBY HOOKUP NOTES
==========================

Dolby units are always setup with their termination switches set to "10k", not 600!!!

There are three types of 361 Dolby hookups:

1. The "surround" hookup where the Dolby is plugged in to the tape positions in the wall and the tape machine is plugged solely into the Dolby unit. The Dolby "surrounds" the tape machine. This is the case for the play machine's Dolby in "Dolby to Non Dolby Copies".

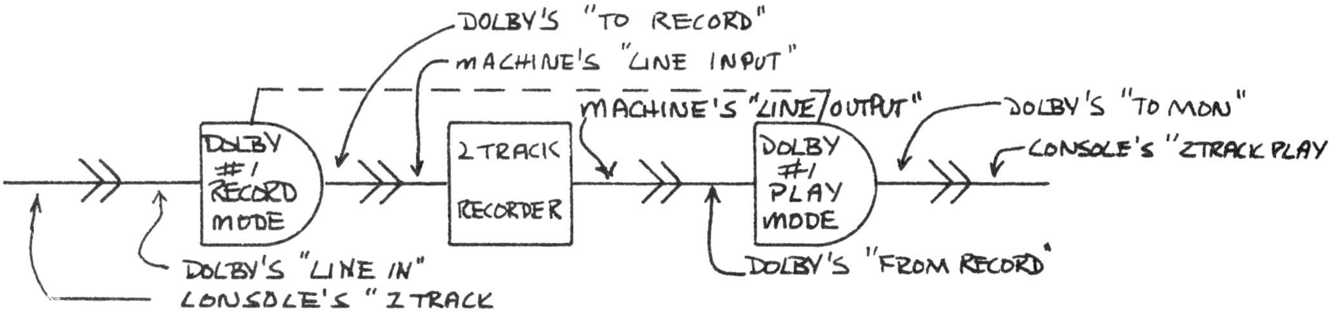

2. The "Encode" Dolby hookup where the Dolby is connected to the input circuit of the recorder but not hooked up to the recorder's outputs ("From Rec" and "To Mon" are not connected). This hookup is only used to encode a signal as in the case of "Non Dolby to Dolby Copies". Here, the #1 Dolby only encodes the non Dolby playback of the play machine on the way to the record machine. The play machine's output is monitored directly into a monitor system without going through the "From Rec/To Mon" circuit of any Dolby unit.

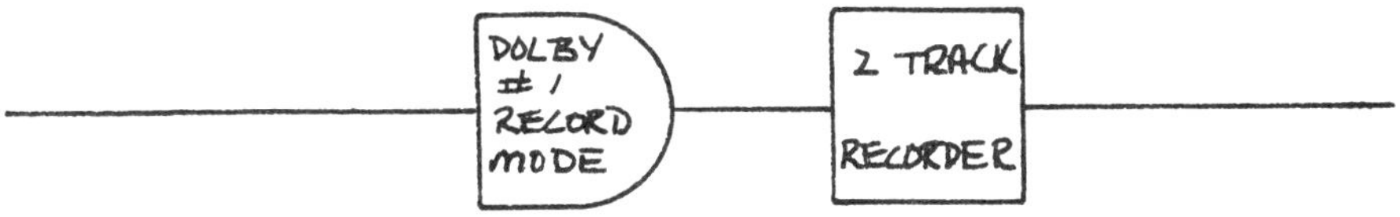

3. The "decode" Dolby hookup where only the "From Rec and To Mon" connectors are used. In this case the record functions (line in and to rec) of the Dolby are not used. This is the case of the #2 Dolby in "non Dolby to Dolby copies" and for both machine in "Dolby to Dolby copies".

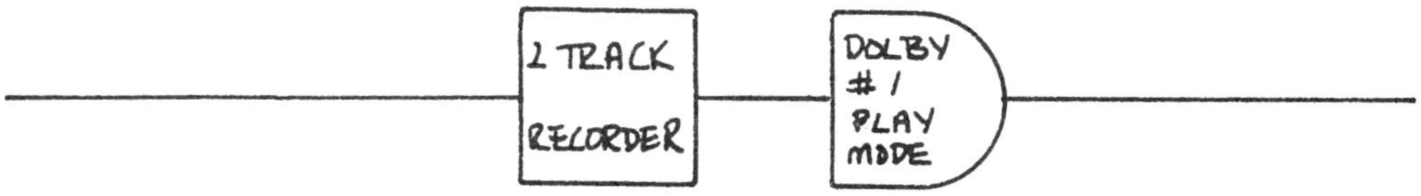

DOLBY TO NON DOLBY COPIES
========================
Use surround Dolby hookup on play hachine

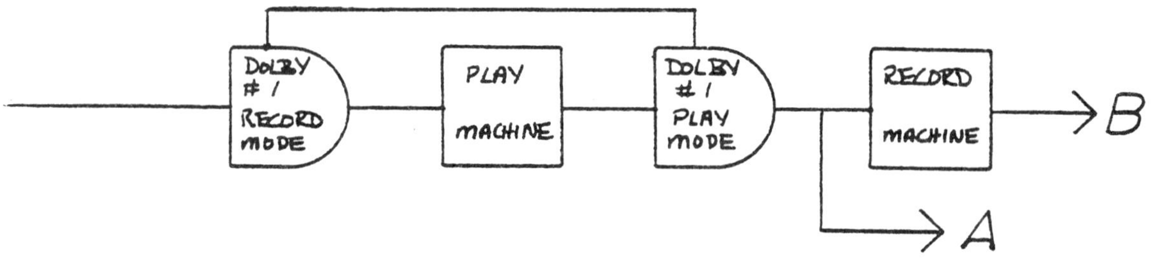

Setup the play machine with the Dolby set to play/normal/NR out. Be sure to include setting the play machine's input mon cal. Switch the Dolby to rec/normal/NR out and adjust the Dolby meters to read on the dot using the Dolby input controls and adjust the play machine's VUs to 0 VU again using the Dolby output controls.

The Dolby is now set to 0 in and out.

To adjust the record machine, set the Dolby back to play/normal/NR out. The console's tone are now relayed through the play machine and through the play function Dolby to the record machine's inputs.

At conclusion of setup, and after checking the alignment fine tuning by playing the nab tones from the play machine to the record machine, play the client's Dolby tone and re-set the play machine's play level controls so that the Dolby tone reads on the Dolby dot (the Dolby is still set to play/normal/NR out. The play machine is no longer set up to its NAB levels, but is now set to Dolby level at which the music is recorded.

Switch the Dolby to NR in and begin copies.

NON DOLBY TO DOLBY COPIES
=========================

Use the encode Dolby hookup on the #1 Dolby preceeding the record machine and use the decode Dolby hookup on the #2 Dolby following the record machine.

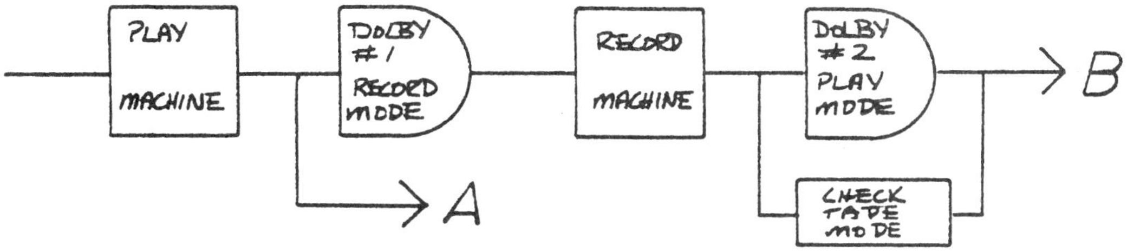

Setup the play machine.

With the #1 Dolby on play/normal/NR out, set up the record machine. When finished (including input cal), adjust the #1 Dolby for 0 in and out by switching it to record mode, setting the #1 Dolby input controls for a reading on the Dolby dot and setting the #1 Dolby output control so that the record machine again shows a zero VU indication.

With the record machine on input, and the #2 Dolby on play/ normal/NR out, press check tape on the #2 Dolby and bring the record machine's outputs up through two faders to two console VU meters to read 0 VU. Now press normal on the #2 Dolby and set the #2 Dolby input controls to the #2 Dolby dot, and the #2 Dolby output controls to again read 0 VU on the console VU meters you used.

Do the final transfer check of the NAB tones.

Now change the #1 Dolby to rec/normal/NR in and the #2 Dolby to play/normal/NR in and make the copies.

Be sure to record a short (15 sec.) Dolby tone at the beginning of each Dolby tape copy you record!

DOLBY TO DOLBY COPIES
=====================

Use the decode Dolby hookup on both Dolbys: Dolby #1 is fed from the mult box following the play machine on the way to the a monitor position, Dolby #2 is fed from the output of the record machine.

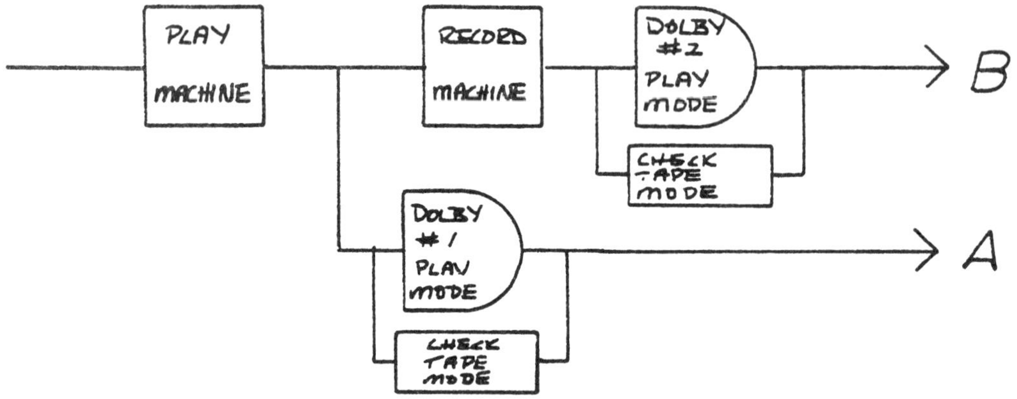

Set both Dolbys to play/normal/NR out. Align both machines without any consideration to the Dolbys at all.

Do the final NAB tone transfer check.

When finished, set the #1 Dolby to check tape and return the play machine to two faders and console VU meters. Zero in and out the #1 Dolby. Repeat the same procedure for the record machine zeroing the #2 Dolby.

Set both Dolbys to play/normal/NR out. Now play the client's Dolby tone on the play machine and set the #1 Dolby's input control so that the Dolby tone reads on the #1 Dolby unit's dot. With the record machine on input, set the input control of the #2 Dolby so that its meters read on the dot. The transfer level between the two machines is unaffected by the Dolby adjustments while both Dolbys will de-Dolby the audio for you to listen to and make a/b tape copy quality decisions. This is so that if the producer or engineer of the original session purposely recorded his music (and therefore Dolby tone) at a reduced level for the purpose of maintaining extra tape headroom, that your copy will not throw his extra headroom away; your copy will be at the same (lower than the alignment tones possibly) level as the original tape.

Be sure to transfer the Dolby tone from the master to each copy!!!! Transfer only the Dolby tone, however.

Now make the copies

MACHINE PLACARD INSTRUCTIONS

Tape Speed Tape Type Dolby (if Dolby)
Client's name tones
Date Initial of maintenance person

Any special information which will be needed (eg. client requsted non-standard bias)

Example:

30 406 Dolby
Sherman's tones
9/19 John Henry

Placard is to be replaced with new alignment data if the setup is for overdub or record.

Old record alignment placard is to be left attached if new alignment is playback only. An additional playback only placard is to be attached. example:

15 Playback Only
Gloria Marshal's Tones
9/20 Sherman

DOLBY PLACARD INSTRUCTIONS

For the original session setup where all tones are recorded:

Dolbys set 0 in, 0 out
studio B
9/19 Sherman

For continuing session where any Dolby tweaking is necessary due to the Dolby indicators not touching the dot (includes removal of original session Dolby alignment sticker):

Dolbys set to training tones
Studio 3
9/19 Sherman Keene

BIASING GUIDELINES

Ampex 351 mono machines

1/4 DB overbias using 1 KHz at 15 IPS
peak bias using 1 KHz at 7 1/2 IPS

Ampex 440 machines

1/2 DB overbias using 1KHz at 15 IPS
peak bias using 1 KHz at 7 1/2 IPS

1 and 2 inch machines except for the Ampex ATR 124:

3 DB overbias using 10 KHz at 15 IPS regardless of user tape speed

Ampex ATR 100 machines:

use master bias adjust - not individual track's bias adjustment controls
2 DB overbias at 15 IPS using 10 KHz tone
2 DB overbias at 30 IPS using 20 KHz tone
1 1/2 DB overbias at 30 IPS using 15 KHz tone

Ampex ATR124 machines:

3 3/4 DB overbias at 15 IPS using a 10 KHz tone
3 3/4 DB overbias at 30 IPS using a 20 KHz tone
3 1/2 DB overbias at 30 IPS using a 15 KHz tone

TAPE COPY BOX LEGENDING

The tape copy box must have certain information inscribed on it by the engineer for the work to be complete. There should be self-stick 7 inch, 5 inch, and cassette box legends supplied by the studio with the studio's logo, address and phone number pre-printed on it along with the blocks where the engineer is to fill in the proper information. Lacking these self stick legends, you must "fill in" the blank box or cassette jacket cover using your own format. This is a list of the information which should be included on the box:

- The date
- Studio's name, phone number
- Engineer's name
- Client's name
- Name of this copy (eg. "radio spots" or "rough mixes")
- Tape manufacturer/type
- Tape speed
- Recording EQ
- Whether Dolby or non-Dolby (say which)
- Track format
- Stereo or mono (or quad???)
- Names of the selections with their timings!!!!!
- Running time of the entire copy (total playing time)

EXAMPLE OF INFORMATION PLACEMENT

Date — Studio
Studio's phone #

Client's name
Copy's title or description

Speed EQ Format Type Dolby/or non

Selection #1	time
Selection #2	time
Selection #3	time
Running time	time

Tape manuf./type — Engineer's name

EXAMPLE BOX LEGEND #1 - RADIO STATION SPOTS

9/12 80 — Sherman Keene Recording
(213) nnn-nnnn

Brickman and Bruister Agency
AM radio spots

15 IPS NAB full track mono non-Dolby

1. Moto-tiller spot #1	:60
2. Moto-tiller spot #2	:30
3. Easy muncher spot #1	:30
4. Easy muncher spot #2	:60
Running time	3:20

Ampex 406 — Engineer: Sherman Keene

EXAMPLE BOX LEGEND #2 - 1/4 TRACK STEREO COPY

9/20/80 Sherman Keene Sound
(213) nnn-nnnn

Rolling Pups
Stereo rough mixes

7 1/2 IPS NAB 1/4 track stereo non Dolby

1.	"Here she comes"	3:44
2.	"There she goes"	12:56
3.	"Another day"	8:22
	Running time	25:14

Ampex 406 Engineer: Mark Marker

RADIO STATION-TYPE COPIES

For media projects, it is much easier to have tones on the beginning of each reel since this makes dubbing much easier. Therefore, do not edit the tones to the tail of the reel.

1. Tapes need not be leadered but there should be 10 seconds of blank tape before and after the program material.
2. All tapes are to be wound onto 5 in. or 7 in. large hub reels - heads out (ready to play)
3. All copies must be listened to for quality control
4. Precise box legends must be provided

SPECIAL AMPEX ATR 124 ALIGNMENT PROCEDURES

General Information

1. Be sure that there is an active EQ for the speed you have selected else you will get err 1
2. Be sure that the NAB button is pressed for any 7 1/2 or 15 IPS setup selected
3. Watch the remote box carefully to make sure that the machine wakes up properly
4. Write and use one of the setups (1 to 4) for your session. Write the set up number on the white paper tape and the information for that setup in pencil!!! on the setup box door
5. Use the master bias control not individual bias controls

15 IPS BIASING

Use only one meter when biasing
Make peak bias result in a machine zero level indication by adjusting the console's oscillator
Overbias 3 3/4 DB using 10 KHz tone

15 IPS RECORD EQ ADJUSTMENT

1. Turn all hi freq rec EQs completely counterclockwise (20 turn pot) until the VU indication stops moving
2. Turn shelf EQ to read -1 db
3. Turn hi freq (peak) rec EQ up past the 1/2 DB dip until the track comes back up to 0 VU

30 IPS BIASING

Overbias 3 3/4 DB using a 20 KHz tone or...
Overbias 3 1/2 DB using 15 KHz tone

30 IPS RECORD EQ ADJUSTMENT

1. Turn all hi freq EQ controls completely countercloskwise (20 turn pot) until the VU indications stop moving
2. Turn shelf rec EQ to +1/4 db
3. Turn hi freq EQ control past the 1/2 DB dip until the track comes back to 0

VU METER STATUS INDICATIONS DURING MALFUNCTION

Meter Indication	Meaning
Meter blinks red and white quickly	You have bias and erase but no record command
Meter blinks white longer than red	You have bias and record command but no erase
Meter blinks red longer than white	You have erase and record command but no bias
Meter blinks red and white slowly	You have record command only but no bias or erase

4. Set up the recorder to perfection using the play machine in input mode to relay the console tones to the record machine (through the play machine)

5. Load the play machine with the master tones and playback the master tones while you record them on blank tape with the record machine. If the levels on both machines are perfect, then continue with the copy of all the audio; if things aren't perfect, then keep adjusting.

6. Listen to the output of the record machine (machine in record/repro mode) for all copying. Listen very carefully. If you hear anything resembling a drop out, investigate immediately. When the copy is finished, play the copy and monitor it track by track to see that all the instruments sound all right, with no noises, humms, buzzes, etc.

7. Carefully mark the box with a duplicate legend or use a Xerox copy of the original legend, being sure to have the word "COPY" significantly displayed on the tape box and the reel itself and include an up-to-date set of track sheets (Xerox copy) from the original master in the copy box. Make sure that the copy's box, reel and duplicate track sheets indicate a copy.

Be sure to copy the tones. If the tape is Dolby system processed, then transfer the whole program including all tones with the Dolby units out of the circuit, unless the instructions explicitly say make a non-Dolby encoded copy. If a non-Dolby copy is required, line up the play machine normally with the Dolby units on bypass, and line up the record machine. Then put the play machine's Dolby units on non-bypass and check all Dolby tone levels and readjust any discrepancies. Record the tones across with the Dolbys in bypass -- do not copy the Dolby tone. Record the actual program material (the music) with the Dolby units in non -- bypass and in play mode.

MAKING THE MASTER REEL

Master reels are those where only the master takes have been edited one after the other with no out takes included.

A master reel is a great help when overdubbing on many songs, one song after the other. The advantage of being able to carry the project tapes to the studio without a fork lift is also obvious.

There will usually be two master reels and hopefully two master reel safety copies. The first master reel should be edited so that there is a length of blank tape (perhaps 1/2 inch of radius), then the songs: first, second, etc. Next to last and very last on the #1 master reel are the master tones and the record pad last. (A record pad is simply one or two minutes of unrecorded tape which is "given" to the master reel to allow the maintenance person to record on while aligning the tape machine. Since this record pad is the same age, type and batch as the project, it gives particularly accurate adlignments.) You don't want the tones and record pad at the head of the reel or the maintenance person aligning the session will have to shuttle the tape through all your tracks (with many edits perhaps) to get to the tones each time he aligns the machine. If there are any splices, then you're asking for trouble.

Note: when you operate the machine always keep one hand within a foot of the reel that is giving up tape during fast winds in case the tape breaks. If you don't and a splice breaks, by the time you get to the machine there will be a nest of tape already on the floor, which you will have to wade across to get to the machine, wrinkling the tape horribly as you you walk on it making it useless for further recording.

So, put the tones second to last and then the setup person won't have to spool through your master takes to get to them. The record pad is edited on the reel last and not left at the head of the reel, so that in case of a distraction during a machine record alignment, some of your tunes won't be erased by the horrified maintenance man, who notices after being distracted that the machine, still recording a 1 KHz tone on all 24 tracks, has wandered off the record pad and is now into what used to be the first song.

The master reel is safety copied after the completion of all tracks, and again at the completion of all overdubs. Some studios have taken this protection to the limit. Management at these studios demands that at the conclusion of all sessions two safety copies be made and the master and the two copies then be taken immediately to three separate fire proof air conditioned security storage areas. Any time the recordings you have just made are irreplaceable, make a safety copy.

TAPE COPIES

*** Please refer to Illus. I7-7 -- Tape Copies ***

Tape copies are usually stereo or mono copies of mixed music masters or interproduction rough mixes. Tape copies are not the same as safety copies because tape copies are made with the recorder lined up to a standard reference level tape. Tape copies are usually made from two track masters or rough mixes on two tracks. They take the form of quarter track (home stereo) tapes or cassettes. In the case of the cassette, the cassette itself should be marked with a felt tip pen as to the contents, and the paper jacket to the Phillips box (a Phillips box is the kind with hinges) should have the tune names and times, Dolby system used or not, and the engineer's name inscribed upon it. In the case of quarter track copies, the tape should always be edited and leadered (unless otherwise told) and the box marked with the tape speed, the recording EQ, which mode (quad, stereo, mono etc.), what kind of heads (quarter track, half track, full track), whether Dolby or not, the tune titles, individual tune times, running time, whether leadered or not, and whether there are any tones (usually not) and lastly and most importantly the studio and engineer's name.

A typical box legend should look like this:

---

7 1/2 ips NAB stereo quarter track non-Dolby

---

1. The Chase Of The Woodpecker 12:33
2. Hairy Henry's Famous Waltz 3:45
3. Jack-Of-All-Trades Polka 6:25

running time 22:55

All selections leadered
No tones
Ace in the Sleeve Recording Studios Inc.
Engineer: Sherman Keene

The reason for the documentation is so that the tape's qualities will be known to the user without any experimentation on his part.

It is rare that the client will ever see your two track or 24 track master tapes and their boxes, which should look great too, but anyone connected with the project who will listen at home will surely see your quarter track tapes and cassettes. You want everyone who comes into contact with your projects to think of your name in the context of great sounds and carefully prepared work. The quality and looks of the record albums are the responsibility of the record company -- the tape copies are yours.

Since many two tracks are set up at +3 compared to "old" Ampex oerating level, most professional tape machines will distort (saturate) the highs if you try to record at 7 1/2 ips at full level. The way to determine this is quite simple -- just set up the machine to record 12,000 Hz so that a console -5 VU reads -5 on the tape machine. Now slowly raise the level to -3, -1, 0, +1, etc., and watch the tape machine's VU levels. When the machine's levels "fall behind" or read lower than those coming from the console, this is evidence of saturation. The slower the tape speed the more pronounced the problem is due to more and more hi frequency boost during recording at slower and slower speeds.

*** Refer to Illus. I6-1 -- The Ampex Record EQ Table ***

A good rule of thumb is to make your 7 1/2 ips tapes 4 dB lower than the level of your 15 ips tapes.

POOR MAN'S AUTOMATION

This is a technique first introduced to me when I was working at the Record Plant in Los Angeles. It is so simple that it sounds silly: always record the output of the monitor busses on tracks 22 and 23 (the 60 hz sync tone is on track 24). Patch monitor booster or monitor out, or whatever your console has for monitor outputs into VU/tape in, and leave tracks 22 and 23 in record at all times. In this way, when there is a new instrument overdubbed onto the tape, the instrument will show up on the rough mix tracks 22 and 23. Monitor reverb and any other monitor effects you use will also be recorded there. In almost all cases the quality of the mix will be much better than the quality of "roughs" that you mix at the client's request. The poor man's automation mixes reflect all the little adjustments you constantly made on the monitor during the several hours it was required to overdub the last instrument, whereas a "quick" rough will reflect only a 15 or 20 minute effort. Secondly, the poor man's automation "rough" mix will sound just like the mix that was in the control room at the last session -- the overdub session. So when the producer asks how much studio time it would require to get "a set of rough mixes" of the album project as it stands so far, tell him the time it will take to copy track 22 and 23 to a quarter track or cassette machine. In the case of the poor man's automation from a live tracking session, the stereo tracks are a first generation mix almost like a live to two track session.

When you begin to overdub, start off the headphone mix using only the poor man's automation tracks sent to cue left and right, while your assistant gets a new cue mix together on two other cue busses, and then patch the new cue mix into service, or just leave the old poor man's automation tracks doing the cue feed with the current overdub sound mixed in too. You can do this until you get a request from the artist for a different mix. The time you save will more than compensate you for the energy used for poor man's automation activities.

AUTOMATION'S LAST PASS

In the automated console, the settings of the faders and in some cases even the pans, EQ's, and reverb sends are all digitized into a recordable sound that, when reproduced, will electronically reset the applicable console devices to the same settings as when the automation track was recorded. The faders, pans etc., don't actually move, the electrical values are set by automation. In one automation type -- Necam by Neve -- the faders actually do move. So when you have a mix you like, record the console's automation "sound" code on one audio track of the tape machine for tape based automation or record the automation data on a computer floppy disk for disk based automation. Realize that if your console only "remembers" the fader settings, then there is a lot of other data you should manually store as well such as EQ, pan, reverb or special effect send level, reverb effect assign, your patches, settings of the effect devices etc. There are many methods of trying to "record" this data. One is by use of console "maps" -- large pieces of paper printed to resemble the controls of the console. You write on the paper where the faders are set, where the sends are set, what EQ is used and so on. Another ploy is the old "overhead high contrast Polaroid camera" trick. It is VERY difficult to make readable pictures of the console settings with enough detail to be actually used. Lastly is the "read the console settings slowly and surely into a cassette machine" trick, pretending that you're in the future trying to reset the console to the sound of your own voice. This technique is simple and it works. If you need to duplicate the mix settings in the future then you can do exactly that.

Spend your energy when you RECORD your tracks. Use all your skill so your tracks won't need hours of mixing. Some mixers go to the extreme of leaving all the monitor system volume controls up all the way and do all the mixing during the recording phases of the project, trying not to touch any monitor system volume controls no matter what during all tracking and overdubbing sessions. They get their balances and keep up a good monitor mix with the monitor system volume pots fixed and do the mixing with the recording faders. This means that for a mix, all that should be necessary is to push the faders up to exactly the same volume (just like the monitor system pots have been all along -- full up and all set the same) and to "just throw in some pan and a little reverb." This showmanship type of recording method although feasible is not recommended.

Almost a lost art in recording studios these days is the ability to provide a finished mix, live to 2 track, at the time of the live session reducing the 24 track machine to only a backup. This is a high form of mixer (or mixer crew) competence. Providing a first generation finished mix at the live session can, of course, only occur at sessions where all the musical parts are played at the session -- no overdubs are planned. This is usually the case with large band or orchestra sessions so these sessions should always go direct to 2 track. This may be too much for a single engineer to handle, so the concept of the second engineer really comes into play: the second engineer operates the mike input to buss output part of the session (and makes the 24 track backup recording) while the first engineer "rides" the 24 track monitor system (operated in PGM or BUSS mode) and adds effects to provide the finished mix, which is recorded live to the 2 track. He, of course, gives helpful hints to the second as the session progresses. This is engineering at its finest and, unfortunately, is becoming a lost art except for a few highly trained, well rounded engineers who are passing the skill on to others.

SEGUES

*** Please refer to Illus. I7-8 -- The Segue Diagram

There are three basic types of segue:
1. The 2 + 2 to 2 method
2. The "organic" live on the 24 track method
3. The 24 + 2 to 2 method

1. The 2 + 2 to 2 method

This is the most often used method. In this method song # 1 and song # 2 need to be segued. Song 1 starts to fade at point A and song 2 starts to fade in at that same point also. Song 1 is fully faded at point B and only song 2 is audible from there on out to its end. After both songs are mixed completely and can stand alone, the producer says, "What would it take to segue song 1 into song 2?" Procede as follows:

1. Set up the playback on three two tracks with the same set of tones -- your project's master tones. Check the record levels and EQ on your record machine by transferring the master tones to the record machine from each of the two play machines, using two faders for each playback two track. Set the fader levels so that the two play 2 tracks make "0" levels on the left and right stereo busses. Keep working on the alignment until the master tones can be played from either of the play machines, through the console faders and stereo buss, and come out as a perfect zero level result on the record machine.

2. Carefully mark the four two track faders on the console so you can set the faders back to their zero-zero transfer settings, as you may need to make several trial runs at the segue. Remember that you must fully fade the faders when making trys at the segue, and it is imperative that you can reset the faders back to their zero transfer level settings so that the transfer function is exact, and that no level jump will occur when you splice the intermixed segue area (point A to point B) part into the end of the master mix of song one and the beginning of the master mix of song two.

3. Position playback machine #1 so that it is ahead of the segue point (point A) and machine 2 so that it is just ahead of the beginning of the mix -- allow for running time to let the machines get up to speed because if you don't the pitch of the music will change at the edit and ruin the segue.

4. Press record on the record machine (using new tape of course) and play mix 1; keep trying various start times and fade in rates on machine 2 (and fade out rates on machine one) until you hit a segue that the artist likes.

5. Sure that the segue is a hit, splice the segue piece just ahead of point A to just after point B into the original two track finished mixes, being careful to save all head and tail ends of the original mixes and marking the pieces (leader them if you want) and the reels of tape so that you can put it all back together if the segue doesn't work.

If you edit carefully, and if your alignments are precise enough, you

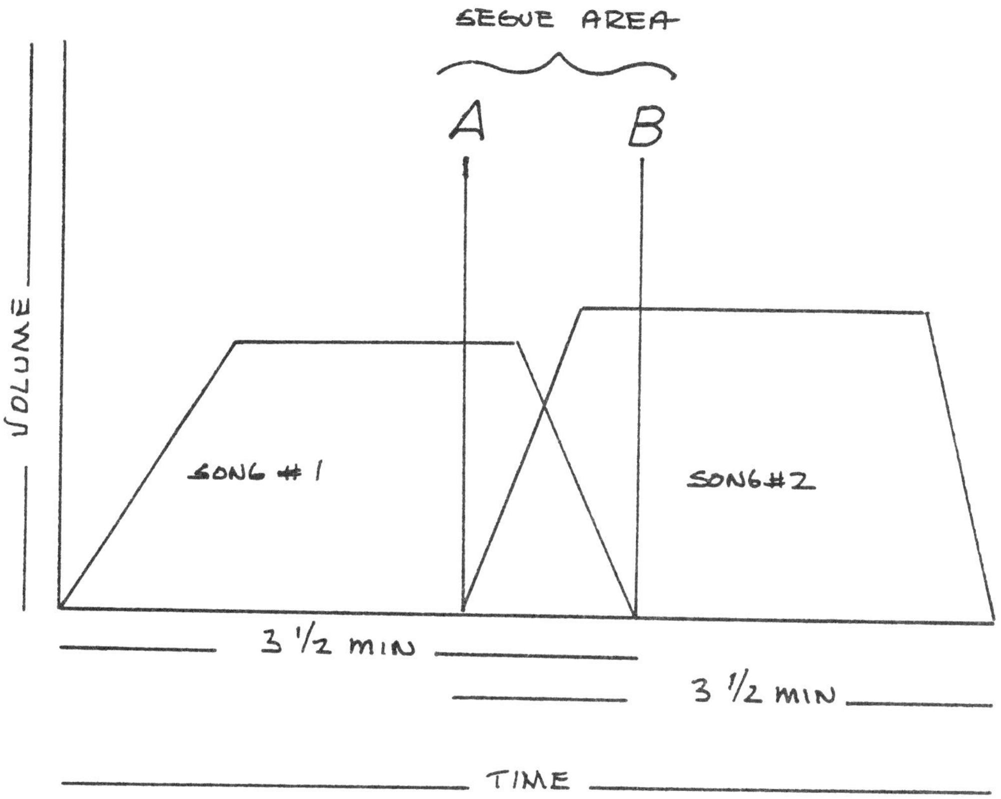
SEGUE AREA
A
B
VOLUME
SONG #1
SONG#2
3 1/2 MIN
3 1/2 MIN
TIME

will hear only a slight loss of quality at the beginning of the segue, but at once your mind's ear will be distracted by the arrival of the next tune. Trained listeners will detect the generation change at segue; the average listener will never be able to hear any technical quality change at all. This fascinating intermixture should more than cover up the fact that the segue piece of tape is third generation audio, not two, like the rest of the piece. When the first tune disappears at the end of the segue (point B), the quality improves back to second generation.

There are demons that plague segue makers. Here are a few samples:

Question: "Why does the pitch of the tune jump down at the start of the segue?"

Answer: Play two track #1 is running slow, or the record machine is running fast, or both. Note: You might not have known that the machines were running at different speeds. Segues that don't work because of pitch jumps indicate speed problems.

Question: "Why does the pitch jump up at the segue and then drift back to normal?"

Answer: You didn't allow enough run-in-time to let the record machine get up to speed.

Question: "Why does the volume jump up just after the segue?"

Answer: The segue fader level alignment was improperly done, leaving play machine #2 with its segue faders too low.

Don't be afraid. Seques are among the most beautiful of effects. They are worth the trouble.

2. The "in" 24 track "organic" method

In this method, the two inch tape isn't "automatically" cut and tail leadered as usual after completing the first of the two songs making up the segue; instead the tail end of the first song is left attached to the remainder of the fresh tape. This is assuming that there is enough tape left on the reel to accomodate the expected second song's length. As the fade would normally begin on song #1, open tracks of the 24 are punched in on, and the opening licks of the second song are recorded. This is an excellent example of when monitor to cue would be extremely handy to use, otherwise as the fade taking place, by manually pulling down the proper faders of the outro of the first song, would ordinarily not be reflected as any change in the cue system mix. Anyway, as you fade the first song at the faders, or it fades by itself, the musicians start recording the opening instrumentation of the second song. After you complete the second song the organic segue is finished. This technique implies that the track allocation will not come out the same as your first song since you couldn't record drums on tracks 1 through 5 for the second if the first song's drums were already on these tracks. With this technique the segue is performed not engineered. This makes for a more musical segue as long as the musicians and engineer are capable enough. As with any function of recording engineering however, if the going gets too rough, use a simpler technique instead.

3. The 24 + 2 to 2 method

Using this method there are several possibilities:

1. The segue area point A to point B of the completed two track mix of the first segue tune is transferred to two open tracks of the 24 track having the as yet un-mixed second tune on it in such a way that the seque

sounds properly timed and placed into the opening of the second tune. If you need more tape to record the end of song #1 onto the multitrack, add some fresh tape. Make sure you erase the tape first so that you won't edit virgin un-biased tape onto your recorded master -- otherwise you will get a thump when the splice goes through.

2. The blank two inch tape on which you will record the second song can have the ending of the first song's finished mix overlaid on it (perhaps on tracks 20 and 21) before your record the track. In this way, the intro of the second song can be played by the band in time with the outro of the first song. When the segue seems properly placed, finish and remix the second tune to a two track along with the closing music of the first song (the segue area).

When the two track part of the segue is recorded on the 24 track, it is imperative that the 24 track alignment be set exactly, so a zero level tone (1KHz) from the two track master tones will be recorded at exactly zero on the 24 track. Not only that, but when the two stereo segue tracks on the 24 track machine are set up to be sent to the 2 track master recorder during the mixing of the second song, it is imperative that they reach the 2 track again at "0" levels left and right. This will assure that the third generation copy of the first song being passed through the 24 track will reappear on the master recording two track at the same levels as the original was recorded, so that when you cut the 24 track borne re-recording (of the first song -- the segue area), which now resides as the "pre intro" stereo recording on the second song's 24 track tape, into the original two track recording from which it came, there won't be a volume mismatch.

To repeat, using the graphic supplied there is an existing, finished, two track mix of song one. Song one from point A to B (the segue area) is transferred to two blank tracks on a 24 track master recorder either already having song 2, which hasn't yet been mixed to the two track, or on which you are ABOUT to record song #2. If laying the segue area over an already existing 24 track master recording, adjust the start time of the playback of the two track, which you are recording on the 24 track, until you find a start time that sounds pleasing or musical. Now that you have the segue area of the first song transferred to the 24 track, do the 24 track mix (or record the song, then mix it) of the second song, which now has the last part of the first song (the segue area) on two tracks. After you finish the second song's mix to your satisfaction, assemble the segue. Cut out of the original complete two track mix of the first tune at point A and edit the tape to the beginning of the mix of song 2, which has the ending of song 1 superimposed on it. This edit at point A will commence the segue. When you tail leader the end of the second song (now connected to the first song) you are done.

Make sure that you take all reasonable precautions, that you have good calibrated levels on the recording and playback machines whenever you are making a tape copy that you intend to cut into another tape. The human ear is very discerning about level jumps and even the most inexperienced listener will be able to hear any sloppiness.

It is O.K. to move the segue area's levels around once you've passed the edit spot as long as you do it gracefully. Only the level mis-match at the cut will sound bad.

BACKWARDS OVERDUBBING

If someone wants to record a backwards part on a tape without wasting a lot of time fumbling around in "backwards land," do the following steps:

1. With the tape moving in the normal direction, for the purpose of locating the places in the song to be overdubbed, temporarily record on a blank track the approximate musical "idea" that is to be recorded backwards with a "ballpark" melody or sound effect idea of the part intended. Without these helpful noises in the backwards cue mix, it becomes very difficult to hear where the part goes, let alone find the right feel and rhythm for it.

2. Turn the tape over and prepare to record on the track number, which is the result of subtracting the track you wanted to record on from the number 25 (if you're working on a 24 -- use the number 17 if you're working on a 16 track). Position the tape at the end of the song, which is now the head of the tape, and record from the end to the beginning of the song. Remember when shuttling around in the song looking for the section you want to record, you must fast forward to get to the beginning of the song and rewind to get to the ending of the song. Another thing to remember is if track 6 is the bass track, track 6 is now at playback position 19. In fact all the playbacks will be reversed in the cue and playback monitor system. To prevent having to redo your monitor and cue mix twice, simply intercept the machine playback and line input positions and cross patch 1 to 24, 2 to 23, etc. Now the cue and monitor positions will be getting the same audio as when the tape was right side up and all the instrument code labels on the monitor system will still be correct even though the tracks have a new sound to them. Another thing is the seemingly hotter level that all your tracks will be playing back at. This is caused by the fact that the dynamics of the music recordings are now coming through the machine backwards. Instead of the snare hitting hard and fading away it now fades in and then hits. This gives the meter more time to move before the musical signal is gone. Consequently when you turn your backwards overdubbed tape right side up, all the levels will go back to normal except the backwards o/d track; if it is at all percussive, it will now be reading considerably hotter than when recorded.

REVERB SYSTEM SETUP

Most reverb systems sound best when the feed or send levels are at a minimum. This holds true both for chamber and EMT type devices. The lowest level that you can use is determined by the amount of noise coming out of the reverb system. To begin, mute the multitrack playback faders and turn up the monitor system to full volume and listen to the noise level of the mixing buss. Move the 2 track master up and down to identify which part of the noise actually belongs to the two track buss. Listen to see that it is mostly high frequency hiss and that no hum or buzzes are present. Turn up the first reverb returns to stereo left and right (assuming it's a stereo device), or a mono device to both busses until the reverb's noise is equal to or slightly lower than the basic mixing buss noise. Turn the monitor volume back to normal and bring up a full range sound, like an acoustic piano from your multitrack into the stereo buss to full level. Send the sound to the reverb device from this module until quite a large amount of reverb is heard. Check that the reverb send meter is not buried (indicating in the red). If not, so far so good. Now check to see that at moderate amounts of reverb the audible reverb balance has each side of the reverb return at equal volumes.

To do this, "jolt" the reverb device with a momentary audio send (use the channel mutes). Turn down the louder return if they're out of balance. Check the stereo balance with the signal plus the reverb. If you want delayed reverb, patch one of your reverb send outputs (at the patch bay) to a spare mono or two track machine and return the delayed (VSO'd?) reverb to a spare fader. (Using a VSO'd reverb delay machine allows you to time the arrival of the delay perfectly in time with the music.) Start the multitrack and re-adjust the mono's record level control upwards, so that the delayed reverb send is recorded at good tape levels in spite of the fact that the actual reverb send level on the console's reverb send meter may be quite low to cater to the reverb device as described above. Now set the reverb delay tape playback fader to the nominal setting point for your console, without sending any of its sound to the mix buss, and adjust the reverb send of that channel so that switching from the reverb send, which goes directly to the chamber, to the reverb send which gets delayed, causes no change in reverb volume in the mix. What's happening is this, say, reverb send one goes to stereo EMT #1. When you send on reverb send 1, you get immediate (un delayed) reverb back through the reverb return network (or spare faders). When you send to reverb send 2, for example, the send (at the same low level that made the EMT sound good) goes to the patch bay and is cross patched to mono machine in and has its volume adjusted at the record level control at the mono to read good tape levels, and returns to a spare fader, which doesn't get sent to the mix, but gets sent to reverb send 1 at such a level (adjusted by the mono machine return fader's reverb send pot) to make a vocal sent to reverb send 2 (with delay) have the same reverb volume as if it were sent to reverb send 1 (no delay). That's all.

Don't forget to see that any stereo audio (including reverb) sounds O.K. when combined in mono.

Balance your effects systems so that switching from one send (and its associated effect) to another keeps a similar amount of effect vs. original sound balance. If you can switch between effects this way without losing the sound of any of the effects, you have what I refer to as a "console keyboard effects" mode: you can "play" the special effects like a piano, changing them (the effect assign switches) at musically opportune times without needing to edit the two track tape due to a level mismatch. If you run out of effects sends, use the console's busses. Consoles having buss pans can make use of the program pans and use the even numbered busses for sends. In this way the pans turn on clockwise just like a send pot would. You can use both the odd and even numbered busses as panned "twin" sends using the pan to blend or select one or the other of the panned effects.

DISTORTION OF MICROPHONES

There are many ways in which microphones distort. Here are some of them:

1. Amplitude distortion -- due to driving the mike element beyond its design limits, where it gets extra resistive to free motion
2. Rub distortion -- by using a mike that has had its element stressed, as in number 1 above, so that now the element has a curled end or is so loose that it contacts other parts of the mike body
3. Capsule amplifier distortion -- which is similar to console distortion, but adds square wave products to the acoustic sound

We know that dynamic mikes resemble little speakers. Just like speakers, the element will move freely up to the point where the "compliance" runs out of stretch, whereupon the element does not move freely anymore. When this happens, the electrical waveforms no longer have properly shaped tops and bottoms. Instead, the top of a triangle wave (coming from a very fine synthesizer speaker) will no longer be triangle-shaped, since the very point of the wave is well beyond the stretch potential of the mike. Instead the top of the recorded wave will be severly rounded. This is distortion. Not only is this sound inaccurate (which is not really so bad in itself since you could consider the new smoother sound as a pleasing special effect under the heading of mikes that have a desirable lack of high frequency accuracy), but continued use of the mike under these circumstancs will further fatigue the "compliance" until the mike becomes unusable even at "normal" volumes. Always use overload immune mikes for high sound pressure work.

If you present a dynamic mike with enough of an overload of sound pressure (this is almost impossible to do with some dynamic mikes) it will "bottom out." This means that the usually razor sharp bottom edge of the bobbin, where the voice coil windings are, will be struck against the magnetic gap, making a curl in the bobbin's form. When the bobbin now moves up and dowm in the gap, it will rub against the sides since the thickness at the bottom (the curl) is now slightly greater than the dimension of the gap. Of course if the bottoming is serious enough, the voice coil will freeze in place and the mike will "die."

When ribbon mikes are seriously overloaded, their little corrugated ribbons get stretched so much that they won't spring back into the proper shape when the loud sound is removed. Since the ribbon is now too long, it will rub or stick to the horns of the magnet body. This ruins the mike.

For condenser mikes, there is a small amplifier located just behind the capsule assembly. The reason for having this little amp is because the condenser capsule output is at very high impedance and cannot survive a long cable run and still retain all the high frequencies. So, at a distance of 1 to 3 inches, the excellent frequency response is maintained. But upon severe overload conditions, the little amplifier is overloaded causing square waves to be added to the electrical audio output. These square waves are harder for the console to deal with if you're using VU meters, as they require more headroom. Also, the tape machine will have a much harder time trying to record the square wave modified audio. The capsule amplifier cannot be damaged by this abuse and will distort for the life of the amp. The capsule can be damaged by long exposure to excess sound pressure levels. When the capsule is overextended, the diaphragm and the backing plate touch an arc of electricity caused by the polarizing voltage arcs across the two and the capsule is ruined. Sometimes the capsule is only stretched greatly, and not ruined, sounding quite normal at first listening, but on any loud sound especially a plosive (the sound of the letter "P" spoken too close to the mike), the capsule will touch the backing plate and and the sound will go away for several seconds while the poor capsule amplifier recovers.

Some condenser mikes have a pad switch which protects the little capsule amplifier from overload on very loud sounds. There is no way to protect the capsule itself except selecting a different mike, moving the mike further from the source (or vice versa), or aiming the mike off axis to the sound power up to 90 degrees. This last trick is best used where there is a strong wind component to the sound such as the kick drum or a plosive singer. The 90 degree trick is also quite useful to get rid of the bass buildup associated with cardoid mikes used at very close range.

TRANSIENT TRACKING ABILITY - HARDNESS

Hardness is my term for the quality that makes microphones sound "bright" or "solid." If a dull mike were used on the snare drum, its sound at loud volumes would be smoother and easier to listen to than a great condenser mike that had a good hard sound; the condenser mike would make your ears hurt and make you cringe, due to the accuracy with which it was moving the tweeters in the control room. The harder the sound of the mike, the more accurate will be the square wave-like material coming through loud and clear. Getting a hard sound to come through the console is fairly easy given a console of excellent design, but trying to get the same clear, hard sound back from magnetic tape recorders is much more difficult. Naturally, the duller (less hard) the sound is from the microphones (during tracking), the easier it is for the tape machine to sound "just like" the original sound when it reproduces it. When you begin to A/B a very hard, real tracking sound (from microphones) to the subsequent tape playback, the more the loss of sound quality will become apparent and the more let down you will feel. (A/B means to compare the machine's sound to the console's sound, a sort of back and forth comparison.) There is no letdown when you hear a digital playback or when you play an acetate during a straight to disk recording. Generally speaking, using 16 track (with its larger record head dimension per track) will yeild a harder, more solid recording than tracking with a 24 track machine. This is why some production outfits like to record their basic tracks on 16 track (along with SMPTE time code) and then copy over the rhythm tracks (and the time code) to tracks of a 24 track, which they then use for the overdubs. The 16 track recording is more solid and is not worn out so much by constant playbacks during overdubbing. When they mix, they synchronize the 24 track to the 16 track and use the original rhythm tracks on the 16 track.

The lighter the element (the less mass it has) the better the mike will track air pressure variations (sound). The mikes with the heaviest elements are the ribbon and dynamic mikes. The lightest element is in the condenser mike. Condenser mikes have the ability to exactly transduce into electrical waveforms the acoustical waveforms they are "listening to."

MANY MIKES OR FEW MIKES

There is a point of diminishing returns when one considers the benefits of using more and more mikes. To some extent the attraction to the novice mixer in using many mikes is to provide more "control" over the sound so that he will be "covered," but of greater importance is: what will be the QUALITY of the sound that you will get, notwithstanding the excellent amount of control you may have over the it?

Consider as an example the mixer who uses a mike on the snare drum and a mike on the hi hat cymbal. Since the snare is much louder than the hi hat, the hi hat mike, being at higher gain to generate a usable console VU reading, will pick up the snare too. The snare on the other hand is so loud, that little or no hi hat can be heard by the snare mike. The engineer solos the snare mike and sets levels and EQ with the drummer playing the snare only, and then solos the hi hat and sets the sound with the drummer playing only the hi hat. Now when the solo is removed and the sounds from both mikes are heard, it is noticed that the snare doesn't sound as clear and crisp as before and it sounds "looser," that is, it doesn't sound as hard as it just did. This is because the hi hat mike can "hear" the snare and this is interfering with its sound. You now have a snare sound made up of one mike at the snare, with another mike (on the hi hat) 2 feet away aimed in the wrong direction. This doesn't sound as hard or solid as just the one mike used on the snare. Multiply this "technique" by 14 mikes used on the drum set and you have what amounts to a real comedy drum sound, at least as far as hardness is concerned. One figure eight pattern mike used between the hi hat and the snare, like a Neumann U-87, would wind up at suitable distances from each and there would still be one mike on each instrument. This would vastly improve the hardness of the sound, but the engineer would lose the facility of having seprate volume and EQ control over the snare and hi hat.

Another aspect of the twin mike example above is the effect of the use of EQ on the two mikes. Since the hi hat mike can hear the snare, any EQ put on it will effect the snare's sound. All the while, the snare will still sound fine when soloed, but when put into a mix with the hi hat mike, the snare sound "goes away" a little.

In addition to the above mentioned considerations is the additional problem of the rest of the drum mikes -- the tom mikes, the cymbal mikes and the overheads all being able to hear and "contaminate" each others sound.

Audio engineers state that when you use one mike on an acoustic source, and when there are no reflections arriving at the mike from nearby surfaces, then the mike is receiving direct sound and will sound as solid as it ever can. But if there is more than one mike, or if a single mike is used and sounds from the source can reflect off a hardwood floor or off a wall and get back to the mike, then the mike is being contaminated with early reflections, along with the direct sound, and weird things happen. If the early reflection sound is of a certain delay period, or if there is another mike at a distance such that the two mikes can hear the same sound source, the distant one having a delay similar in time to an early reflection, large and undesirable peaks and dips will occur in the response of the recording channel, which would not be there if there was only one mike in use. The more reflections (or mikes) there are, the more these peaks and dips will occur, causing "lumpy" or "peaky" recorded sounds. These holes in the frequency response are sometimes called notch filtering and ruin the solidness or hardness of the sound. Phase cancellations also ruin those qualities of the original sound that made it blend well in a mix with other sounds. You should expect phase cancellations whenever you have mikes in reflective surroundings or when more than one live mike is in the sound field of an acoustic source. The fact that the mikes having phase cancellations get recorded on different tracks does not eradicate the problem, the sounds from the tape tracks will get mixed sooner or later. Whether the sounds get mixed sooner (during tracking recordings) or later (during re-mix), cancellations will eventually occur.

Other considerations for using less mikes include the time forgotton recording theory: "The best mixer of multiple sounds is air; no console can mix sounds as well as air can." Under this heading come all studio and remote recording phenomenon whereby more and more mikes come up with less and less sound quality. Good examples of this can be heard with large vocal groups. When a large vocal group, already having a great blend and kept in an excellent mix by its conductor (in a pretty sounding acoustical environment of course), is miked at some distance (hopefully near to the conductor) with a single stereo mike, or with two mikes used as a stereo mike, you have a sound that is superior to multiple mike usage. If you should ever find yourself in a great hall with a remote recording session on your hands, don't forget to walk around to find a place where the mix and sound are in excellent proportions and to put a stereo mike there. Using air to do the mixing in the studio can best be heard when recording multiple vocals (backing vocals) by using one mike on circular pattern or a PZM mike taped against the wall or control room window compared to, say, five mikes on five different vocalists. Sometimes you will find a vocal group that sounds great in the room but terrible on five different mikes. When this happens, try using one mike and get the mix by moving towards and away from the mike. When you're having the singers "do" the mix, be sure to advise them of how sensitive the least amount of movement is. This is especially true if some of the singers move closer to the mike while others move away. If great difficulties aries, use a PZM microphone taped to the outside of the control room window, as this mike helps the singers to tell whose voice is whose in the phones during actual overdubbing. Wonderful horn sounds can be had by recording with two mikes in stereo and getting the mix by moving the players around as you stand near the mikes. The vibrancy and excitement of horns recorded this way cannot be matched by the multiple mike method.

Lastly, there is the phenomenon of a boomy floor. Say you are miking several instruments with single mikes, each on a mike stand, each mike being assigned to a separate track, then the recording on each track will reflect one instrument's worth of audio and one serving of bass boom from the floor (transmitted from the bass amp speaker bottom or kick drum through the floor and up each mike stand). It doesn't matter to the mike whether the mike is standing still and the air is moving, or whether the mike is moving (along with the floor) and the air is standing still -- both produce audio output, which is recorded. When the song is mixed, each track will only have one instrument on it but every track will have on it a little floorboom, which all add up (these are coherent noises, meaning they are almost identical in waveform and in phase). As you push up more and more faders at mix time, more and more boom will be heard since the booms recorded on the different tracks are all in phase if they were recorded on the same pass. To detect floor booms, very good speaker systems are required. Many studios have monitor systems that will not reveal the boom problem, since their monitors don't go down that far. If this is the case, the disk cutter will have the unfortunate job of informing you of their presence in your mixes.

UNUSUAL MICROPHONE TECHNIQUES

There are many unusual microphone techniques. You will probably wind up knowing more than I do but here are some that I feel are worth passing on:

1. The "twin spinning mikes trick" for the flying lead guitar sound...

This effect sounds like the guitar amp was suspended from a small silent airplane flying around in front of a stereo mike during the overdub. Sometimes the sound goes straight from the rear of the room towards the listener, sometimes the sound flys away. The sound can cross the speakers from left to right, or vice versa. It can cross stereo near to the observer or cross the room far away from the listener.

The method is easy, simply put the amp in the middle of the floor and have one of your two assistants lie on his back and spin a light weight dynamic mike with a wind screen from his raised arm. Have him twirl the mike cable, with the mike at the end acting as a weight. The cone shaped pattern caused by the mike revolving around the assistant's arm will be two or three feet across at the bottom. Have the other assistant, holding an 8 foot mike boom, stand to one side and twirl the longer cable above and around the smaller circle made by the first assistant on the floor. Position the instrument amp so that the front panel is tangent to the cones so that for half the swing the mike faces the amp and for the other half the mike faces away from the amp and picks up mainly room sound and reflected sound from the nearest wall.

*** Please refer to Illus. I8-1 ***

Record the two mikes in stereo and get levels as best you can. Pan the stereo sound hard left and hard right in the mix. Do not try to do away with all the wind noise from the mikes since these noises aid in the effect. Compress the microphone channels to pick up more "room" sound if you want to.

2. The "put the mike in a container to get strange EQ and filtering sounds trick."

Any slender mike with a good all-around frequency response will do for this operation. Simply put the mike into the opening of some vase or container (best if the container has hard walls) and put the vase mike by the instrument you want to record. A piano is a good instrument to use since it has such a good frequency content. Remember that a closed cylinder resonates at a different octave than an open one, so wah-wah tone control can be effected by carefully closing off one end of a hard surfaced tube. Diameter, volume and aiming of the "thing" are important to the sound you get.

3. Using a pencil to control a singer's incurable popping problems:

For condenser mikes tape a pencil or pencil shaped object to the front of the mike, directly over the center of the capsule. When a large "pop" sound comes along it gets split like a bridge support in a river does

Illus. I8-1

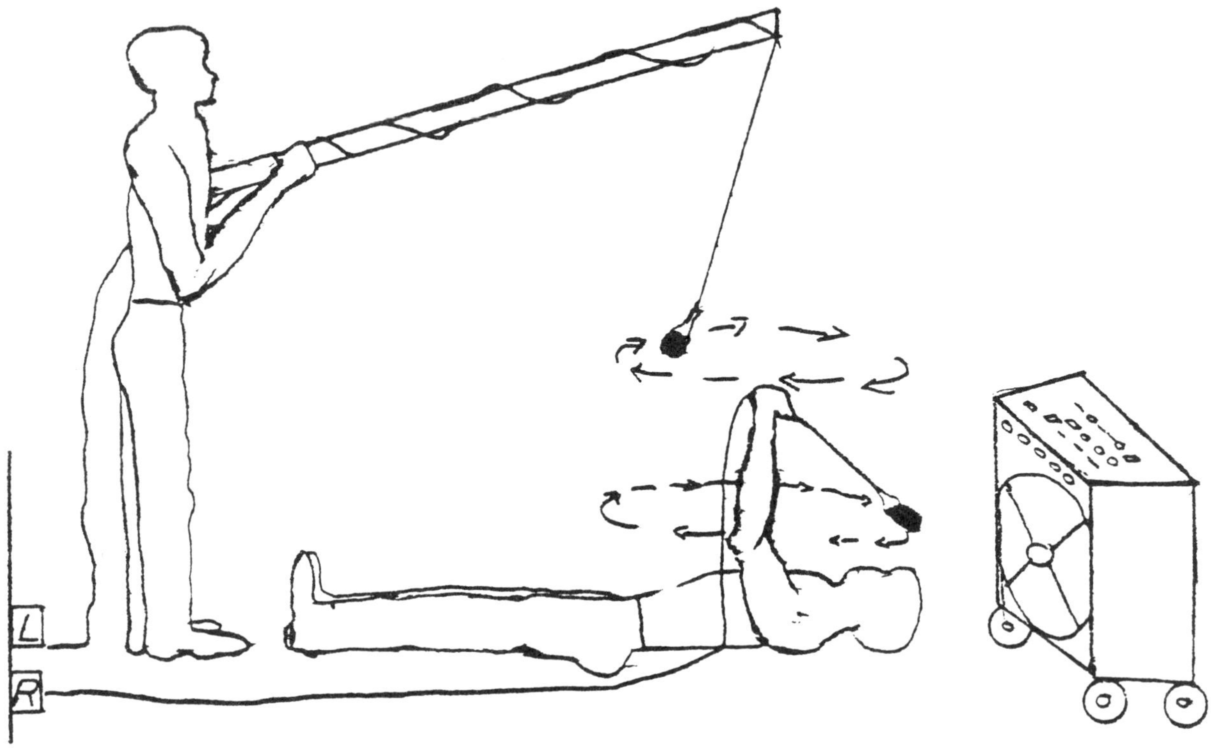

THE FLYING GUITAR TRICK

to the water, and the pop misses the capsule. It doesn't deflect the interesting parts of the audio however.

4. Using the wrong side of the mike:
This is a technique for doing away with the highs and tuning in to the truly huge and pompous low end, which will flow around the contours of the mike body and into the back of the mike, where the element is listening. This low, smooth sound can be mixed with a mike used normally faced for deepening effects.

PRACTICAL MICROPHONE CONSIDERATIONS

If you were getting started as a studio owner, you might be faced with difficult choices selecting mikes, considering the constraints of the budget. The best mikes go for $800 and up so money goes fast. One great mike compared to 10 O.K. mikes is a hard choice especially given the demands of today's engineers.

Don't put your highest quality mikes on the snare drum with an inexperienced drummer at the sticks. One clean hit on the mike instead of the snare will put an end to the mike. Always remember that ordinary maintenance persons cannot repair microphones with broken elements. Mike repair takes a long time; it sometimes requires a year and many, many phone calls to recover a mike from a factory repair department. Sometimes you call about the progress on your mike, only to be told that they can't find it. Instead of these problems attendant to using a rare and exotic mike, use a sturdier, more expendable, easily repaired mike on sessions involving rowdy rock & rollers, or inexperienced players especially drummers. Set up with care and forethought to protect the valuable mikes.

ACOUSTICAL DELAY

Acoustical delay is obtained by placing a mike at a considerable distance from an acoustical source of interest. The best results begin once you get your delay mike more than about 15 feet away from the source. Acoustical delay is completely unpredictable and the sound of the "effect" is different for each note played. One acoustical delay setup is to put a uni-directional mike way up over the drums and mix the sound into the "drum overhead" recording signals during tracking. Another is to run a bass tunnel made of two walls of baffles running away from the bass amp and putting a mike in the tunnel with a baffle roof at some distance. The purpose of the roof at the mike end of the tunnel is to prevent vertical sound energy from getting into the mike from above. This acoustically delayed bass audio gets mixed in with the other bass sounds such as direct bass, miked bass, post effects bass direct, etc. This kind of setup greatly expands the required number of faders since a full bass treatment would then be:

DB1 (direct bass)
DB2 (post bass special effects pedals direct)
MB (mike bass)
MB@20 (mike bass at 20 foot distance)

The number of different bass sounds available just by changing fader settings alone at the console without even any change in the bass' EQ settings or other effects is enormous.

The use of two room mikes in real stereo during the tracking session to listen to the "whole band" makes a very handy stereo recording (especially if PZM's are used); if there is any audience reaction noises in the room, so much the better. The stereo room tracks generate beautiful reverb sounds, for overall reverb without actually bringing up the stereo room into the mix itself. Use of the rough vocal performance track for the vocal reverb send with the master vocal, gives an unpredictable and therefore fascinating sound to the vocal's reverb (this is a subliminal audio engineering technique). It is important to remember that the ear/brain combination is able to achieve recognition of an effect when used on a sound in a mix very quickly if there are no random qualities to the "effect" being used. The more unusual and unpredictable the effect is, the more fascination is produced in the listener. This fascination more or less commands the listener to pay attention, that is if the effect is musical and not ugly or mechanical (technical) sounding. Since acoustical delay lines, when used properly, give a different sound note by note from all other notes in the instrument's range, the ear/brain never tires of attempting to "figure it out." This attraction to a sound is not wholly on a conscious level and many times the attraction of a listener to a song is as much on sound as the musical or lyrical content. All engineers should acquire the technical and musical expertise and judgement necessary to "improve" the sound of the artist by understanding and using the exotic and random qualities of acoustics to make their recordings more listenable and interesting.

ISOLATION BOOTHS

Isolation booths are small rooms that have

1. Overhead lighting
2. AC outlets
3. A mike input panel with at least 4 mike inputs
4. A window of sufficient size so that the artist(s) on the inside can see the rest of the session happening outside
5. Sound treatment so that the room is "dead" -- there is no real liveness to the sound inside
6. A music stand, which has been carpeted, so that no metallic bounce or resonances will be "heard" by the mike
7. An air conditioning outlet, which produces no wind noise or rushing sounds

Isolation booths have two basic uses:

1. To keep loud sessions from leaking into the recording of a soft source, like a gently played acoustic guitar, or a soft spoken singer that needs to be recorded live with the band
2. To keep a loud instrument from leaking into recordings of soft instrument tracking sessions, like an all acoustic guitar basic track with no drums but with a very loud electric guitar

In the first case the soft instrument is isolated from the main room, where the crash and thunder band will be playing; in the second, the gentle session is not disturbed by the Marshall amp because it is put in the isolation booth. Remember that the small dimensions of the vocal booth will be recorded and that if the engineer doesn't use close miking techniques, the ear/brain will be able to detect that the recording was made in a small room. Small acoustic environments make for small sounding recordings.

For large studios, isolation booths should be movable, with the best arrangement being the one where the booth is really two separate half rooms held together by clamps, which can be left hooked together for one large booth, or pulled apart and put with their open ends to the walls making two half sized vocal booths. If the booths are immovable, it is best to put one of the booths's windows facing the control room so that silent cues can pass back and forth between the artist and the engineer and/or producer.

While on the subject of cues, there are a set of excellent cues that I have worked out during my recording career (for an artist who can see into the control room):

1. Points to top of head - go to the top
2. Artist points from his right to left - rewind
3. Artist points from his left to right - fast forward
4. Traffic cop stop sign - stop
5. Twirls fingers in vertical plane - play tape from there
6. Scissors motion - record from where you are on the same track
7. Finger pointed straight up - get ready to punch in (out)
8. Finger jabbed towards engineer gun like - punch! (in or out)
9. Make motions like playing the instrument that is out of balance, point to phones, point up or down -- please adjust this instrument in the phones (a very handy sign language command during vocal overdubs since any spoken word means that the tape will have to be stopped and rewound to erase the words spoken)

EXTREME ACOUSTICAL DYNAMIC RANGE

Each classical instrument has a pretty well defined dynamic range potential, except for the pipe organ and perhaps the cannon used for the 1812 overture. Strings can only be played so loud. Horns can only be played so soft and no softer or the sound stops. In the case of electric instruments, however, anything is possible if the artist is wealthy enough and can assemble enough amps and speakers. An electric guitarist can play in the dynamic range from the threshold of hearing to the threshold of pain and beyond.

At some theoretical sound pressure, the air itself comes to a point where no more volume can be produced, since each rarefaction wave would constitute a vacuum in the atmosphere. I believe that this astronomical sound volume is produced by an explosive force of 10 sticks of dynamite at a distance of 50 feet.

Percussive instruments require a large dynamic range capability be set aside for them when you choose a mike to use. A snare drum is a very loud instrument. If you want to prove this to yourself, get a good, loud, solid cue system drum mix for yourself in the control room and then go out and hit a few solid snare hits with the phones on and then a few with the phones off. One adventure along these lines will forever assist you to understand drummers who always request "killer" drum volumes in the cue system. This is the best reason to always use a separate cue feed for the drummer, since other session players will agree that the drum levels needed in the cue system by the drummer are too great for them.

When you are faced with recording dynamic ranges that are impossible to accomodate with the equipment at hand, use the technique of double recording at two different levels and stacking the tracks to one track later:

As an example let's say that we have to record a lead guitar, which will start out as just picking noises from the strings with the amp turned off, and then at the push of a foot switch, three huge guitar amps will light up with the other part of the sound for this track. As the engineer, you will never know when the foot switch will be pushed and so cannot possibly "ride the fader."

First, adjust a limiter so that there will be no gain reduction happening while the guitar is played in its soft "voice." When the guitar amps lights up, the limiter will prevent any fires from breaking out on the tape due to the extreme levels. At the "godzilla" volumes adjust a level capable mike (Sennheiser 421 for example) so that usable console readings result. Record these two sounds on two different tracks and combine them to a single track later, since once you know where the changeovers are in the guitar performance, riding the faders at the proper moment becomes easy. Stack the two tracks to one and erase the 2 contributer tracks. These tracks are now free for further recording work.

Remember anytime you "chase" the volume of a recording up to the softest point in its dynamic range and you hear console noise, you need a more sensitive mike.

HUMAN HEARING

The ear can hear atmospheric disturbance waveforms of as little as .0000000002 times the usual atmospheric pressure. In addition to this amazing sensitivity the ear/brain combination can hear with some degree of accuracy over a 1,000,000,000 to one range of volumes.

In order not to be driven batty by the extreme range of loudness of some sounds in life, the brain compresses the range of sound volumes, which it otherwise would hear from the ear, to a reduced dynamics version that it can withstand.

All in all the total motion excursion range of the eardrum from the smallest sound audible to the largest sound before physical pain is from approximately 10 to 40 millionths of an inch.

The intensity of sound is nothing more than the measure of the sound's pushing power on a surface of one square centimeter. As the distance of a listener from a source of acoustic energy decreases, the intensity increases and vice versa. The reason for this is quite simple if you only remember your geometry formula for the area of the surface of a sphere

Area = 4 x pi x the radius squared

If you were four feet from an acoustic source, the area of the sphere of sound (supposing the sound energy sphere didn't hit a wall, floor or ceiling on the way to you) would be

4 x pi x radius (squared)
4 x pi x 4 squared or
201.1 square feet

But if you moved in to two feet, the area the sound power would be spread out over would only be

4 x pi x 2 squared or
50.3 square feet.

This means that the acoustic energy at 2 ft. distance would be spread out over only 1/4 the area it was at four feet distance. So the intensity at 2 feet is four times that at 4 feet. This relationship is known as the inverse square law, which can be written as:

Intensity = 1/distance squared

So as the distance increases by a factor of 2, the intensity decreases (inverse law) by a factor of 4 (to 1/4). Unfortunately this relationship only holds true for even amounts of distance change; it won't work for a distance change of 4 feet to 3 feet for example, only from one distance to another distance that is an even number of times larger or smaller. Three times further from a source = 1/9 the intensity, four times closer is 16 times the intensity.

Intensity is also related to watts of power coming from a speaker. Intensity is related to watts in that to get four times the intensity (6 dB) requires four times the watts (6 dB).

Intensity is interesting, but sound pressure level measurements using SPL meters are what is used to work with sound and volume measurements in recording studios. Sound pressure is analogous to electrical pressure or volts. This means that twice as many volts is the same thing as twice as much sound pressure. All microphones measure sound pressure. When a mike is moved closer to the source (in a sound field where reflections are not significant), the console VU meter shows twice as loud by showing 6 dB more level each time the mike/source distance is halved.

The console VU meter says four basic things. (1) Each 3 dB means twice the watts or half the watts depending on which way the VU is going of course; (2) Each 6 dB means twice or half as many console output volts; (3) each 10 dB is approximately twice as loud or half as loud, which requires about 10 times more or less watts; (4) each 3 dB means 1/2 or 2 times the intensity.

We will work with these numbers again when we do the calculations in Advanced 3.

*** Please see "Sherman's Ohm's Law and the dB" -- Illus. A1-1 ***

Practically speaking, to measure acoustic sound pressure level, use a sound pressure meter. A very inexpensive one is usually available through Radio Shack. When the sound pressure level changes by 10 dB, then the volume is twice or half what it was and this requires an increase or reduction of about 10 times the power (watts) from the monitor system power amplifier.

FLETCHER MUNSON PHENOMENON OF INTEREST

1. When the sound pressure or volume increases, the highs and lows sound louder in proportion to the midrange. As the sound pressure decreases, the amounts of relative highs and lows appear to decrease.

2. As the volume of a sound is increased in volume, it appears to decrease in pitch slightly.

3. Once you have listened to loud volumes for a given time, your hearing will be "accurate" only at or above that loud volume from then on or until a sizable rest period is taken in relative silence.

These simple laws bring into audio engineering more than their share of misery. Here is how:

When you record at loud volumes and then mix at soft volumes, you will have to boost the highs and lows that you won't hear mixing at low volumes (shall we say "realistic" volumes) due to law #1. This addition of mixing EQ may be equal to and the opposite of the highs and lows you removed at the recording session, where you were hearing the highs and lows to "excess." If you were to come full circle with this volume foolishness, you would then monitor at high volume again at disk mastering time and therefore hear highs and lows to "excess" again. Naturally you would EQ the two track

Each 3 db =	2 times or 1/2 the watts (no relationship to volume)
Each 3 db =	2 times or 1/2 the sound intensity
Each 6 db =	2 times or 1/2 the volts, ohms, amps or distance from mike to sound source
Each 6 db =	4 times or 1/4 the intensity
Each 6 db =	4 times or 1/4 the watts
Each 10 db =	approximately 2 times or 1/2 the volume and about 10 times or 1/10 the watts

Electromotive force is the name for the electrical pushing or activating energy provided by a battery or a battery substitute.

Current is the measure of the number of electrons passing a given circuit point in a given time. It is the quantity of electron flow. the speed of the electrons is always the same: the speed of light. It is the quantity which is measured as the current.

Resistance is the opposition to the flow of electrons. any time there is work done by electricity, there must be resistance to the flow of electrons. Poor conductors or even good conductors over a very long cable run will have resistance without producing any work - this is wasted electrical energy known as loss.

Electromotive Force = E = volts = 6 DB steps for 2 times or 1/2 as much

Current = I = amps = 6 DB steps for 2 times or 1/2 as much

Resistance = R = ohms = 6 DB steps for 2 times or 1/2 as much

Distance = D = feet = 6 DB steps for each 2 times or 1/2 the distance

Power = P = watts = 3 DB steps for each 2 times or 1/2 as much

I = E divided by R

R = E divided by I

E = I times R

P = E squared divided by R

P = E times I

D = rate times time

master to get rid of the unwanted highs and lows. Thus, you would have taken three EQ passes on the project just because you didn't listen at the same volume throughout the project.

If you listen to a playback of a song that you just finished mixing (at low, living room volumes) once or twice at high volume, you will probably put your ears through the volume barrier in rule 3. Once having done this you will have to mix further tunes at this higher volume or quit for the night. There is no way to regain your former accruacy except by getting rest and quiet.

When the phones are turned up too loud, the artist will perceive the pitch as being lower. This will make him sing flat. There is also the phenomenon of having something covering your ear (like an earphone cup) that further confuses the artist's ability to perceive pitch correctly. It is for this reason that I am always amused when bassists show up at sessions without amps to tune their basses with. You will notice the pitch shift (due to volume changes) phenomenon the most during vocal overdubs when going from a quiet studio with only a few headphones operating to the control room, where the music is playing loudly, at which time it will sound as though the tape speed has slowed slightly as you enter the door.

ECHOS AND REVERB

An echo is one discrete repeat of the initial sound. Echo is NOT reverb.

Reverb is the sound of millions of discrete echos that blend together into an indistinguishable array of sound, which is similar to the original sound in volume and pitch but much longer in duration. Reverb comes from three major sources: reverb chambers, reverb plates, and reverb springs.

Reverb chambers are made of shiny, hard walls, which support sound reflections. Millions of echos bounce around in the reverb chamber to their hearts content blending together into reverb. These chambers are best built with heavy, hard, smooth walls that are at least parallel on two sides.

To reduce the reverb time, rugs and small absorbent items can be placed in the chamber. For maximum reverb time the chamber should be dry and dust free.

Reverb plates are large steel panels that are fed the audio signal by a small motion transducer, which vibrates the plate with the same motion as is represented by the input audio waveform. The plate's vibrations, which reverberate long after the input signal stops, are picked up by one or two pickups attached to the plate. The plate's signals are amplified and returned to the control room at line level as reverb.

To reduce the reverb time, a parallel soft fiber panel is moved near the steel plate. The closer the fiber panel is moved, the less is the reverb time. For maximum reverb time, the absorbent panel should be at maximum distance and the springs that suspend the plate should be as tight as possible without breaking.

Spring reverbs are sets of springs, which are shaken at one end and activate pickups at the other. They almost always sound "sproingy" especially with percussion like program material and are not really as good a substitute for live chambers as plate reverb devices are. Actually spring and plate reverb are not supposed to be substitutes for a live chamber -- they each have a special sound quality that works well with certain sounds.

Echoes are obtained in the studio by the use of digital, tape, or other delay devices. Tape machines are very useful because they can easily produce long delay times and are usually already available. Although they do not produce long delay times easily or inexpensively, digital delay devices provide short delay times with extreme repeatability, accuracy and handiness.

Echoes and reverb are both available in location recording by miking the hall where the musical event is happening. The "sound" of a great hall has very pleasing musical potential when mixed in with the recording's direct microphone sounds. Always record the "room" when recording live performances.

Echoes from a music hall have acoustical qualities that make the echoes and reverberations different from note to note and from one volume level to another, whereas the echoes from a digital delay or tape delay always sound exactly like the originating sound. The ear/brain is attracted far more by a subtle randomness than it is by a carbon copy. Remember this benefit of incorporating acoustic random variances in your echo treatments and you will have better results.

Lexicon has a device called the model 224 digital reverberator that will digitally emulate up to seven room and reverb functions including music halls, reverb plates, and reverb chambers. Complex digital data is fed to the microcomputer by means of EPROMS or ultraviolet light reprogrammable memory chips. By changing the EPROMS, you can change the programs, which dictate the possible sounds the 224 has. Since the sound comes from EPROMS, which are easily replaced, amazing new halls and effects will be available in the future -- I'm sure of it.

WOW AND FLUTTER

Wow is slow flutter, flutter is fast wow. Both are tape speed inaccuracies.

The frequency at which a wow becomes a flutter is approximately at 2 events per second.

Wow is caused by slow or large diameter irregularities in the tape path such as:

A lump in the tape pack

A sticky portion in a large rotating tape guide

A faulty (and varying) power supply voltage to a motor

A faulty reel which is pinching the tape once each time it goes around

Flutter is caused by fast or small diameter irregularities in the tape path such as:

Small diameter tape guides

A faulty capstan motor

Faulty scrape flutter filter guides (between the heads)

A capstan or other small diameter tape element that has been severly bumped and is now out of round

Faulty tape tension servo action causing rapid tape holdback fluctuations

Very high speed flutter sometimes sounds like distortion. Some types of servo machines are subject to an insidious type of servo flutter, which can be absent when you start a long day of live tracking sessions. During the first recordings, while you are still doing playbacks regularly to check the tape and the machine, everything sounds O.K. It's after you stop checking to "save time" (probably at the insistance of the producer) that the machine reaches the high operating temperature, at which time the servo may begin to flutter.

Only slow (wow) type disturbances or steady off speed problems can be resolved with a resolver; flutter is not fixable with a resolver. Always record the 60 hertz line frequency.

To check for the presence of flutter use a flutter meter or record a 3 KHz tone and then play back a section of it. Or record the three KHz tone and then play it back while listening to both the console's 3 KHz tone and the machine's played back version of it. Any machine speed fluctuations will be apparent as audible beats.

RECORDER HEAD ALIGNMENT

Tape machines have four adjustments for alignment of the heads:

1. Azimuth -- the perpendicularity of the vertical axis of the head compared to that of the tape

2. Height -- the altitude of the head above the deck compared to the altitude of the tape above the deck

3. Rotation -- the twist of the head as viewed from above, which determines if the gap area of the head touches the tape dead center in the head/tape contact area

4. Zenith -- the perpendicularity of the face of the head to the plane of the deck as viewed from the side. If the top of the head is tilted back, for example, then the tape will want to ride down towards the bottom of the head with increased holdback tension from the supply reel. If this seems unusual, think of tape as a belt and the head as a "pulley"; a belt will always seek the highest point on a pulley. This is why belt pulleys are convex (raised in the center).

To check a two track's azimuth, play the azimuth setting band of a reproduce alignment tape in satisfactory condition and first adjust the azimuth adjusting nut (check the instruction manual for the tape machine in question) so that a peaking of the playback levels on both tracks reaches a maximum. Next, patch one track into the left and one into the right jacks of the phase scope and then check the phase scope to achieve an in phase picture. For multitrack machines the first half of the tracks should be summed to one buss and the other half of the tracks should be assigned to another buss and these two busses plugged into the phase scope.

To check height, record a 1 KHz tone at high level (perhaps at +10 VU), and then "develop" the tape with Magna See. Magna See is a solution of metal particles suspended in a solvent. When the solvent evaporates, the metal particles are left on the tape wherever there is any magnetic flux. Measure the guard bands to see that the record head is sitting at the proper height. Look at the "developed" tape with a jewelers loupe, graduated in mils or 1/1000ths of an inch. Procede to check the play head by recording normal level 1 KHz tone on all the tracks and then adjust the playback head height until the tape plays at the same VU reading with the tape upside down as it does with the tape playing right side up. (P.S. This is very difficult to get perfect).

*** Please refer to Illus. A2-1 -- The Ampex Track Formats Diagram ***

To check rotation, mark the face of the head with a sharpie or other dark felt tip pen. Play enough tape to buff off the ink and check with a magnifying glass to see that the wear pattern is directly over the gap.

To check zenith set the holdback tension as light as it will go small reel setting). Now by using your hand as a clutch, steadily increase the holdback tension and watch the tape's altitude at each tape head. If the tape rides up or down even a little it means that there is a zenith problem. (P.S. This adjustment is really difficult to perfect taking sometimes as much as a week to get perfect). Very important to the eventual success of aligning the heads is the perpendicularity of the deck components, including the capstan motor (adjusted by shimming) and the alignment of the basic tape guides, not counting any guides associated with the head block. A complete zenith alignment means taking off all the heads and guides and having the deck consist of nothing but the flywheel idler and the capstan and pinch roller. You then put one deck component in at a time and check zenith constantly as you go.

Bumping or jarring a machine (especially a two track) can have a catastrophic effect on the head alignment, so these machines must be moved with extreme caution.

Without accurate two track head alignment, each area on a roll of tape will get recorded with a different height or azimuth. There would be no

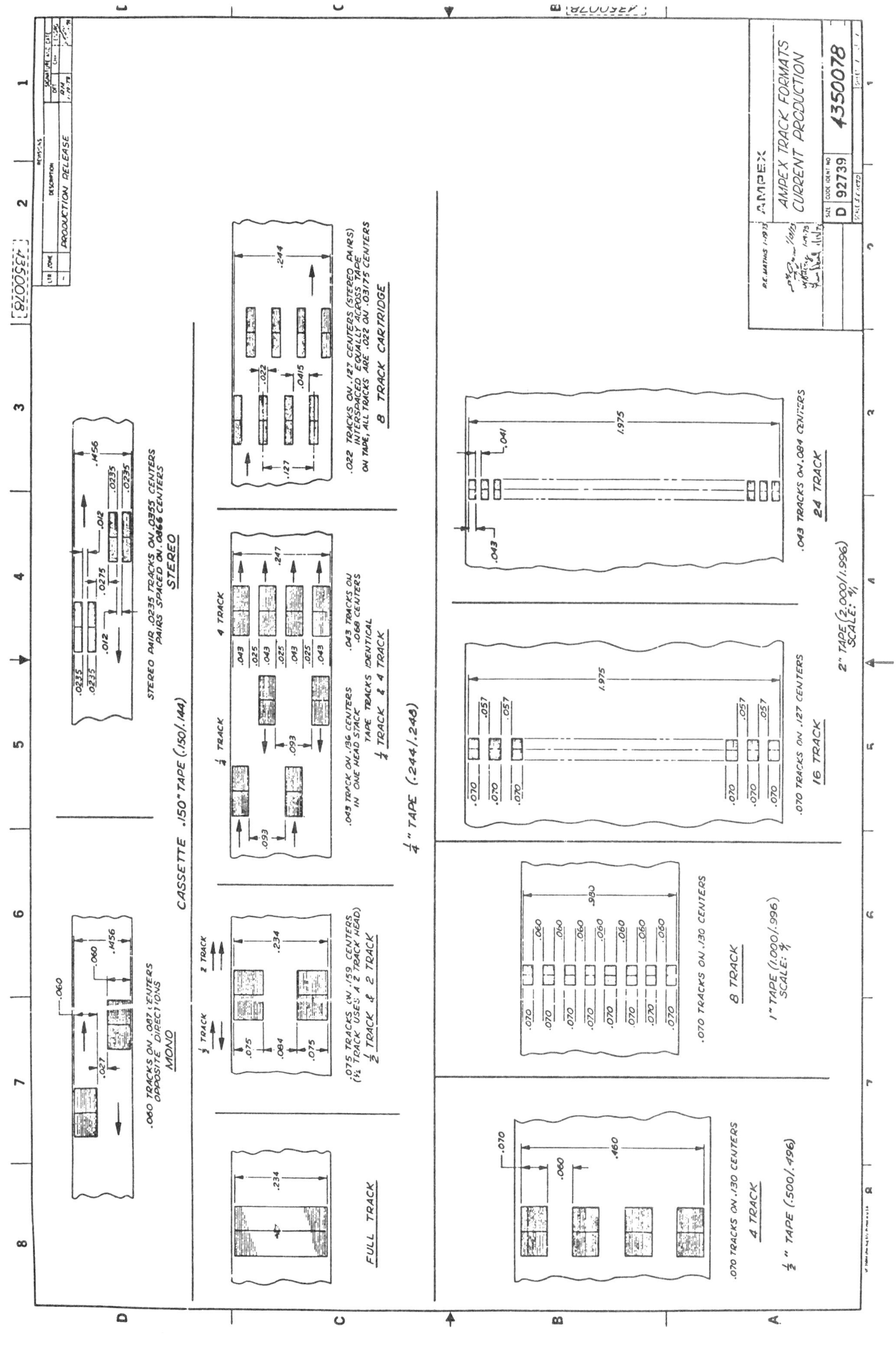

.060 TRACKS ON .087 CENTERS
OPPOSITE DIRECTIONS
MONO
STEREO PAIR .0235 TRACKS ON .0355 CENTERS
PAIRS SPACED ON .0866 CENTERS
STEREO
CASSETTE .150" TAPE (.150/.144)
FULL TRACK
.075 TRACKS ON .159 CENTERS
(½ TRACK USES A 2 TRACK HEAD)
½ TRACK & 2 TRACK
.043 TRACK ON .136 CENTERS
IN ONE HEAD STACK
.043 TRACKS ON
.068 CENTERS
TAPE TRACKS IDENTICAL
¼ TRACK & 4 TRACK
.022 TRACKS ON .127 CENTERS (STEREO PAIRS)
INTERSPACED EQUALLY ACROSS TAPE
ON TAPE, ALL TRACKS ARE .022 ON .03175 CENTERS
8 TRACK CARTRIDGE
¼" TAPE (.244/.248)
.070 TRACKS ON .130 CENTERS
4 TRACK
½" TAPE (.500/.496)
.070 TRACKS ON .130 CENTERS
8 TRACK
1" TAPE (1.000/.996)
SCALE: 4/1
.070 TRACKS ON .127 CENTERS
16 TRACK
.043 TRACKS ON .084 CENTERS
24 TRACK
2" TAPE (2.000/1.996)
SCALE: 4/1
PRODUCTION RELEASE
AMPEX
AMPEX TRACK FORMATS
CURRENT PRODUCTION
D 92739
4350078

way to change the tunes from their originally recorded positions onto a master reel without the sound being affected by the mismatch of the heads of the mastering two track to the shifting azimuth of the individual cuts. The two track, which made the out of height and azimuth recordings, could not even play its own recordings once the tunes were edited onto new reel locations.

THE VSO

The term VSO is actually a misnomer. It stands for Variable Speed Oscillator. Actually a VSO is a "Capstan Motor Driving Oscillator And Power Amplifier For Variable Tape Speed," otherwise known as a CMDOAPAFVTS. Since CMDOAPAFVTS is an unwieldy term, we will use the term VSO like everyone else. Ampex manufacturers one which they call the "Varispeed."

A VSO can be made out of any oscillator that can produce a steady 60 hertz sine wave and any power amp (like for speakers) that can deliver 110 volts of the 60 hz sine wave without distortion. This 110 volt signal is used to power a synchronous capstan motor instead of the 110 volt 60 hz line voltage coming from the wall. As you vary the oscillator around the 60 Hz frequency, the tape speed varies around the normal 15 or 30 ips. It is best to consider the VSO as a special effect device for overdubbing and mixing only; the tape machine should never be used for tracking or mixing with the VSO in operation. The 60 Hz coming from the power company is much more stable than you get from a studio's oscillator set "by eye" to 60 Hz on the dial. You can set the VSO oscillator to exactly the frequency you want using a calibrated frequency counter if you have one.

Ampex tach lock systems VSO uses a line level VSO signal at about 3/4 volt and at a frequency of 9600 Hz. To VSO an Ampex 1200, tune the oscillator to 9600 and patch the sine or square wave directly to the Ampex speed accessory jack (a Jones plug) at the rear of the machine. Look in the instruction manual for which pins to wire the hi and ground of the oscillator to.

There are many ways to use the VSO. Here are some:

1. To change the tape speed during mixing so that the tempo is more to the liking of the artist or producer.

Example: by speeding up the tune the song sounds "brighter" (less sleepy) and therefore "more commercial."

2. To change the tape speed during overdubbing so that the dynamic range and timbre of the instrument being overdubbed is more "interesting" during subsequent playback and mixing at normal tape speed.

Example: by recording some drum overdubs at higher than normal speed, when the drums are played back at mix (normal) speed, they sound huger and beefier. This technique works especially well doubling the snare. By overdubbing the entire set of drums at higher than normal speed, the drummer sounds much more funky and relaxed at normal speed later. In addition it sounds as if his drums are bigger than they really are.

3. To put your studio piano in tune with the piano on a master tape recorded somewhere else on a different machine.

Example: the artist didn't tune the guitars to the piano carefully (or not at all) before he recorded a track consisting of three acoustic guitars, bass and drums. The guitars and bass are perfectly tuned to each other, but not to the studio's piano. Rather than tune the piano to the tape (very expensive and time consuming), VSO the tape to the piano temporarily for the overdub. Better yet -- leave the VSO at the setting that puts the tape "in pitch" and leave it there. This way people having perfect pitch won't cringe and change stations when they hear your song on the radio. This consideration might amount to naught however since radio stations have been known to speed up their turntables in order to get more records (and commercials) squeezed into each hour of air time. Therefore, if you speed up your record and they do too...

4. To add "precision flutter" to an overdub.

Example: acoustic guitar in the Beatle's song "Here Comes the Sun."

5. To speed up a recording so that it conforms to a timing criteria.

Example: you find that your 30 second commercial production has come out being 34 seconds long. Using the VSO at 68 hertz instead of at 60 will make the playback occur in 30 seconds instead of in 34 seconds.

*** Details of VSO and other mathematical calculations sometimes required of audio engineers appears in Advanced 3 and 5 ***

6. To enrich the sound of doubling an instrument while overdubbing.

Example: after overdubbing a picking style guitar part you are asked to "double it." If you just go ahead and make a second recording you'll have one type of sound. If you change the tape speed ever so slightly from normal, you will have a much warmer and interesting type of sound. This is caused by the displacement of one overdubbed instrument's overtone series into a slightly different pitch than the other's (due to the pitch change by using the VSO) so that the overtones don't sit on one another but are in different places, letting both instrument's sound come out.

NAB AND ELEVATED LEVEL

Once upon a time there was only Ampex operating level. When high output tape came along however, recorder setup procedures, which had resulted in zero levels showing up on the tape machines, began returning +3 dB levels due to the more energetic tape. Since the tape was getting recorded 3 dB hotter, the industry came out with a test tape that was 3 dB hotter as well. This is what elevated level is all about. The new level is expressed in magnetic recording strength units called Nano Webbers per meter, which is abbreviated NW/M. We will work with these units in Advanced 3.

Ampex operating level is still the only real standard. All other standard level tapes are specified in reference to the old Ampex operating level.

PURC

Purc stands for Pick Up Record Capability. What purc does is to modify the timing normally used by a recorders's switching circuits when they activate the record and erase heads during punching in and out. Ordinary recorders activate their erase and record heads the instant you press record. Purc machines do not.

The purpose is to be able to punch in with another selection on an existing musical recording and not wind up with a momentary dead zone where there is no signal.

The way it works is that the machine upon getting a record command begins to erase immediately, but does not begin to record anything until the exact moment when the erased area on the tape has had time to get from the erase head to the record head. If it did not wait to begin to record, there would be an area of tape that had been recorded twice, and there would be a jump in the music level or a "glitch" until the erased portion of tape had reached the record head, at which point there would only be one musical recording again.

During punching out (going back to the music originally recorded on the tape, which is now being erased to make way for the replacement music now being recorded over it), the erase head stops erasing immediately upon getting the punch-out command, but the record head continues to record until the erased but not yet recorded portion of tape makes it over to the record head. It then stops recording the new music. This prevents a blank area of unrecorded tape. If the record head stopped recording immediately, it would not record anything on the already erased but not yet recorded tape that was between the erase head and the record head at the time the punch out command was given.

XLR AND RTS CONFIGURATION

*** Please refer to Illus. I7-9 -- The Yamaha Wiring Diagram ***

It would have been nice if all recording studios and recording equipment companies would have agreed on which pin of the audio cable connector -- the XLR -- was the signal + , the signal - , and the shield, but they didn't. In addition there is no real standard for using RTS (ring, tip, sleeve -- sometimes known as TRS for tip ring sleeve). Instead there are two different "standards" in use.

The "almost world-wide standard" is used in the eastern parts of the U.S., the South, in almost all U.S. film companies, in most of Europe and for most microphones throughout the world. In this standard the hi wire goes to pin 2 and to the ring of the RTS. The low wire goes to pin 3, and the tip. The shield goes to pin 1 and to the sleeve.

*** Please refer to Illus. A3-1 -- The XLR and RTS Format Drawing ***

The "other standard" is used in the western part of the U.S. Except in some film companies and some of Europe's studios, which has the hi wire going to pin 3 and the tip of the RTS, and has the low going to pin 2 and the ring. The shield is the same.

This lack of standardization causes more trouble than you can believe and hopefully one day will be changed. As it stands now, West Coast studios use pin 2 hi for mikes and pin 3 hi for line level signals like tape in and out.

SIMPLE WIRING PROCEDURES

*** Please see Illus. A3-1 ***

Starting from the top of the page we have the XLR and RTS format drawing. The pin numbers at the left are stamped on the actual XLR cable or chassis connector and there can be no doubt of which pin is which because of this. The RTS cable connector has screw down terminals to which the wires (+ and -) are attached by means of a small ring solder fastener, which gets soldered to the wire end and then screwed to the plug.

The signal polarities are the important feature of this drawing. The West Coast (where I work) is shown first with the signal positive (hi) wire soldered to pin 3 of the RTS and to the tip of the RTS. The East Coast and the format for most of the rest of the world is shown to the right.

Balanced wiring.

In the middle of the page is the diagram of how the cable is wired. The circular disk on each end of the wire drawings is the shield, which is braided around the actual signal carrying wires. To the left is the balanced variety of cable format where the signal hi and lo wires are in the shield. The shield does not carry any signal at all, as it only supplies the shielding action against hum and RF interference. (RF stands for Radio Frequency and is associated with radio transmitter caused interference).

XLR AND RTS FORMAT DRAWING Illus. A3-1

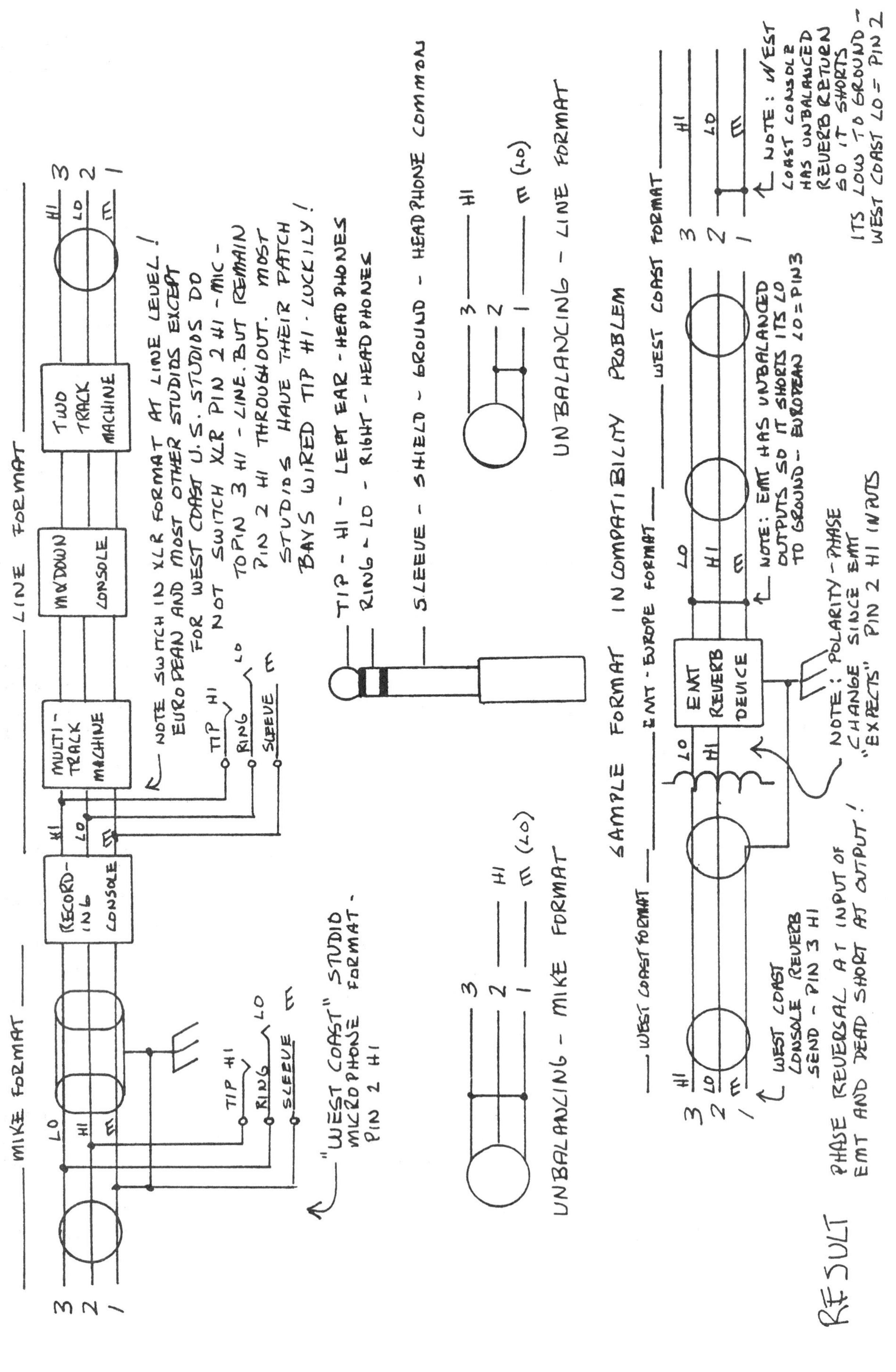

Unbalanced wiring.

Here there are only two wires since the equipment doesn't have a balanced output. The signal lo or return wire is attached to the shield. Now the shield is carrying the lo part of the signal along with its proper job of supplying RF and hum protection. This means that on the West Coast the XLR cable plugged into an unbalanced device would have its pin 2 and its pin 1 wires "strapped" together. On the East Coast, however, it would be pin 3 and pin 1 that would be strapped together. This is the basis for some difficulties.

Unbalanced to unbalanced.

The third drawing shows how this difficulty shows up. What we have pictured is a European EMT reverb device that has an unbalanced output with pin 2 hi mating with a West Coast console that has an unbalanced (pin 3 hi) reverb return system. This means that while the EMT is sending it's reverb signal down the wire on pin 2 and shorting pin 3 to ground (per its format in Europe) the American console is unknowingly listening in on pin 3 and shorting its end of the same wire pin 2 to ground (per its format in Hollywood). Because of this, both units are shorted out and inoperative. This little problem may exist whenever any new piece of "foreign" outboard equipment comes into a West Coast studio.

Balanced to balanced.

Here we see what happens if a balanced East Coast limiter is used in a balanced situation in a West Coast console. The signal comes from the console where pin 3 is hi. This signal is processed by the limiter but as an inverse phase signal. The signal is handed to the limiter, which expects the signal hi to be on pin 2 (+). Since it isn't, the signal is out of phase in the limiter, which hands the signal "out of phase" back to the console on what the limiter thinks is the wrong wires (pin 3 hi). But the console expects the signal being hi on pin 3 to be proper so the signal is no longer "out of phase." Amongst my audio engineering companions comes up the esoteric point of whether a limiter would limit a bass drum the same, seeing as how the bass drum head would be retreating from the mike (from the viewpoint of the limiter, which finds the signal out of phase) when the drum is kicked by the drummer, and whether it would limit this upside down event the same as it would if the drum were to move towards the mike as it is supposed to. For now it will be best to leave these types of considerations alone. Remember though, that when the drummer kicks the bass drum (with a mike in front of the drum head) a pressure wave is caused, which will produce a positive voltage in pin 2 and a negative voltage in pin 3 of the mike cable, that is if the mike's wired correctly. All mikes for a West Coast studio should be pin 2 hi, pin 3 lo.

OHM'S LAW AND THE OHM METER

Ohm's law describes the basics of electrical activity. The formula is extremely easy to work if you have your calculator. Use one with a 1/X (inverse) key, an X squared key, a square root key and most importantly a log and antilog key (log X and 10 to the power X).

The electricity comes from a battery (or a battery substitute). The electrons "pile up" at the + terminal and want nothing more than to get to the - terminal. If you put any conductor between the + and - of the battery, the electrons will flow.

The measure of the motivating force -- the push of electrical pressure -- is the EMF, which stands for electromotive force. The more volts, the more push. The unit for EMF is the volt. The symbol for the volt is "E." Why they picked "E" is anybody's guess.

The measure of the number of electrons that flow past a given point in a given time is the measure of the current or the electrical activity. The more push, volts, the more current. The unit for current is the amp. The symbol for the amp is "I." Again, why "I?"

Any restriction to the flow of electrons from the + of the battery to the - causes the current to lessen. This restriction to the flow is called resistance. The unit of resistance is the ohm. The symbol for the ohm is R. This symbol makes sense.

Power is the ability of a battery to push against a load (resistance) and still hold up its proper volts. If a battery drops its voltage when given a load, then it lacks the power to do the job. The unit of electrical power is the watt. The symbol for power is "P."

The Ohm's Law Formula is:

$$I = E/R$$

The current equals the push divided by the resistance.

Stated in units it means the amps equal the volts divided by the ohms.

With 10 volts and 2 ohms there will be 5 amps flowing.

To measure volts use a voltmeter which you attach across the battery with the + of the meter to the + of the battery and so forth.

To measure ohms use an ohm meter and disconnect the unit to be measured and adjust the meter until the needle reads on scale.

To measure amps use an ammeter, which you connect to the circuit by first disconnecting one of the leads connected to the battery and then reconnecting the wire to the battery using the ammeter as a jumper. The + of the ammeter goes to the wire leading to the - of the battery and the - of the ammeter goes to the wire that leads to the + of the battery.

Power is usually not measured directly. Use either of these two formulas:

P = E squared / R
P = E x I

The power equals the volts squared divided by the ohms, or power equals the volts times the amps.

If you play a steady tone from your cassette machine through the speakers and measure 10 volts with your ohm meter across the speaker terminals, and the speakers say they are 10 ohms on the decal near the terminals, then you have:

P = E squared / R
P = 10 squared / 10
P = 100/10 = 10 Watts

THE dB IN MATHEMATICAL TERMS

The dB as a technical unit does two things. First it expresses a ratio. Secondly, it uses an unusual numerical system know as logarithms. The study of logarithms is interesting for those who possess scientific curiosity, but for the purpose of this book, I will deal in logarithms only in terms of push button technology. In other words when you want to go into logarithm mode, push the log x button on the calculator; to get out of log mode, push the 10 to the x power button.

When we say the word ratio it means to divide. Using the calculator you can find the ratio between 25 and 13 as a ratio of 25/13 = 1.92 or 13/25 = .52

Since all ratios can be expressed as the two numbers divided either way, we have to decide if the ratio is supposed to be greater than or less than 1 in each case. Since we have less now than before, the ratio we should use is the .52 since this is less than 1.

Going from 30 volts to 5 volts is a ratio of .17 change. Going from 5 to 30 volts is a ratio of 6 change.

Having the concept of change in ratio numbers understood, we will now begin some log pushbutton adventures.

The log of 1 = 0. The log of 10 = 1. The log of 100 = 2. The log of 1000 = 3. Check these with your calculator.

From this short listing we can see that the change of the log from 0 to 1 equals a change in real numbers of 10 numbers (0-10). But the change of one log's worth from 2 to 3 is a change in real numbers of 900. It is precisely because of this relationship that logs are used. By relatively small changes in the logs of numbers, the real number can change by tremendous leaps and bounds.

To calculate the dB change, first find the ratio of the change in whatever unit you are to calculate, find the log of the ratio, and mulitply this log by 20.

Using the example above, to find out the number of dB change from 5 volts to 30 volts, first remember that each doubling of the volts is 6 dB. So from 5 to 10 volts is +6 dB; from 10 to 20 is +12 dB; and from 20 to 40 volts is +18 dB. But none of these increments end at 30 volts so we cannot use this simple relationship. Instead we find the ratio:

$$30 / 5 = 6$$

With the log button we find that the log of 6 = .778

$$.778 \times 20 = 15.563$$

So the dB change from 5 to 30 volts is 15.563 dB.

The formula we just worked out is written:

$$\text{change in dB} = 20 \times \log \text{change in E (volts)}$$

This same formula is used for finding the change in dB for volts (E), ohms (R), amps (I), impedance (Z), distance of the mike from the source (D), and Nano-Webbers of recorded energy on the tape (NW/M).

Let's move from volts to ohms

For a transformer we want to use as a direct box transformer

We read from the transformer case that the input impedance is 15,000 ohms and the output impedance is 250 ohms. What is the change in dB? Well, to start out with, there is going to be a loss anytime the signal is going from a large number of windings to a small number as in this case. There is exactly the same amount of gain if the transformer is hooked up the other way, with the 250 ohm side as the input (primary) and the 15000 ohm side as the output (secondary). For our direct box we will be trading many volts with little power (hi impedance) to few volts with much power (low impedance), which the console mike input channel can use.

The ratio is 250/15,000 = .017

The log of .017 = - 1.778

20 x - 1.778 = -35.563 dB

The loss of this transformer is 35.563 dB. In calculator push button jargon this would be: 250 divide 15000 = log x times 20 = (35.563)

If we hook the transformer the other way, we have a ratio of 15,000/250 = 60

The log of the ratio 60 = 1.778

20 x 1.778 = 35.563 dB

So the dB change is the same, only it is positive this time meaning a gain of 35 dB.

Now in the matter of distance...

If you move a mike back four times its original distance this would lessen the volume by 12 dB (each twice the distance = 6 dB). But going from 2 to 5 feet is a ratio of 2/5 = .400 and is not an even change of distance, so the 2 times the distance = 6 dB formula cannot be used; the new log formula must be used:

Change in dB = 20 x the log of the change in distance
= 20 x log (2/5)
= 20 x log .400
= 20 x -.398
= -7.959 dB

So from 2 to 4 feet is -6 dB, but from 2 to 5 feet is 7.959 dB.

Now for you Nano Webber fans....

If one tape box had the notation "This tape is recorded at 185 Nano-Webbers per meter (NW/M)" and another box said, "This tape is recorded at 300 NW/M," what would be the dB difference between the two?

Well, as you record "hotter" you are recording with more NW/M. So the second tape is "hotter" than the first.

Ratio of the NW/M = 300/185 = 1.622

The formula:

The change in the dB = 20 x the log of the change in the NW/M
20 x log 300/185
20 x log 1.622
20 x .210 = 4.199 dB

The second tape is 4.199 dB hotter than the first.

There is one audio relationship that uses a different formula, and that is the relationship of one wattage to another. The watt formula uses a 10 instead of a 20. Any time you calculate for watts or power use:

The change in dB = 10 x the change in watts (power)

From 50 watts to 175 watts in dB is:

10 x log 175/50

10 x log 3.5

10 x .544 = 5.441 dB

So the dB change from 50 to 175 watts is + 5.441 dB.

REVERB

Reverberation is the most widely used special effect. The most common use of reverb is to add some acousitc flavoring on a sound that has none. Basic studio sound is dry because it is recorded in a dead studio. By adding reverb you can make certain sounds in the mix, or the whole mix for that matter, sound like it happened in a huge hall.

Reverb makes things sound farther away. If the mix is generally "dry" (without reverb), then any instrument having reverb will tend to stand out. In a generally "wet" mix however, a dry instrument will stand out.

If reverb is delayed by a tape delay channel, the reverb will happen after the actual sound is finished and will seem longer in duration, as well as making the actual instrument sound clearer than if real time (without delay) reverb is used. If the delay is too long, the reverb sounds artificial, silly, and un-musical.

If you turn the tape upside down and print reverb from a recorded track on another un-used track and then turn the tape right side up again, you will have pre-reverb. This reverb starts out soft and grows in volume ending abruptly at the sounding of the track that generated it. This is a very eerie sound that is well used on open (un-cluttered) drum parts to make an individual drum sound more interesting. Reverb used in this way calls so much attention to the sound it's attached to that it enables you to turn down the instrument and still hear it easily in a mix.

Reverb can be mixed with a short duration sound before sending the mix of both to a phaser, flanger or other signal processing device. In essence, the presence of the reverb stretches out the basic sound giving the special effect device more audio to work with.

Using delay to fit the arrival time of the delayed reverb into a place in the beat where there is no rhythm part can bring out a strong rhythm influence, which would other wise be lacking in the rhythm structure. Sometimes this effect, called "slap", will make the beat stronger than overdubbing an actual instrument part in the same place in the beat.

Reverb will make a part seem even more out of tune when used on an instrument or vocal track that is slightly out of tune. If you return reverb to the phones (cue system), and too much is put on the vocalist's voice during vocal overdubbing, the vocalist will tend to sing flat.

"Columbia Reverb" is the name I give to the type of reverb that was used extensively on Simon and Garfunkel records. This special reverb technique and particular sound is obtained by using a Langevin or other sound effect filter on the reverb send channel after the 15 ips tape delay so that all frequecies below 5 KHz are removed. Thus all low notes and sounds will generate no reverb, but all the sung consonants (T's, S's) will get a brief splash of bright, delayed reverb. This really enhances the diction of the vocalist and makes the record sound brighter and clearer.

Some interesting psychoacoustic effects can be made with reverb. If there are two lead vocal recordings on the tape, try using the proper one for the up front vocal and the other (unused) one for a reverb send only (not bussed to the mix at all) so that there will be some subtle differences in the actual voice and it's reverb counterpart. This will probably cause subliminal fascination and a more interesting sounding vocal.

DYNAMIC PROCESSING

Dynamic processing is used for several purposes:

1. To reduce the dynamics of the final product for practical considerations of user medium limitations (cassette, in flight airline music, etc.). If the dynamics were too great, the music would appear only at the hi level passages and then disappear again into the noise background. If the listener tuned in during a low passage, he would have to turn up the level greatly to hear anything above the noise and soon would be jarred and disturbed by the louder volume passages when the next one occurred.

2. To reduce the dynamics of the final product to make the product sound louder on the radio so that it will compete and be commercial. Some radio station engineers feel that their station is a favorite with the listening audience because of the sheer volume at which they transmit the music. Since approximately each 10 dB on the stations VU meter is half as loud, they never want to see any real drops in level, so they limit the sound.

3. To increase the dynamic range of a signal, so that unwanted lower level signals fall below audibility, due to the masking effects of the rest of the sound (like using a Keypex on the drum overhead mikes because otherwise you would hear weird room noises coming in from the mikes). To increase separation from already recorded tracks, learn to use a Keypex.

4. To decrease or increase one instrument's audio so that it will "track" or keep "in volume" with the dynamics of another instrument having different dynamic characteristics. Example: if you have two guitars that are supposed to play in harmony together, and one of them gets too loud or too soft compared to the other, it could be compressed to achieve a similar dynamic range to the steadier guitar and in this way the harmonies between them will hold together.

5. To cross limit the instruments in a mix so that when one instrument gets louder, others get softer to make way for it so that proper on scale VU meter readings can be maintained.

*** Please see Illus. A4-1 -- The Cross Limiting Diagram ***

In cross limiting one instrument is "ducked" out of the way to allow more room for another instrument. In the graphic, the kick is to have the most important sound, so we are going to "back up" the bass guitar at those brief instants when the kick hits. We won't really hear the bass dropping away since the powerful kick sound will be right there masking the effect, so to speak. The kick goes straight to the mix and is thus not affected in any way. The kick also goes to the triggering limiter. The triggering limiter triggers the triggered limiter that has the bass running through it. When the controls are set up properly, the bass "backs up" each time a kick drum sound comes through, allowing the kick to have a tremendous amount of authority since it doesn't have to "compete" in any way with its lo end counterpart -- the bass.

For triggered limiting, the concept is the same except that you're controlling the whole of the mix and causing the whole band to move out of the way for the "featured" instrument of vocal.

CROSS LIMITING AND TRIGGERED LIMITING DIAGRAM

Illus. A4-1

CROSS LIMITING

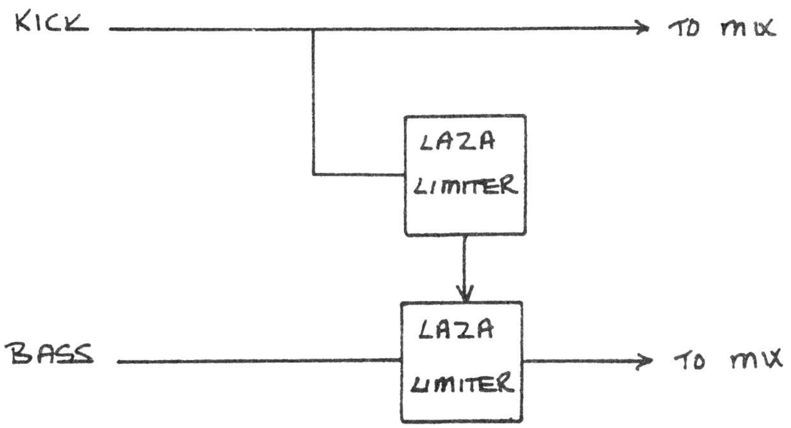

TRIGGERED LIMITING

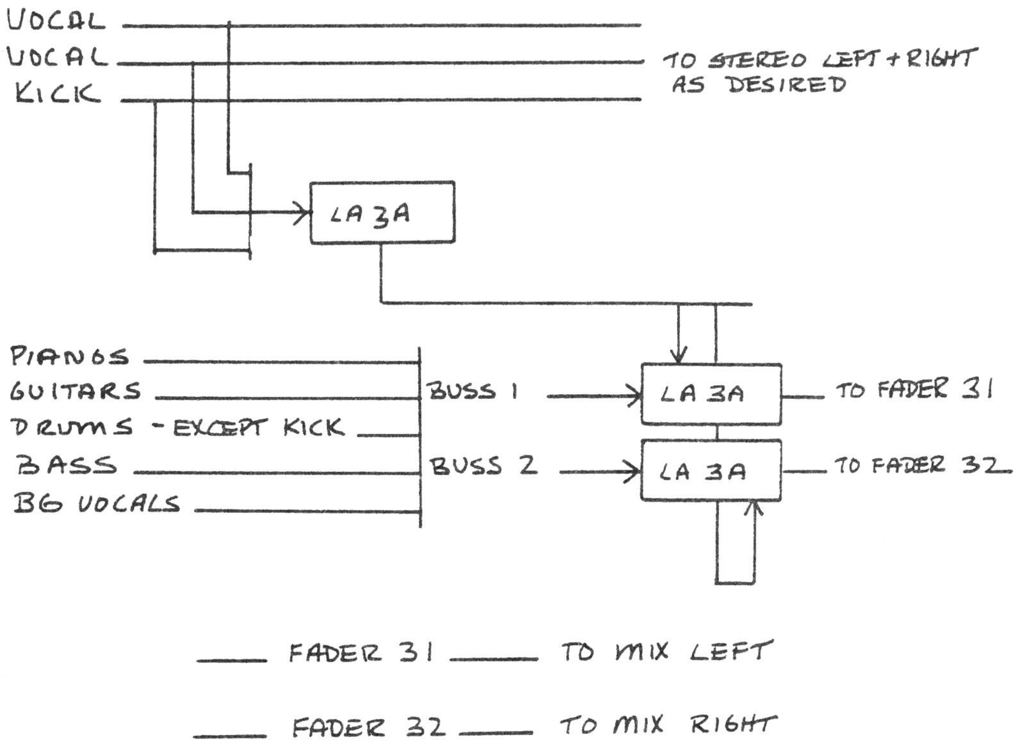

A reverb send from the module having the vocal is used to send a vocal signal to the input of the "trigger limiter" in our example. The output of the trigger limiter is not used. Instead, the gain reduction control of the trigger limiter is adjusted to make the music (running through the triggered limiters) "duck" down in level enough to allow for the presence of the vocal. When the music is playing, it occurs at full VU leaving no room for the voice. When the voice comes in, it causes the music to "back up" just the right amount (by the setting of the reverb send off the voice channel going to the trigger limiter). The more the reverb send is turned up (or the higher the vocal fader is pushed up), the more the music backs up. This technique is very useful for bringing up "difficult" instruments to extreme loudness in the mix, like bass drum, low toms, etc.

6. To increase the sustain of an instrument (the volume is increased by a compressor in inverse proportion to the instrument's natural tendency to fade out).

7. To prohibit extreme transient spikes, which would distort the tape, since these transients are usually not necessary to the musical enjoyment of the work.

8. To hold down the violently loud parts of a musical performance so that the intriguing quieter moments (otherwise too soft to hear) become audible.

9. To "imprint" the dynamic envelope of one instrument onto another instrument's sound quality (this is performed by envelope generators and Vocorders). See Advanced 5.

10. To limit the program while in the remix situation so that harmful changes to the sound due to limiting can be optimized with the fader settings and EQ at mix time, instead of waiting until it "comes out" on AM radio, where they limit the music tremendously, and then hearing how the mix is altered or perhaps ruined.

The devices that produce dynamic processing are:

1. Compressors
2. Limiters
3. Expanders
4. De-essers
5. Bass-x's
6. Manual fader motion by the engineer
7. Dolby noise reduction units used as special effect devices
8. envelope followers
9. Vocorders

SLOPE AND THRESHOLD

Compressors are gentle dynamic processers. They have a slope that is generally less than 4 to 1. The slope of a device is the ratio of input dB change to the output DB change. If the input to a compressor went up 12 dB and its output goes up 3 dB, the device has a slope of 12 / 3 or 4 to 1, which is written 4:1.

Stronger slopes (higher numbers) are caused by the action of limiters. A limiter is therefore a strong compressor. Slopes from 4 to 1 all the way to 100 to 1 are obtainable. With a slope of 100 to 1, a 100 dB increase at the input would be reflected at the output of the device as an increase of only 1 dB!! Now that's limiting! This kind of limiting is referred to as "hard limiting." This is useful for performing function 7 in the first list above, regarding limiting extreme volume excursions not musically required.

Expanders are devices that make the instrument fade away more quickly once the instrument fades down to a given volume. As it passes through this (operator set) volume level, the expander expands the dynamic range of the sound by fading it in addition to its natural fade characteristics.

De-essers are limiters that only act on hi frequency audio. They do not alter the mid or low frequency ranges. The signal is split up by a filter network so that the highs go through one path and the mids and lows go through another. The highs are limited and rejoined with the mids and lows later at the final stages of the device and mixed back together for the output signal.

Bass-X's, which are very old devices, are the bass equivalent of the de-esser. The Bass-x is a bass only limiter. The highs are not affected. Theoretically, if you connected a Bass-X in series with a de-esser, you would be able to limit the highs and lows and leave the midrange alone. Doing this would sound most unusual.

Since the VU is generally faster than the eye, manual fader motion by the engineer is the most exciting and risky form of dynamic processing. If you know your artist well and are feeling "on top of it" on a given evening, you might try "hand limiting" a vocal overdub. In the early days of recording, "limited" vocals were performed by the performer watching his own VU readings on an oversized VU meter set before him as he sang. He would work at his recording performance until he could keep the VU needle in the full modulation area.

361 Dolby units can be used to "beef up" some kinds of sounds because they act in unusual ways on the dynamics of the audio. If the signal is run into the Dolby unit at about -20 VU, the effect can be heard quite nicely on snare drum tracks. Note: The Dolby unit should be on record, NR in, normal mode and the audio patched from the console reverb send to the Dolby unit "line in" and the Dolby's "to record" patched back to a console channel for mixing as needed.

Envelope followers are described in Advanced 5.
The Vocorder is covered in Advanced 5.

The ear hears music differently than compressors and other dynamic processors do. Electronic devices act on the peak program material at the speed of light. This causes most of the mysteries between what the eye sees (on the device readouts and controls) and what the ear hears.

*** Please look at Illus. A4-2 -- The Dynamic Processing Graph ***

Starting at the bottom left and going along to the right we have the axis of increasing input (to the device). Starting at the same point and going upwards we have the resulting output from the device. Along the 1:1 line we have the "curve" of a wire or an amplifier operating normally: no gain or loss through the device -- the volume or level at the output exactly equals the volume or level at the input to the device. This makes the 45

DYNAMIC PROCESSING GRAPH Illus. A4-2

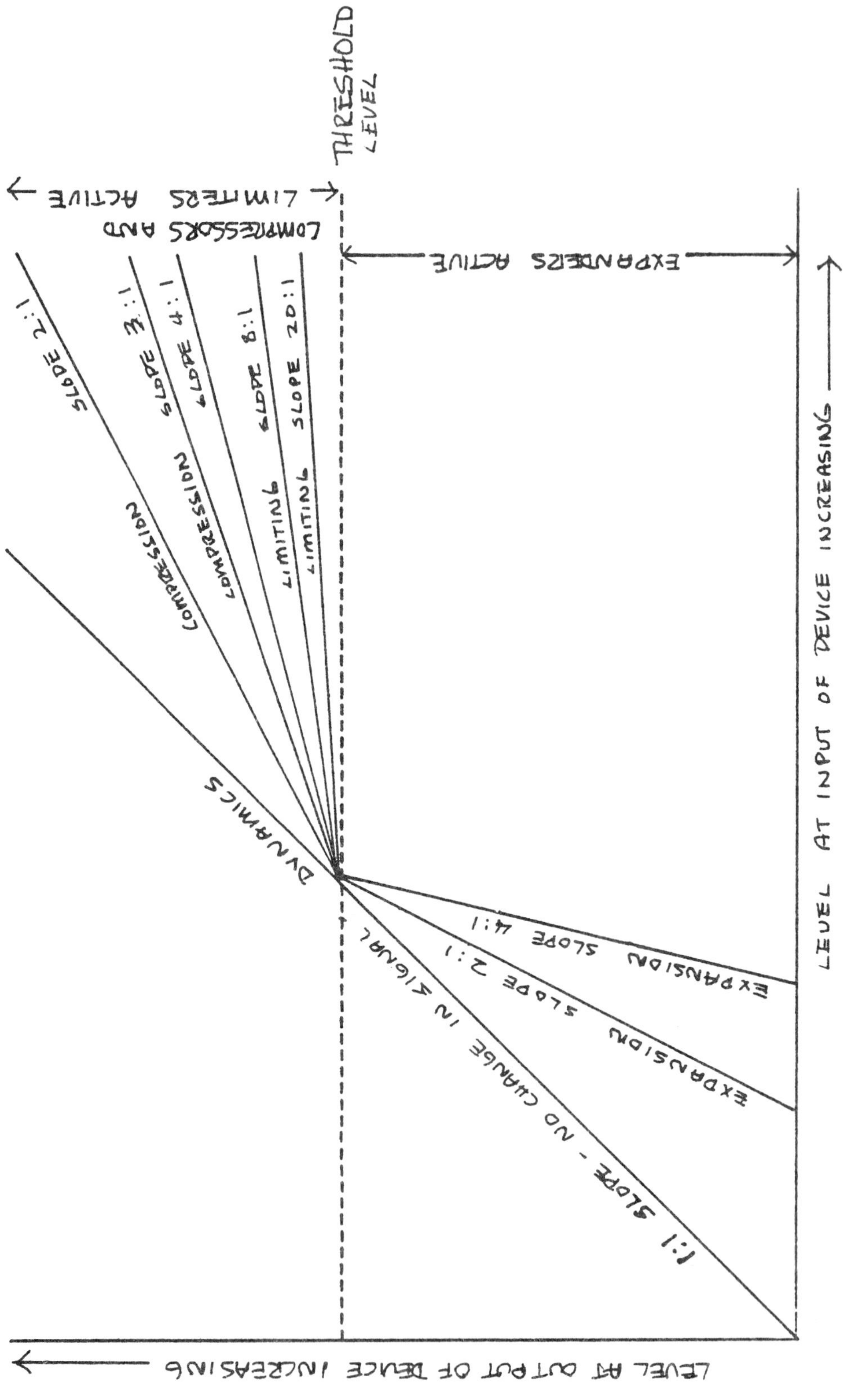

degree line.

The horizontal black dashed line at mid graph indicates the threshold of the compressor, limiter or expander. Above this level, the relationship changes for compressors and limiters. At 2 to 1 slope there is a 1 dB increase in output for each 2 dB of input. At 8:1 there is a 1 dB increase in output for each 8 dB of input.

As the instrument begins to fade, its dynamics start to trace downwards along the graph. Instead of fading 8 dB (using the 1:1 line), it traces down the 8:1 line and only gets 1 dB softer. This is where the increase in the instrument's sustain function comes into play. If this amount of fading brought the instrument down to the threshold line, then the instrument's fading dynamics would return to normal again and continued fading would see the instrument fade normally.

Essentially The expander does the same thing the compresser does, but in reverse. As the large volume passage traces down the line from the top of the graph, no change in the dynamics occurs, but when the instrument drops into the threshold area (below the line for expanders), the rate of fading of the instrument increases causing the instrument to fade with greater rapidity. While the input drops 5 dB - moving right to left at 2:1 slope -- the output drops by 10 dB -- moving top to bottom.

For compressors, expanders, and limiters, the level for the threshold is adjustable by the user. If you want just the peaks attacked by a limiter or compressor, adjust the threshold to a higher level so that the meter just barely indicates any activity. This means that only the peak material in the audio is through the threshold and not the main body of the sound (this is assuming that you're dealing with a percussive sound having peaks and not a smooth tone).

As you adjust the threshold of an expander to be more and more sensitive (higher on the graph), more and more of the lower volume musical events will disappear until you reach a point where even the most noticeable parts of the instrument's sound will be affected. At this point the device turns into a special effect device since you can hear its effect clearly. Before you can hear anything happening, the device is merely getting rid of audio that might muddy up or clutter the sound, and the loss of such clutter is both unnoticeable and beneficial to the listener. It is very instructive however to periodically deactivate all the devices you are using to show yourself what you gained (or lost) by working on the sound's dynamics.

dB over threshold

The gain reduction meter on a dynamic processing device does not indicate how many dB over threshold the audio is. The relationship of the dB over threshold to gain reduction (GR) readings on limiters and compressors is according to this simple formula:

dB over threshold = GR x slope / slope-1

This means that if a 1.3:1 compressor's meter was reading 12 dB in the GR (gain reduction) mode then

dB over threshold = gr x slope / slope - 1
= 12 x 1.3 / 12-1
= 15.6 / .3
= 52 dB

This means that for 12 dB of gain reduction to have occurred at a slope of only 1.3:1, then 52 dB of audio must have passed through the threshold. This 52 dB of audio would have been acted on by the slope of 1.3:1, making the actual increase only 52 DB / 1.3 = 40 dB. Since with no gain reduction, the increase would have been the full 52 dB, the increase of only 40 is a reduction of gain of 12 dB (52-40).

This large difference between the small amount of gain reducion shown compared to the large amount of signal passing the threshold is much less for higher slopes.

For a gain reduction reading of 12 dB with a slope of 20:1 the dB over threshold would be:

dB over threshold = GR x slope / slope-1
= 12 x 20 / 20-1
= 240 / 19
= 12.6 dB over threshold

This is only .6 dB off the indicated gain reduction reading of 12.

OPERATION OF DYNAMIC PROCESSING DEVICES

Now that we have a little background of the terms and varieties of the devices at hand, let's cover the uses as previously listed in the dynamic processing section.

1. For practical considerations such as noisy environments, the dynamic range is reduced. This is done with the use of a gentle compressor (2:1 slope perhaps), which is adjusted to have much of the audio passing through its threshold. This differs from the use of a limiter where the slope is much more drastic, but very little of the actual dynamic range is through the threshold. When compressing the aim is to come out with a new dynamic range. When the softer parts of a piece occur, they aren't as soft. The primary factor for deciding how many dB of indicated compression gain reduction to use, is by listening carefully to the music itself and to the weird noises that will crop up if too much of the low level material is raised into audibility by the action of the compressor.

Much attention should be paid to release times. No releasing (audio becomes louder) should occur during momentary pauses in the music. This is called "pumping" or "breathing" and sounds terrible.

To avoid momentary compressor distortion when loud sounds suddenly come through a limiter that has had no audio coming through it for several seconds and which is set to high amounts of limiting, use a very low frequency tone to "pre trigger the limiter." This will prevent a noise fade in and the momentary burst of distortion that would occur as soon as the audio hit the limiter. This "warm up" tone can always be edited out at the two track editing and assembly stage of production.

Remember to follow the twin compressors used in stereo mode with twin EQs. This is the most often overlooked and important consideration when dealing with compressors and limiters -- use a following EQ to put back some of the lost sparkle and thump that the sound hopefully had before you comressed it.

I feel free to use "audio slang" to discuss strange sounds and audio effects. Using homemade terms is far easier than the verbosity that results from attempting to describe unusual sounds using "scientific" terminology each time one of these concepts comes up in a discussion.

"Sparkle" is "high end," also known as the "super highs," which come right out and tickle your ears.

"Thump" is low fequency power you can hear with your chest as well as your ears.

"Moo" is really rich bass that seems to fill the room like a giant water balloon and doesn't seem to be coming from anywhere in particular. It requires the musician to play with legato notes and without any real attack noises from the instrument.

"Honk" is overly accentuated mid range sound. It is what a voice sounds like with cupped hands in front of the mouth.

"Squashed" means limited so much that there is really no difference in recorded levels: whether you play hard or very quietly, everything comes out at the same level. In some cases, like in fuzz lead guitar parts, noises like hums, rasps, buzzes, and strange overtones from the ringing strings on the guitar make a very unusual "fill in" between the notes the player plays. Too many of these oddball sounds are distracting and unpleasant of course.

"Open" means the opposite of squashed -- the sound of no limiting or compression at all. Open sounding tracks sound like they were recorded outdoors -- no room sounds are included.

2. Commercial music for AM radio get a separate kind of mix, different than the album mixes. The AM radio play tunes should be mixed at the same session as the album cut. After the album mix, the console should be re bussed to just one buss and the same effects and balances, EQ, reverb, and "moves" should be made to make a semi-squashed version of the open version just mixed in stereo. The amount of squash depends on the artist, the type of music, and how "loud" they want their song to "come in" over the radio.

3. Increasing the dynamic range requires an expander. One expander is the Keypex.

The Keypex is first adjusted to the desired range. This is done by guesswork at first and later tuned in for best results. Setting the range to 10 dB is a good starting point. The threshold adjusts the point at which the sound will begin its more rapid fading. The more threshold, the more audio will be below the "begin expanding here" volume point. The less threshold (a counter clockwise turn) the more music will reside above the threshold. Any music that falls below the threshold will be quieted by the expander, according to the settings of the range and release controls; any music above the threshold will not be acted on at all.

The range control will effect the amount of expanding because the more range, the more the distance the Keypex will follow the sound expanding all the way. With only a little range, the Keypex will only expand a little, making only a short ride for the expanding area, and then it won't expand any more even as the sound continues to get softer. With the range set to full (fully clockwise), the Keypex follows the sound and expands it down to inaudibility.

The release control affects what's going on too because if the release is set too slow, the Keypex will just start to expand when the next sound comes along causing the Keypex to open up again. Each time a strong sound comes through the Keypex, the release control is set to its starting point again and no expanding begins until the release time passes. After the waiting period, expanding happens. If the release time is too fast however, the Keypex will release after each low frequency wave peak passes (if there is any low frequecy material), thereby adding distortion, because it will be trying to expand each descending low frequency wave front as it passes. This type of distortion also occurs when limiters are set with their release times too fast.

If you juggle the range, threshold and release times just right (it isn't easy), you can get rid of unwanted sounds without hearing anything strange happening to the wanted sounds. You can get rid of:

A. Tape hiss

B. Room boom (leakage of room boominess being picked up by some of the mikes in the room)

C. Snare leakage coming in on the hi hat mike

D. Drum kit leakage coming in on the tom-tom mikes

If you want, you can press the key switches on two Keypexs, with the release controls set to maximum time and with the range controls set to max as well. If you have the threshold set high enough so that no expanding will happen to your stereo mix when the key in buttons are not pressed, then pressing the key buttons will begin an automatic fade of sorts. The one really useful thing about these auto fades: if started on a particular beat of the song, fade endings can be intercut since any mix of the ending should be equally faded at any given point due to the precision of the Keypex's fades. Remember this technique for fancy fades. When you want the players to jump around in stereo, for example, or where your hands are already busy enough without giving up one hand just to do the fade -- consider using the Kepex for fades. A clever maintenance person can modify two Kepexes to track very well together, as well as slow down the fades to very, very slow speeds to be used especially for auto fades. Realize that any product guarantees will probably be voided by any modification.

4. To match the dynamics of two instuments, watch the VU meters carefully. If the rowdy one has a VU dynamics of 10 dB and the steady one has dynamics of 3 dB, then you could use a slope of

10 / 3 = 3.3 to one

With the threshold set so that the rowdy instrument has indicated gain reductions at -10. So when the rowdy instrument attempted to go from -10 to 0 it would only go up

10 / 3 = 3.3 dB

instead. This would make the dynamics of the two instruments the same.

5. To cross limit you need two limiters that can be hooked up in stereo.

*** Please refer to Illus. A4-1 -- The Cross Limiting Diagram ***

In the example, the kick is going to cross limit the bass so that when the kick needs some extra VU room, the bass will "back up" to supply the room. As the kick becomes louder, the bass moves back farther. In between kicks, however, the bass is at full volume. The bass only moves "back" at the instant a kick is happening. The softening of the bass will be unnoticeable because the bass's volume loss will be hidden by the presence of the loud kick.

The kick goes directly to the mix but also goes to the input of one Teletronics LA-2A limiter, which work especially well. The bass goes to the mix but has a LA-2A patched into its pre EQ patch point (whatever that point is called in your console installation). The #6 barrier strip terminals are wired together. The kick's limiter output is not used. The bass is set up with no gain reduction control action on its limiter. The gain reduction control is turned up only on the kick's limiter, not the bass's limiter -- the bass isn't causing any limiting of itself, only the kick can limit the bass. Since the audio output of the kick's limiter doesn't get used, this level control isn't turned up. As the kick gain reduction control is turned up, the bass begins to be limited. The kick's gain reduction is adjusted so that the bass backs up just the right amount to accomodate the amount of kick desired. The more gain reduction you dial in on the kick's limiter, the further the bass will back away during loud kicks. If you adjust everything correctly, the more kick you use, the more the bass will back away to give the kick VU room.

This cross limiting technique works excellently with any combination of ingrediants where different types of sounds are competing for being loudest. You can make the whole mix sound much louder before you actually hear any "pumping" of the cross limited instruments. This is because there is always something loud taking over for the disappearing instrument: the loud instrument whose presence is causing the cross limiting.

A really nice cross limiting trick is to "fill in" the undesired pauses in a lead solo. Have the nuttiest player do a freaky noise track (guitar screechs, grunts, etc.) during the length of an instrument solo. After overdubbing the fill in noise part, set up a cross limiting situation so that when the lead solo instrument pauses, the cross limiter opens up and allows the noises to come through. It will sound as if the player played the collection of sounds as part of his solo.

Triggered limiting is included here although it is nothing more than an elaborate version of cross limiting.

For triggered limiting use three LA-3A's (or any three limiters that can be interconnected in the fashion shown). In the case of UREI 1176 limiters, for instance, a stereo interconnecting box would be used with an RCA connector mult cord from the box to the two controlled limiters. In the mix there is a small band with two lead vocals and a requirement that the kick sound as huge as possible for, say, disco. The band wants the music as loud as it can go for the whole record, while not burying the music by having the vocal "too far out front". The singer on the other hand doesn't want to be "buried in the track" either. When you hear these kinds of comments, you will probably save as much time as you spend setting up triggered limiting as you would trying to please everybody with conventional techniques.

The vocals are mixed directly to the stereo buss, but the band is mixed (pans and everything) to console busses 1 and 2 instead. These busses are channeled through two LA-3A's and then to two faders, which are in turn sent to stereo left and right. The control mix elements (the two vocals and the kick) are sent to a third LA-3A, possibly by assigning them to a reverb send buss that controls the other two LA-3A's (with the left and right stereo band running through them). The EQs at faders (in this case) 31 and 32 are used to restore any sparkle and thump lost due to the use of the limiters.

The band is first mixed up to full VU range leaving no room at all to put any vocals on top. Then the vocal is set up without the band to read full VU as well. Now when both the vocal and band are turned on, the VU's should pin. You now correct this problem by slowly increasing the level of the reverb sends coming from the two vocals so that when the vocals hit loud notes, the trigger limiter shows gain reduction happening to the proper extent to move the track back the correct amount to not pin the meters. Now both the vocal and the band are at maximum volume. The kick can be pushed until it really pumps, by adjusting the kick's trigger send level so that it gets as much VU room as it needs.

6. There are several ways too increase sustain. One way is to turn up a speaker with the instrument's sound so that slight feedback occurs. Another is to record the instrument (tuned up of course) at higher than normal tape speed, which will give it more sustain when mixed (played) at normal speed. Another is by using limiters.

To increase sustain using a limiter, adjust the limiter to hard limiting (a slope like 20:1), adjust for mid scale gain reduction, and carefully adjust the release time to be as close as you can get it to the natural decay time of the instrument. A guitar, for example, takes about 1.2 to 2.5 seconds to become louder by the amount shown as half scale (6 to 8 dB) gain reduction on a limiter. If you set the release time so that the limiter lets go of the gain reduction at the same speed at which the instrument is trying to fade away, the down fading guitar and the up "fading" limiter will cancel each other out, making for no fade for the 1.5 to 2.5 seconds during which time the limiter is releasing and still has audio above threshold -- the limiter's gain reduction meter still shows gain reduction decreasing. Adjust the limiter by ear as well as by eye so that the sustain seems prolonged. If the highs of the guitar get munched by the action of the limiter, use a little audio from a contact pickup or a mike very close to the strings mixed with the sustain sound to bring back the sparkle.

Another even more bizarre trick is to take an instrument recording (like drums for instance) and add lots of bass to it, then limit it greatly, and then remove lots of bass past the limiter. What this does is give lots of moo to the limiter to which it can apply lots of limiting. Not having such a rotund sound will have the limiter limiting on the brilliance of the sound to a greater extent. Anyway, when you de-bass the sound, the result is a strange stretched drum sound much different than the run of the mill limited drum sound. This technique sounds amazing with certain kinds of drummers who can play long phrase drum parts, as the releasing action of the limiter causes the current drum hit to sustain until the next drum is hit, which then takes over drowning out the previous drum.

7. Prohibiting spikes is accomplished by adjusting a large slope limiter to very quick attack and very quick release times. This way, the spikes (transient attack noises above the program level that drive the tape crazy but do not improve the sound very much) are attacked and held to a maximum level while the program is not affected. If the needle moves at all, you have adjusted for too low a threshold. The only way to adjust the limiter to attack spikes, since you won't be able to see them on the limiter gain reduction (GR) meter, or see them being removed from the console's buss VU meter, is by watching an oscilloscope or PPM (peak program meter) connected to the buss that you are working on. Adjust the limiter's threshold so that you see a reduction of the peak level but so that the limiter's gain reduction VU doesn't move. This will give you a starting point from which to work. If the limiter's gain reduction meter does move very much, it means that considerably more than the peak material is being acted on.

8. Some instruments produce intriguing sounds that are rarely noticed. These I call subliminal sounds.

*** Please refer to Illus. B3-1 ***

A typical subliminal sound is the brilliant snap that would come from the electric bass's electrical output, if it weren't for the physics of the instrument's pickups, which prohibits these sounds. If you want to record these sounds, you won't be able to pick them up using the bass's output cable. The attack noises from a bass do a lot towards enabling the listener to hear the bassist's rhythm ideas and where each note was played. A bassist playing with a pick can hear these attack noises when he isn't wearing phones because the sound travels straight up into his ear from the string. When hearing the playback, these subliminal noises are missing. If the bass is compressed, the sublminal sounds are reduced even more since they're percussive and get attacked by the limiter more since their volts are usually greater than the large round tone signals coming from the bass.

To accentuate small noises coming from instruments that add to the brightness and interestingness of the sound, use a limiter that is adjusted to very slow attack and very fast release times. This gives the leading edge noises a chance to get through the gate, so to speak, before the limiter begins to work. The fast release pertains to the desire of the engineer to have the gain wide open by the time the next note comes along so the small opening noises will be as loud as they can be. If the notes are far apart, or if distortion occurs, use longer release times.

9. To imprint the dynamics (envelope) of one instrument on another instrument's sound use a Keypex.

*** Please see Illus. A4-3 -- The diagram of a Keypex used as an envelope follower ***

In this setup, the idea is to take a really dull sounding set of toms and put some musical differences in the sound of each hit, instead of having a set of drums in which each tom sounds really dead and without a tone of its own. To do this, set up a bass overdub to a spare track and then patch a Keypex into the bass's recording channel on the console. Buss the drum overhead mike (or a mix of the individual tom tracks) and send this buss to the Keypex's Key input.

Set the Keypex for maximum range, 3/4 release time to start, and then play back the tom tracks mixed and patched to the Key inputs, and adjust the Keypex's threshold (press Key in) so that the entirety of the tom's main sound keeps the light indicator on the "0" (top) position, with the light falling immediately after the main tom event passes (this is set by the release control). Drum leakage should be low enough so that there is no light indications on the Keypex during snare hits. If there isn't enough separation on the overhead or tom tracks, try using another Keypex to increase the tom compared to other drum separation so that only tom hits will drive the envelope Keypex to the "0" light. With this setup complete, have the bassist listen to the tom parts and prepare some notes to play in time with the tom hits. Hopefully these notes will be musically in tune with the song itself. Record the Keypexed bass at normal levels on the tape, but adjust the monitor system so that just enough of the keyed bass comes through to flavor the tom sound. If the bassist plays at the wrong moment, the Keypex will block the bass from getting to the monitors and tape. In fact,

THE KEYPEX AS AN ENVELOPE FOLLOWER Illus. A4-3

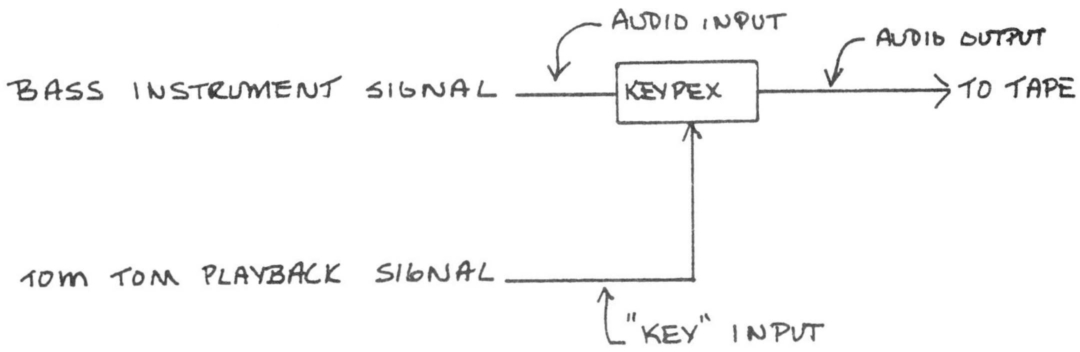

if the tape is stopped, the bass cannot get through the Keypex at all.

Use this technique with very fast release times to double the snare on a record. The agreement of the two snares will be heightened greatly if the engineering and musicianship are really up to doing the gating carefully.

Put up some lo pass filtered pink noise (you can probably borrow some from the technician in the shop) through a Keypex, EQ it and key it with a voice for an alien sound.

If the snare reverb is too long in duration, put it through the Keypex with the snare keying it. By adjusting the release time you can custom fit the decay time of the snare reverb.

To make a really tight overdub of a tambourine with a snare or other percussion sound, make a 10 second mono recording of someone shaking a tambourine continuously. Make a loop of the recorded tape and pass the looped sound through a gating Keypex triggered by the snare (or whatever) and re-record the gated tambourine sound on a new track. This is a very powerful technique -- making a loop and using a Keypex to gate the sound into the envelope of another instrument. Using the Keypex in this way turns the recording studio's console and accessories into a kind of synthesizer.

For automatic fading (very, very useful on projects where you have to do mixing tricks during a long fade requiring) just set the release to the longest possible time (to start with anyway), using the Keypex set to maximum range and with the threshold set so that none of the music falls out of the zero top window. When you push the key in button on a cue, like a tag drum fill (so all subsequent fades start at the same point), the automatic fade will begin. All completed fades will then start at the same point in the song, and will fade at the same rate as any other fade done at the current release time setting. Sometimes it is necessary to trim the release time pots so that the stereo fade is even left and right all during the fade.

10. You should limit the program while you are mixing it, rather than letting a future user like an AM radio station limit it and thereby change your mix significantly. If you are going to make a mix for a user who wants a small dynamic range, due to the conditions under which the music will be played (high background noise, etc), or because of limitations in the medium that will be used (cassette), then try to compress the product at mix time and thereby give yourself an opportunity to optimize the sound in consideration of the changes that will be caused by the compression. Compressing a mix changes the balance between the instruments. In addition, the compression invariably removes some of the brightness and roundness from the sound. Therefore you should always follow a limiter or compressor with an equalizer.

If a user compresses a mix that you have already compressed, there will be much less alteration of your basic sound than if the user compresses a mix that you provided at full dynamic range.

Projects that you should release at full dynamic range:

A. Product destined for record album release

B. Product destined for magnetic film sound tracks (first run movie) release

C. Live (mikes to tape) to digital recording

Projects that you should use peak limiting on as it leaves your studio (by tape or by phone patch):

A. Direct to disk

B. Direct to live FM broadcast

Projects that you should use compression on while mixing:

A. Light compression for cartridge duplicating master tapes -- for use in cars, etc.

B. Heavier compression for cassette duplicating master tapes -- for the noise problem

c. For optical sound tracks for movies

d. Duplicating master tapes for high background noise environments such as background music, trains, planes, etc.

PARAMETRIC EQ

Parametric EQ is the most powerful type of EQ. Filtering is the only equalization that is stronger. A parametric EQ has both variable amounts of EQ and variable frequencies at which the boost or cut is located. With a given frequency, you can smoothly go from full boost, to no activity at all (flat - no EQ), to full cut. Other parametrics have amount knobs, which go from minimum (off) to full (fully clockwise). For these types of EQs, a push button that makes the amount control either boost or cut is provided.

If you hear a boominess, for example, you can tune in the boomy range with a parametric EQ set for full boost, and upon finding the proper frequency range, convert the boost to cut and remove the boom. Remember that whenever using parametric EQs, there is a tendency to nullify the bumps and dips in the monitor system with the program EQs. This should not be done. This EQ, though making the audio sound good in one room -- the one you're working in now -- will make the program material sound lumpy in another room that has different monitoring lumps and valleys. Another thing is that if the control room where you are working has any serious problems with one particular frequency being absorbed by the room, any amount of EQ you pour into this absorbtive range will be eaten by the room acoustics. You will never hear enough of this frequency. If you find yourself adding lots of boost at a frequency where you know there should be some audio for the EQ to work on, but there is no apparent change to the sound, then you have found an acoustical well. Any boost poured into this well will be recorded on your tape, but will only be heard (to your dismay) in other listening environments, never in the present control room with its acoustic well.

The general rule of thumb "the mike requiring the least amounts of EQ is the best mike" really holds true for parametric EQ's. This is because as EQ's go, the parametric rumples the waveform the most.

The "natural" waveforms of five vocalists may blend quite nicely without EQ. If five mikes are used and each is parametrically EQ'ed to "sound great" by itself, the resulting "blend" when all the mikes are mixed may well be unusable.

If one engineer (perhaps over zealous with his console full of parametric EQ's) records a project, and another engineer at a studio with no parametrics mixes the project, there will be no way for the second engineer to undo the EQ damage done by the first engineer. Even if the second engineer had parametrics, the result to the music of being first wrongly EQ'ed and then "correctively" un EQ'ed later would be disastrous.

EQing a stereo pair (from stereo mikes) with two paramtric EQ's using different EQ on each mike channel can cause a stereo piano to sound horrible when listened to in mono, due to the phase cancellations that occur.

DIGITAL DELAY

A digital delay device is used to get echo -- precision repeats of an electrical signal that exactly resemble the original sound. The most important use of a digital delay is to get very short delays. Long delay times are easily gotten with tape delay. Digital delays are useful in that the delay time can be changed suddenly to a precise new delay time.

Phasing can be accomplished with digital delays in the same way as flanging is accomplished with tape machines.

*** Please see Advanced 5 -- Flanging ***

Use of some room sound with digital delay devices makes the sound much more unpredictable and therefore more interesting.

Entire "musical environments" can be "created" by the use of delay.

TAPE DELAY

Tape delay is most handy for long delay times. The almost universal use of tape delay is for making a delayed reverb send. To do this patch one of your reverb sends to a tape machine instead of letting it go to the chamber. Run the send levels a little hotter than normal, since you want to get good levels on the delay tape to prevent noise. The output of the tape delay machine is returned to a spare EQ in position, and that position is operated with the reverb send set to about 1 O'clock and is assigned to the chamber or the EMT, but with the stereo buss not assigned. As the fader is pushed up, the delayed send goes out to the reverb device. A VSO is used to adjust the time delay (slower tape speeds make for longer delays) so that the reverb "comes back" on or off the beat as desired. If more delay is needed, a two track can be used with track one being sent directly to track two and then the playback of track two is returned to the console.

"Straight delay" is also used where the tape delay is used in the mix, without sending any reverb. There are combinations of these two uses. When tape delay is used in the mix and for delaying reverb, use the pre function for the reverb send so that the reverb send level is independent of the fader's contribution to the mix of the "straight delay," or bring the delay back to 2 faders; one for the level of the delay into the mix buss and the other the amount of delay to be sent to the reverb device. It helps a lot when the reverb send for an effect is thus remoted to a straight line fader, which can be carefully marked at different positions on its scale, rather than using a rotary pot.

ECHO LOOPS

This is the term for intentionally causing feedback in a tape delay channel. There are two basic types of setups for doing this: returning the delay machine to one fader or to two faders. When two faders are used, they follow a mult; one is sent to stereo (the effect level) and the other is only sent back to the effect (the regeneration amount).

In the one fader method, the console channel used for the echo return, uses its reverb send to send back to the input of the delay machine (for regeneration). Set the echo return fader for the desired amount of echo return level, and then the reverb send (set to post) is set so that the proper amount of repeats occur at the effect's mix voulume. Now when a "rev" is required (this is antiquated terminology for a regenerative tape echo event), the tape echo channel on the console is pushed up to the mark (you marked the spot right?) and the delay will be heard along with the repeats (multiple echos). If you pull the fader down slightly from the mark, the level of the effect will be reduced, but the number of repeats will be reduced twice as much. By pushing the fader slightly above the mark, more repeats will occur than originally set. Of course, if you are willing to adjust the reverb send control on the echo machine's console return channel "on the fly" (during an actual mix pass), any number of repeats is possible. To change the number of repeats simply adjust the reverb send control on the delay return channel (which can be marked also). On "pre" the number of repeats and the volume of the effect will be unrelated. Having them related (on post), gives a rowdier effect. Pulling down the fader cancels the sound that was trapped in the regeneration.

TAPE LOOPS

A tape loop is usually performed with a quarter inch tape machine. A recording is made which is then connected head to tail so that the emulsion is continuous. The tape is then threaded through the two track (or mono). Have someone hold a plastic reel horizontally on a pencil so that proper tape tension is maintained. If another two track of the same type is handy, thread the tape around the plastic reel mounted on the other machine. The distance between the machines is then the adjustment for the tape tension. In this way one recorded sound can be made to repeat over and over as long as your patience holds out.

TUNNEL ECHO

Tunnel echo is a particular type of tape echo, which sounds something like the sound that would come back from a long tunnel with shiny walls. To make stereo tunnel echo you need a two track (a four track for quad).

*** Please refer to Illus. A4-4 -- The Tunnel Echo Diagram ***

The band goes to left and right in stereo. The lead guitar gets sent to the tunnel echo system via reverb send #1. Reverb send #1 goes to the left track; reverb send #2 goes to the right track of the two track. All reverb sends are selected to post (not pre) and are operated at "0" VU for good tape delay signal to noise.

As the lead guitar plays a "hot lick," which you want to send down the tunnel, push up the two tunnel echo faders; to clean up the lead guitar (remove the tunnel) just pull down the tunnel faders.

What happens is this:

1. The lead guitar signal comes from the console channel #20 reverb send #1 and goes to the mix and to the tunnel two track track (track 1)

2. The tunnel two track plays back the signal late as an echo to fader #25, which is being sent to the left side of the stereo mix

3. Fader #25 sends the late two track signal to reverb send #2, which sends the echo back to the two track (track two) for additional delay

4. The two track (track 2) plays back the signal late and sends the echo to fader #26, which sends it to the right side of the stereo mix

5. Fader #26 also sends this echo back to the two track (track #1) via it's reverb send #1 which starts the loop

The reverb send on module 20 should be set for "0" VU on reverb send VU #1. The reverb send on module #25 should be set for "0" VU on reverb send meter #2 when the tunnel echo is in proper perspective to the rest of the mix. The reverb send on module #26 should be set for the desired amount of repeats while the two faders #25 and #26 are in position for the desired amount of effect in perspective with the mix.

When the twin tunnel faders are pushed up, the note currently being played will be "trapped" in the tunnel by the action of fader #25's reverb send coming alive, and the number of repeats -- the tunnel time duration --will be determined by the setting of reverb send #1 from module #26.

For added weirdness of course, simply change the tape speed of the

TUNNEL ECHO Illus. A4-4

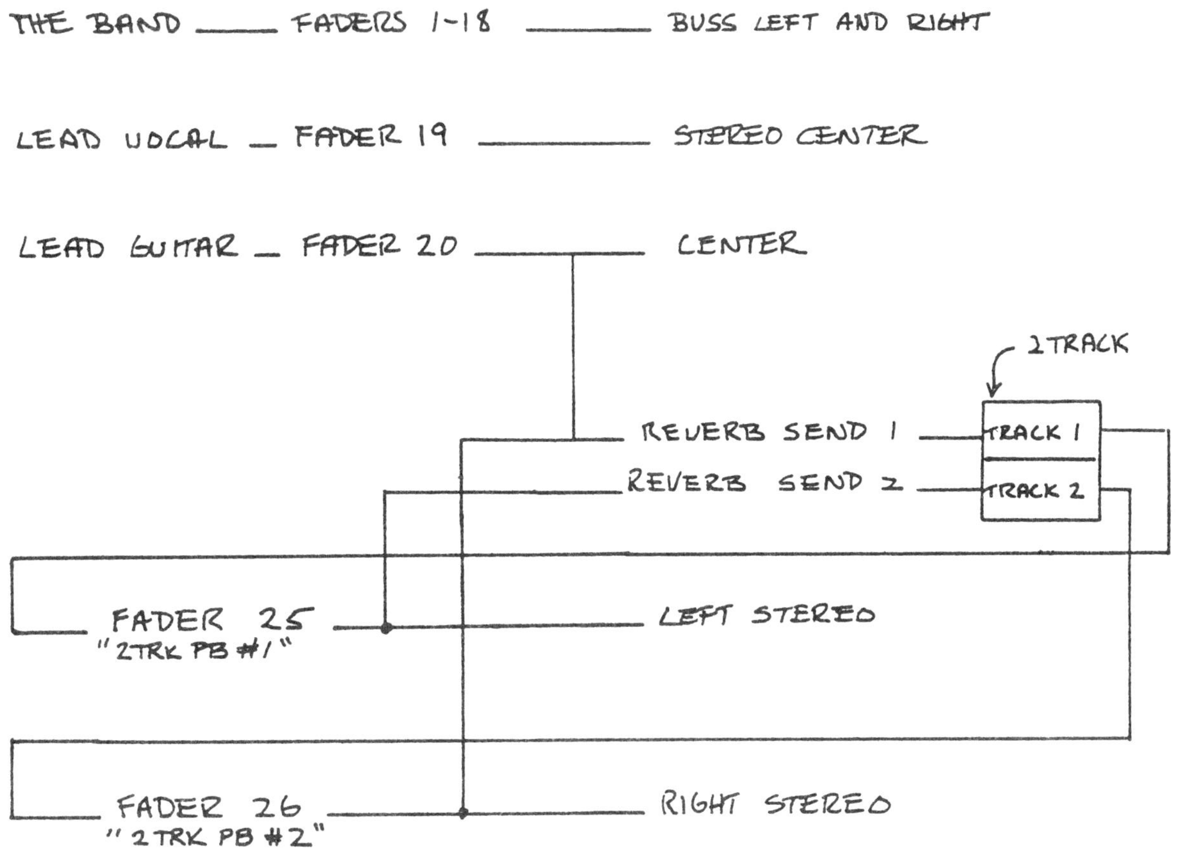

tunnel two track slightly (or greatly) with a VSO while there is lead guitar (or something else) trapped in the tunnel. Also patch a digital or tape delay into the return of the tunnel two track (track 2) (module #26) and add and delete delay while the mix is running.

THE EMT 250

This is a digital reverberation unit, which can perform delay, reverb, echo, echo loops, delayed reverb, delay with regeneration for comb filter effects, and space echo (reverb). The strong point of this device is that the reverb quality can be altered. The decay times for hi and low frequencies can be set differently, which can make the sound of several types of chambers. Space echo (reverb) is the sound of a mathematically perfect reverberation space in which all signals decay as slowly as possible; even though high frequencies should decay rapidly, they do not. It is a startling sound.

THE LEXICON DIGITAL REVERBERATOR MODEL 224

This amazing device provides (with proper options) the sounds of being in many different size rooms and reverb chambers, and the sounds of almost any imaginable reverb device. Many astonishing rooms can be "constructed" by use of the reverb time, crossover, treble, and initial delay controls.

DIGITAL PITCH CHANGING

Evantide's primary pitch shifting device is called the Harmonizer. This device provides delay, pitch shifting with regeneration, many voltage control possibilities and an anti-feedback function. Pitch shifts of an octave are possible, but the musical quality is not good at one octave up or down. At lesser intervals however, the device works well with good musical results. Here are some of the things you can do:

Generate parallel musical "harmonies" from single note recordings already recorded without having the artist "figure out" a harmony -- something that is sometimes very difficult for the artist to do, especially if he is not an accomplished musician and the original part just "came out." In this case, the artist may not be able to play the same notes ever again, let alone harmonize them for the record. This is especially true for lead guitarists, a prime reason to NEVER, NEVER do a "run down" for a lead guitar solo without recording the take. Even if you record way too hot, the performance may be irreplaceable, and even if the tape distorts, the guitar instrument amp is distorted to begin with. Lowell George always had his twenty four track machines set up with track #21 adjusted to +14 instead of to merely +3 just to have available a "guaranted tape saturation track" for the lead guitar.

The Harmonizer will always harmonize the same interval regardless of whether this is a musical interval or not. It is partly for this reason that the designer made arrangements to have a synthesizer keyboard input to the rear of the Harmonizer.

By setting up this synthesizer keyboard input system, a skilled keyboardist can set the Harmonizer to proper musical intervals as the solo progresses.

Some of the things the Harmonizer sounds great on are:

1. Electric keyboards, with the Harmonizer set nearly to 1.00 and a little regeneration added, makes the keyboard sound very present. If used to excess, this makes an excellent swirling special effect.

2. Vocals, with the harmonizer set to different delays on the two outputs makes the sound of many more persons singing than there are. The same holds true for strings.

3. Generating precision and musical wow by moving the pitch shift control as a part plays through the device, like the acoustic guitar on "Here Comes The Sun."

4. Lead vocal doubling, with the harmonizer set to nearly 1.00 and without any regeneration or delay makes the vocal sound fatter and clearer. This is similar to the technique of changing the tape speed slightly for a double but less effective, since the electronic Harmonizer "artist" is still singing exactly the same notes and therefore has relatively the same overtone series. In order to get the exact same benefit of VSO alteration of tape speed for a double with the harmonizer, you would have to pitch shift the cue send so that the artist heard the band at the "altered" pitch and sang at the slightly different pitch for the second track. This reminds me of one of the truly horrible tricks you can pull on a singer with the Harmonizer: patch the harmonizer into his vocal channel and fool around with the pitch as he attempts to get a hold of the song's pitch. This bit of pure devilment is reserved for your closest friends. Not for anybody with whom you wish to continue a working relationship.

DIGITAL OPERATIONS WITH MUSICAL NOTES

At this time I have to re-introduce you to logarithmic expressions. A logarithm, as you may recall, is simply the power that you raise the base to in order to get the desired number. The base number in audio is 10. If the desired number is 100, then the log would be 2 since you have to raise 10 (the base) to the second power to get 100 (the number). Logarithms are very powerful in the study of audio, acoustics, and music since all three of these studies relate to human hearing, and humans hear logarithmically, so to speak. The log notation formula is simply:

log (to the base 10) of the number N = X
10 (to the X power) = the number N

This means that for each number N, there is a log N = X. If you raise the base (10) to the power X, you get the number N.

O.K., now forgetting the formal way of dealing with logs, please get your calculator which has an inverse key (1/X), a log key (logX), an anti-log key leave log number system and return to regular numbers (10 to the X power). To get the log of a number, enter the number and press the logX key. To un-do a log, press the 10 to the X button. That's all there is to it. Here is the one formula that covers almost every case:

Ratio of dB = 20 x log of the ratio of volts or
ohms or
amps or
impedance or
NW/M or
distance

And one additional formula for watts (power):

Ratio of dB = 10 x log of the ratio of watts

These formulas bring to light exactly what the dB is: a ratio of one quantity to another. Here are some examples...

Volts

If your hi fi puts out 3 volts at the first "dot" on the volume control, and puts out 5 volts on the next dot up, how many dB is that?

Ans: ratio always means divide so we must divide the two numbers 5 and 3 to get an answer greater than 1 (we're going up in volts right?)

5/3 = 1.67

The equation says to find the dB change we have to multiply 20 times the log of the ratio of volts so

log 1.67 = .22
.22 x 20 = 4.44 dB

So you should mark the next dot as +4.44 dB.

Ohms

If you have a transformer that says 50,000 ohms primary and 250 ohms secondary, how much change in dB is there?

Well, there will be a loss since the volts follow the windings and there are less windings on the secondary so the ratio of ohms will be

250/50,000 = .005

The formula says 20 x the log so

the log of .005 = -2.301

20 x -2.301 = -46.021 dB

So there will be a loss of 46 dB through the transformer. This is a typical type of direct box calculation by the way.

NW/M

NW/M ("Nano Webbers per Meter") is the measurement of tape recording signal strength (magnetic strength). "Old" Ampex zero level is 185 NW/M. If a client from Amsterdam or some other studio brings you a tape with no tones on it, but the box says, "Recorded at 300 NW/M," will you know what level to set up your recorder to? Here's how to figure it out:

The change in the NW/M is

300/185 = 1.622 times "hotter recording level"

20 x the log of the change is

log of 1.622 = .210

20 x .210 = 4.199 dB

This means that your visitor's tape is recorded at plus 4.2 dB over old Ampex zero. If you have a zero level tape at 185 NW/M, then you can procede to align for their tape by setting the playback tones at -4.2 dB throughout the alignment and then set the record circuitry for zero on the machine's VU meters.

Distance

How much would you expect the level to drop if you moved a mike back from a distance of 1 foot to 3.5 feet? You cannot use the inverse square law explained earlier since this is not a distance movement of increments of twice as far. Instead you can now use the new formula:

First find the ratio number. The distance is increasing, so the volume will be less

1/3.5 = .286

now find the log

log .286 = -.544

and now the dB

20 x -.544 = -10.881 dB

So the loss in volume will be -11 dB on the VU meter.

Watts

Finding the dB related to watts or power is the same procedure except that you have to remember to multiply by 10, not 20.

If your transmitter goes from 300 watts to 112 watts, how many dB would that be?

The watts are reduced so you have to divide the two readings to get a number less than 1

112/300 = .373

The formula says 10 x the log of the ratio so

log .373 = -.428

10 x -.428 = -4.279 dB

The loss in watts expressed in dB = 4.3

DIGITAL OPERATIONS WITH MUSICAL NOTES

Now on to the logarithmic expression of pitch. The distance from the note A 440 to A 880 is not evenly divided up with the 12 tones of our musical scale. The first few notes take up only a small portion of the frequency distance, with the last few notes being much larger steps. This is the same way that logarithms work. So we can relate musical intervals to pitch shifts by converting musical intervals into logarithmic intervals. The Harmonizer reads out in pitch ratio numbers, so we must convert back to non-log ratio numbers before we dial in our answers to the readout of the harmonizer. Since we have our handy dandy calculator, this is quite easy to do and really is only a matter of remembering a series of button pushes.

The musical octave is divided into twelve notes. The ratio of one note to the octave note above it is 2:1 or two to one. The octave above A=440 is A=880. The log of the ratio of a one octave jump is

log 2 = .301

So a single note's worth of pitch jump (in log notation form) is one twelfth of that

.301 / 12 = .025

If one note's worth of jump is .025, five notes jump for example would be

5 x .025 or .125

Now to use this information to get from the frequency of one note (A=440) to the frequency of 5 notes higher

log 440 = 2.643

log of a pitch change of 5 notes up is

5 x .025 = .125

So the log of the note that is five notes above A = 440 is

2.643 + .125 = 2.768

Now to find out what note (frequency) this is we must exit from log notation with the 10 to the x key

10 to the power 2.768 = 586.138 Hz

586.138 Hz is the frequency five notes above a = 440.

Here's the important point. To find the new frequency add the pitch change (in log notation) to the original frequency (also in log). But to find the Harmonizer setting, add the pitch change (in log) to the Harmonizer reading (in log form) when the Harmonizer is at normal pitch (at normal pitch the Harmonizer reads 1.00).

Working the above example for the Harmonizer we first calculate the log of the pitch shift:

5 notes at .025 = .125

log of starting number (1.00) = 0

0 + .125 =.125

The antilog of .125 (just use the 10 to the x key) is 1.334.

To make the Harmonizer shift pitch five notes up, use 1.334 in the Harmonizer window.

To make a 60 Hz VSO shift 5 notes up where 60 Hz = normal tape speed:

log of 60 = 1.778

1.778 + .125 = 1.903

antilog of 1.903 = 79.983

To raise the pitch of the tape by five notes use the VSO oscillator set to 79.983 (80) Hz (remember 60 Hz is normal tape speed).

To go down in pitch just take the log of one note's pitch shift x the number of notes desired and subtract this log from the log of the starting unit (whether it's log is based on 1.00 or 60 or 100% or 15 ips or whatever).

Eg. what should I set the harmonizer to to get 7 notes lower in pitch?

7 x .025 = .175
log 1.00 = 0.000
0 - .175 = -.175
antilog of -.175 = .668

To get 7 notes lower using a Harmonizer use a window setting of .668

The Ampex VS-10 variable tape speed accessory indicates 100.0 when the tape is at normal speed. To get 3 notes lower tape speed you would set:

3 x .025 = .075
log of 100 = 2
2 - .075 = 1.925
antilog of 1.925 = 84.140 (84.1)

To slow the tape down 3 notes on the piano (to be played three chromatic keys lower) use 84.1 on the Ampex VS-10.

How many ips is 4 notes up from 15 ips?

4 x .025 = 0.1
log of 15 (ips) = 1.176
1.176 + .1 = 1.276
antilog of 1.276 = 18.880

Four notes above 15 ips is 18.880 ips.

PHASING WITH TAPE MACHINES

In the days before digital phasers (the first Evantide Phaser was called a Flanger) phasing had to be done with tape machines. The sound of tape machine phasing is still the best as far as I'm concerned. Tape machine phasing can be in mono, stereo, or quad. You must use a mono, two track, or four track respectively to get the effect.

For example, let's use stereo phasing on the stereo room tracks of a live 24 track studio session tape.

1. Turn the tape over and play the tracks (now backwards) in sync mode

2. Patch the two stereo room mike tracks (subtract the track you want from the number 25 to find the tracks while upside down) to a recording two track machine, which is properly lined up of course

3. Set up to record on two open tracks on the 24 track from the two track. The two track will be delaying the sound due to the two track being in the record and reproduce mode

4. Record the two track delayed audio back onto the 24 track on the new tracks (count the tracks carefully)

5. Turn the 24 track tape right side up and patch the playbacks of the newly recorded advance tracks (delay with the tape recording upside down makes advance tracks) to the same recording two track for re-delaying

6. Balance the two left signals (the original room left and the advanced now delayed room left) and the two right signals so that they are at exactly the same volume

7. Just before the phasing begins in the mix, bring up or un-mute the advance tracks and have your assistant put just the smallest amount of drag on the supply reel of the two track (note: you cannot use an ATR-100 for this due to this machine's refusal to allow the operator to slow the tape down by

dragging on the reel flanges). For the luxury method, use a VSO on the two track to "tamper" with the tape speed.

8. Correlate the amount of drag (and resulting phasing) with the musical events taking place.

The technique is the same with a four track or a mono.

It is important that the amounts of the phasing and original tracks be exactly equal, for only then will startling amounts of phasing occur. To be certain of this, calibrate the console submasters, and use the dummy buss technique.

Phasing with an electronic phaser is usually done by an internal lo frequency oscillator, by foot pedal (quite useful if your hands are very busy mixing), and by a rotary knob at the front panel. All phasing is in mono.

FLANGING

Flanging is the same thing as phasing except that it is caused by someone dragging their finger on the flange of a reel. The Evantide Instant Phaser has replaced the Flanger. Instead of just having swish/wish phasing at a constant rate (with the oscillator), or a pedal or knob controlled phase, the phaser is trigged by peaks in the volume. You can adjust the threshold and release controls so that at a given volume a complete flange cycle commences. The release time sets the speed at which the cycle of phase "in" and "out" occurs. If you adjust these two controls properly, you can get the phaser to phase at optimum musical events without having to supervise it. With some modification it should be possible to make the envelope follower "follow" a different input signal than the signal being phased by the phaser (like a key in). If you work this out (or have an electronics shop work it out), you can get precise timings on the phasing since the presence of any kind of signal at the "key in" will make the phaser cycle. A 400 Hz console tone burst recorded on a spare track can be used or a recording of the word "now" or anything so long as it is patched to the "envelope input" or special modified "trigger" input.

Be aware that this modification will probably void any warranty on the device

ENVELOPE GENERATORS

*** Please see Illus. A5-1 -- The ADSR graphic supplied ***

There are six controls on some types of ADSRs (attack, decay, sustain, release). Two are for volumes, and four are for timing the events.

Volume:

1. The volume sets the overall volume of the whole envelope.

2. The sustain volume sets the volume of the sustain portion. If the sustain is set wide open, the sustain is as loud as the attack itself.

Timing:

1. The attack adjusts the time required for the waveform to rise from dead soft (or the current volume if it hadn't released yet) to full volume as set by the volume knob.

2. The decay time adjusts the time from the moment when the envelope

A D S R DIAGRAM

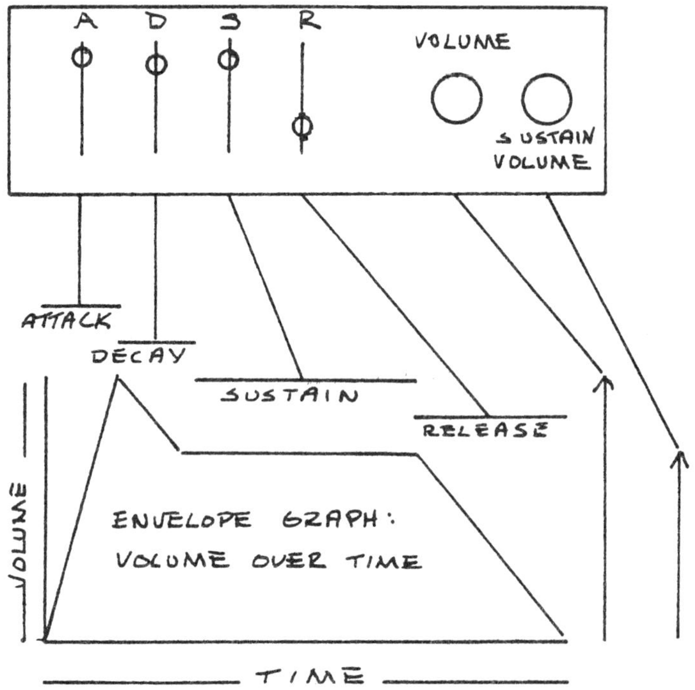

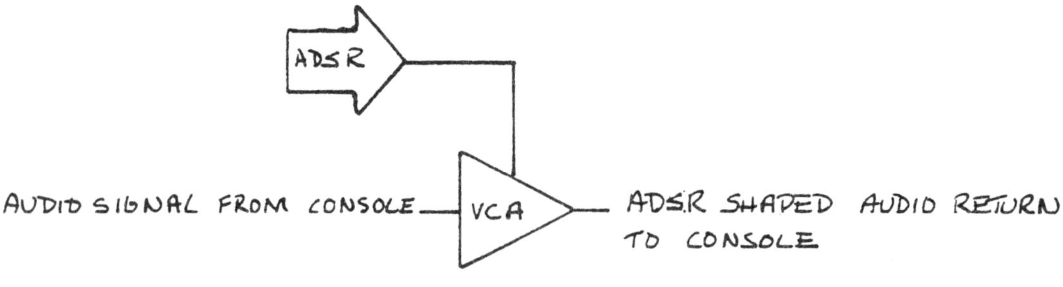

reaches its peak level (at the conclusion of the attack phase) to the time when the envelope fades to meet the sustain volume level (set by the sustain volume).

3. The sustain time sets the time during which the level rides along relatively steady. If this is set to the maximum short setting, the sustain ends the instant the keyboard note is released. If it is set too long, the note will "sustain" for perhaps 2 minutes or so.

4. The release time is the time it will take for the sound to fade away after the sustain time has passed.

There are a few other methods of using these controls and there is no real convention between synthesizer makers. Any careful reading of the synthesizer directions will yield satisfactory results from any major synthesizer maker.

The hookup for a synthesizer/console use of the ADSR is shown. The ways to tell the ADSR when to begin to create an envelope are shown to the left of the ADSR block.

1. Sample-and-hold -- a device that puts out a pulse of DC whenever a signal goes by that the sample-and-hold can "hear." The sample-and-hold has controls of its own but this is, afterall, a book on audio engineering, not synthesizers....

2. Keyboard control voltage -- anytime you press any key on the keyboard the keyboard trigger voltage will start an ADSR event.

3. A simple switch -- just a one volt battery with a momentary contact switch (push button) will trigger the ADSR.

When the ADSR gets a "begin" signal, it pours its control voltage into the control voltage ports of a voltage controlled amplifier, whose gain for the audio passing through it will then be varied to resemble the shape of the envelpoe delivered to it by the ADSR. When the ADSR gets through its release, there will be no more control voltage coming from it and the VCA will be fully off. The VCA will go to full gain whenever the ADSR gives out its full voltage.

Now just patch some audio to the VCA and it will come out of the VCA with an envelope "imprinted" on it by the settings of the ADSR.

RING MODULATORS

A ring modulator is similar to the ADSR hookup as described above except that instead of an ADSR to adjust the gain of the VCA, a tone is used. The tone is selected to produce "ringing" or bell-like sounds. Ringing sounds are produced when the tone doing the volume "gating" has a musical relationship to the key of the music going through the VCA. Especially pleasing to me is what happens when you put a major third tone (four piano notes higher) than the key of the music into the control voltage input of the VCA, and then put an instrument's sound through the VCA. The more sustain the musical instrument has (like a piano), the more ringing happens. For each different note played, different overtones and harmonics come forth due to the action of the ring modulator.

Some ring modulators (not associated with complete synthesizers) come with an internal oscillator. This makes it handy to use. In any case, if the key changes, the frequency of the oscillator must be changed to preserve the current musical effect.

THE APHEX

The Aphex is a misunderstood and secretive device. Basically, the Aphex is supposed to clip, distort, and phase shift the music in such a way that the music "sounds sweeter" to the ear. It is supposed to have come about accidentally while the "inventor" was trying to build a phono pre amp. If well used the Aphex does improve upon music recorded with little or no ambience, as is commonly done today through the over use of "separation" techniques and lots and lots of baffles. Music that is recorded with little or no sound baffles, and with careful aiming and careful use of mikes in a reasonably good sounding room, should sound Aphex-like to begin with.

VOCORDERS

The Vocorder is like the envelope generator except that the envelope shaping is done by another sound, not an electronic timing device. The audio goes through the device in what is known as the "carrier." The controlling input (which has no output) is called the voice input. The voice dynamics (envelope) is then imprinted on the carrier instrument. A guitar plays the melody of the song while the "singer" reads (not sings) his part. The guitar is patched to the carrier input and the carrier output goes back to the console. The spoken word audio is patched to the voice input. The controls are then set up so that the resulting carrier output is a "singing guitar."

The simple ADSR-VCA-console hookup diagram is useful in thinking about the Vocorder. In addition to the envelope following functions of the Vocorder, there is also a hiss and consonant detecting system in the voice input. Whenever there is a hissing or consonant (guttural) sound present in the voice input, the Vocorder will supply a similar sounding hiss or consonant sound at the output mixing stage of the Vocorder. The relative levels of the hisses and consonants in relation to the carrier audio is set by means of mixing controls.

Vocorders are wonderful devices and produce amazing sounds. Some of my favorites are:

1. The singing synthesizer trick
2. The bass flavored kick drum trick
3. The electric piano flavored kick drum trick
4. The bass flavored tom-tom mix trick

PHANTOM SUPPLY

The phantom supply works between the mike input transfomer and the microphones.

*** Please refer to Illus. A6-1 -- The Phantom Supply Block Diagram ***

The phantom "power" can come from a battery or more likely, a power supply (with a spare phantom supply backup). The minus of the battery goes to the ground of the mike input system and thereby goes to all the mike cable shields by way of all the pin 1's on the XLR. The positive volts go to each and every mike signal wire (pins 2 and 3 of each mike input XLR). The +48 volts go through a 6,800 ohm resistor on their way to each mike signal wire. This is partially because if you plug in one bad mike (with its cables shorted pin 1 to 2 to 3), the short will not cause the phantom power supply to "fold" or malfunction.

If you will recall the discussion on phantom supply given earlier in Intermediate 7, the presence of the DC volts on the mike wires will pose no problem since the volts will be pushing on both hi and low wires of the microphone and console input transformers equally, thus causing no electron flow in the console or microphone transformers. If the resistors used (the two 6,800 ohm phantom resistors) are not exactly equal in value however, then there will be a flow of DC in the mike signal wires, which will "eat up" some of the headroom in both the mike and console transformers. The presence of DC "trickle" in the phantom supply voltage, due to an unbalance in the volts applied to the hi and low wires (by there being different value resistors delivering the volts), will cause loss of headroom and a discoloration of the sound. If either of the signal wires is shorted to the shield, then the full 48 volts will flow around and around the mike input wires for XLR pins 2 and 3 causing gross distortion problems. It is partly for this reason that no unbalanced microphones should ever be used in the studio (because this kind of mike will likely short mike pin 3 to the shield).

From the phantom microphone's point of view, there is a free supply of DC coming in on its hot wires. All the mike need do is to put a pair of resistors calculated to reduce the 48 volt supply to the required number of volts needed by the little mike amplifier. These mike resistors are chosen carefully to be a perfect match, like the resistors you put in your console for supplying the phantom volts. So, the mike picks up its volts and turns on its little amplifier, which then processes the mike element's signal and gives it to the mike output transformer. The fact that there is DC on the mike's hot wires is no problem since many, many kinds of electronic signals can be run on the same physical wires.

From the console's point of view, the DC it is sending out causes no interference to its ability to process the push-pull audio coming in from the mike. Both transformers do their jobs of blocking all electrical signals common to both wires (common mode rejection) and act as a magnetic barrier so that no outside signal wires are ever actually attached to the console -- they only can transmit autio through the action of the magnetic barrier.

One thing though, you will want to include a way to connect the phantom volts only to those mike inputs that need it. For ribbons and dynamics, the phantom volts should really be switched off. Very few consoles

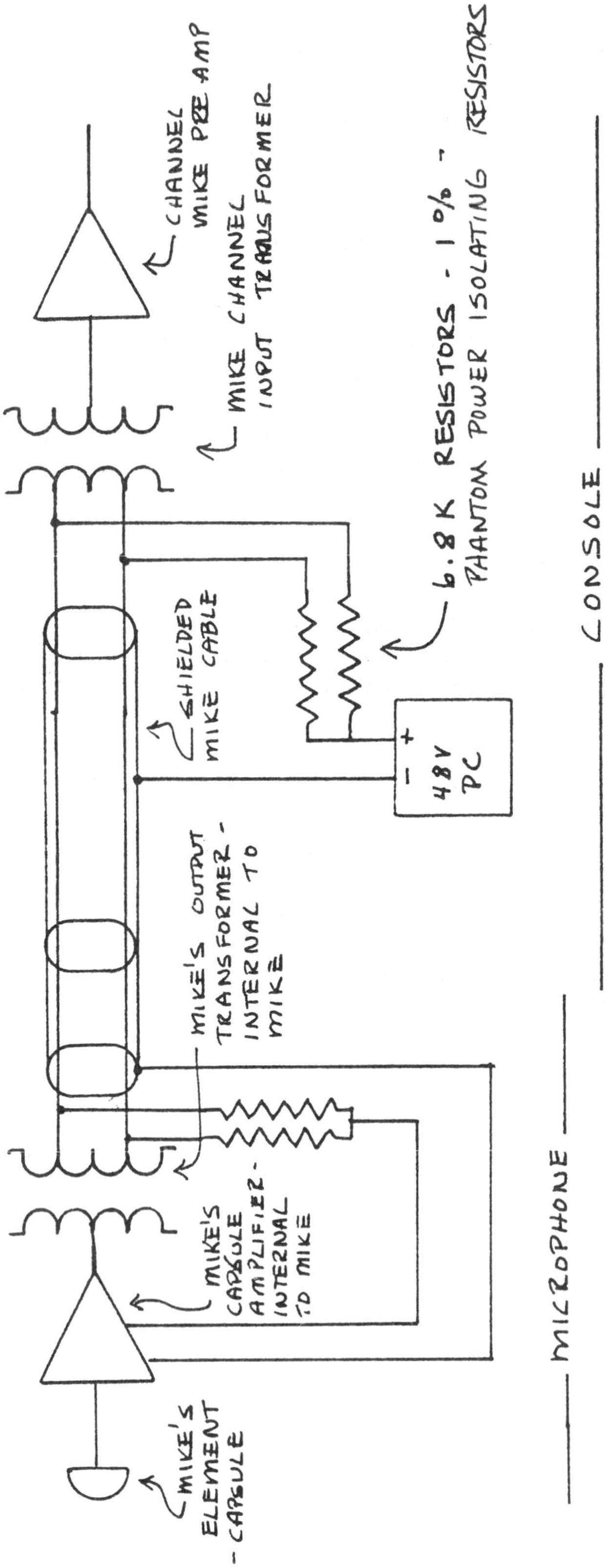
CHANNEL MIKE PRE AMP
MIKE CHANNEL INPUT TRANSFORMER
6.8 K RESISTORS - 1% - PHANTOM POWER ISOLATING RESISTORS
CONSOLE
SHIELDED MIKE CABLE
+
-
48V DC
MIKE'S OUTPUT TRANSFORMER - INTERNAL TO MIKE
MIKE'S CAPSULE AMPLIFIER - INTERNAL TO MIKE
MIKE'S ELEMENT - CAPSULE
MICROPHONE

provide this fine point however.

TACH SERVO SYSTEMS

This is a system of comparing an ideal event rate with the actual event rate and by the comparison, adjusting the actual event so that it occurs on time. Many audio tape machines, both higher quality hi-fi as well as professional audio varieties, use tach servo systems to monitor their capstan velocities to assure good tape speed accuracy.

A tach servo system is really only a very accurate variety of cruise control as is used in automobiles. If the car begins to slow down for any reason, the cruise control gives the car more gas; a servo system gives the capstan motor more volts.

Associated with all tach servo systems is some kind of tach. A tach is short for tachometer, which in this case is a strobe wheel that has stripes around its perimiter so as to stand still when the medium (tape, disk, etc). is moving at the proper speed. In the case of the tach for a servo system, you do not observe the strobe wheel. An electronic counter continously counts the strobe markings and compares the number counted with the reference frequency coming from a very reliable frequency source. If the number of strobes counted is correct, the capstan motor gets the same amount of voltage as it did in the last counting period; if the number of strobe counts is too few, the capstan is starting to run slow and the servo system gives the capstan motor more volts to speed it up. All this happens very rapidly; it must or else the human ear would detect the speed changes. Tach servo speed changes are minute because the speed at which the system makes samples is so extreme.

To adjust the tach servo system on the Ampex 1200 perform these steps:

1. Put the tach servo card (the middle card at the near left side of the machine's top panel) on its extender
2. Set the duty cycle with the proper pot (see owner's manual)
3. Now set the overall gain for minimum tape speed flutter and minimum jitter of the square wave on the oscilloscope. Measure the flutter with a suitable flutter meter.

DISTORTION OF MICROPHONES

Microphones usually distort because they are in sound fields that are too strong (music too loud), or because they have been abused. Abuse of microphones is really to be frowned upon since it is possible to do session work without damaging any equipment especially microphones.

Microphones are especially critical since even the tiniest noises coming from the mike will take the ride up in level provided by the mike pre amp and become painfully loud. A mike that is partially defective or intermittent is worse than having none since intermittant microphone problems always seem to show up whenever the "heat's on" with your client listening.

In a tracking session, the only thing worse than interruptions due to bad mikes or cables is an interruption due to poorly maintained headphones. Any problem is bad enough, but discovering that your mike supply was semi - destroyed by the visiting producer / mixer who was working in your studio a week ago is really the worst. Find mike and head phone problems before the client gets to the studio using the set up procedure given in the first few sections of Basic engineering.

Mikes get damaged in these ways:

1. Being hit with a stick -- drummers
2. Being dropped -- carelessness or rushing
3. Being towed or pushed over while on a mike boom and hitting something -- poor mike cable routing or mike setup
4. Being used in excessive sound pressure areas, which stress (and damage) the mike element -- incorrect mike choice or placement
5. Getting rained on -- mother nature at outdoor remotes
6. Getting a chemical goo infestation - allowing smokers to direct their smoke at the mikes along with atomized tobacco infested saliva while singing

Overloading condensers can damage them partially or completely ruin them. Sometimes the polarizing voltage can arc between the diaphragm and the backing plate, which ruins the capsule.

Overloading dynamics can damage the little bobbin of voice coil wires. If the little element is driven down into the "gap" far enough, the bottom of the voice coil may impact on the bottom of the gap, and thus distort the edge of the bobbin making it rub the sides of the gap, causing horrible distortion or perhaps a complete freeze up.

Overloading ribbons can cause them to become overstressed. If they get stretched, the ribbon will not have its proper springiness and will distort sound or just go dead.

DISTORTION OF CONSOLES

There are numerous ways for consoles to distort:

1. Incorrect voltages coming from the power supply -- everything is distorting; all the active components of each system are distorting early (at surprisingly low levels) because there isn't enough voltage to drive a full sized waveform -- sometimes this problem shows up because the console indicator lights dim
2. Low phantom supply voltage -- non-phantom mikes sound O.K.; only

phantom mikes sound weird

3. A short between a mike input hi wire and a console shield -- the particular mike input is horribly distorted while other phantom mikes are O.K.

4. Mike transformer distortion -- using a really hot mike, or a supposed "mike" level output from a synthesizer that is hot enough to drive the mike input transformer into distortion

5. Mike pre distorted and not gain adjustable -- gain/pad switch is broken so that the poor mike pre is stuck on full or near full gain even though the gain/pad switch clicks and turns O.K.

6. All audio sounds smeared -- possibility of high frequency oscillation in the console that is using up some or most of the console's headroom

The list could go on and on. The point is, that you must check out various possibilities before deciding on the cause, extent, or nature of the difficulty. Your reaction, once properly oriented to the trouble, must be swift and in proper proportion to the size of the problem.

There are unlimited problems and solutions. It would not be proper to try to imagine them all here. Once you start doing sessions experience will be your true teacher. Before then, take the best advantage of your apprenticeshp by studying mixers doing their work. Ask as many questions as you can get away with for now, while your prestige is not at stake.

DISTORTION OF THE TAPE MACHINE

There are many problems that can cause you to have distortion introduced by the tape machines. Here are some of them:

1. Running into a batch of bad tape -- you can expect to find a piece of bad stock once in a while; the maintenance man may discover it while doing the alignment, but if he doesn't, you will. Whenever you can, always listen to playback while recording tracks. If you can't, always check your recordings by doing check playbacks (doing this requires at least twice as much studio time).

2. Using the wrong bias setting -- this can occur only if the previous tape machine users have not left an alignment sticker on the deck clearly showing what tape-type the machine is set for or if a person were to change a bias setting and not tell anyone. In rare cases, the master bias amp may drift to a higher or lower gain. If the amp goes higher, high frequencies will be reduced; if it goes lower, recordings will be brighter and distortion will occur.

3. There are magnetic particles in the record or playback head gap -- the only way to get the gap really clean is to lay a Kim-Wipe soaked in Ampex head cleaner (not to be used on certain types of heads -- use denatured alcohol instead) over the face of the head and then demagnetize the head with a strong head degausser. This "knocks" the particles out of the gap, where they get stuck to the Kim-Wipe. This technique should be done at least once a day and several times after having the heads relapped.

4. Attempting to record too hot -- recording "hot" is O.K. as long as you know where your peak levels are and monitor the machine in repro mode while recording. Using peak meters is necessary to know what your peak levels are. When using VU meters, you will have to guess where the peak

levels are, and sometimes you will guess wrong.

5. Lining up the tape machine without noticing that it and another machine are plugged into the same console playback system. Two machines to one playback system - this will cause the alignment to be performed at the wrong levels and EQ.

DISTORTION TYPES

1. There is magnetic distortion, which is called saturation. Saturation happens in transformers and in magnetic tape recordings. It softens the transient quality of the audio and makes the sound duller. Saturation does not produce square wave material the way solid state devices do when they distort.

2. Solid state distortion is very simple to explain: when the waveform attempts to rise above the maximum level, the solid state device simply shears off the top and bottom of the waveform. Sometimes spikes are added to the leading edge of the upward and downward waveslopes.

3. Tube distortion sounds quite beautiful if not overdone. Tubes distort differently than do transistors. Tubes distort by making the waveform have round lumps rather than square tops and bottoms. This type of roly-poly distortion is called soft distortion. Slight amounts of soft tube distortion sound better than the undistorted signal sometimes. This phenomenon is true especially on sibilant vocalists, horns, bass drum, and bass guitar. It is worth every good engineer's while to get a portable rack made consisting of perhaps 6 old professional tube mike preamplifiers powered by a really old tube power supply. The "soft supply" is almost as much as a requirement as the tube preamps themselves. You will get a nice surprise if you use tubes on instruments that seem to require mandatory EQ to sound full or beefy. After trying to equalize some types of bass guitar to sound huge and giving up, it always gives me a thrill to patch the mike to a tube preamp, and just bring it up through a fader with no EQ and have just the sound I was trying to get and more. For the full effect of the "tube experience", patch: mike out to the tube mike pre in to a tube limiter/line driver in (LA-2A) to a tube EQ (Pultec or Lang), then direct to tape using the machine's VU meter for the level meter. LA-2A's make excellent "line amps" for the mixing buss if your project can stand a dB or so more noise due to the tubes in exchange for the tube sound.

4. There is phase distortion by equalizers (including the pre-emphasis EQ of most tape recorders). Most EQs distort the phase; that's how they change the relative levels between the highs, mids or lows. When EQ is applied to audio signals, the tone parts of the waveform do not become distorted. They only get larger or smaller if the frequency of the tone falls within the effective frequency range of the EQ. The more ragged (or square wave like) the audio waveform becomes, the more distortion is caused by the effects of EQ. The square wave gets altered the most with EQ. Using very powerful EQ such as 1/3 octave room EQ or parametric EQ can easily change a square wave into a sine-like wave.

There are two basic types of distortion: harmonic and

intermodulation.

5. Harmonic distortion is caused by the distortion of the waveform by amplifiers. Tube amplifiers produce even harmonic distortion; transistor amps produce odd harmonic distortion. Of the two the even harmonic distortion is the most musical.

6. Intermodulation distortion is caused by poor amplifier design. What happens is that the presence of one audio signal in the amplifier's input will distort the waveform of another audio signal also passing through the amplifier. This means that if you were to pass a simple tone type waveform (like an electric organ playing only one note) through an amplifier with high intermodulation distortion, it would sound O.K.; but if you put a full range instrument with many notes present at once (like the piano) through the same amp, it would have the smeared intermodulation distortion sound. Odd harmonic distortion will be heard as a frizziness; intermodulation distortion will sound cloudy and dull as well as smeared.

Professional audio equipment that is working properly will not distort noticeably in any of these ways.

7. Polarity distortion is caused by a phase reversal in the audio system somewhere. What polarity distortion does is make the final product out of phase with reality.

An example is when the kick drum is played, a positive or pressure wave front is created in front of the drum head. If a mike is placed in this position, the mike should detect a positive or pressure wave. The mike should therefore send a positive voltage down the XLR connector on wire #2 and pull voltage (negative) on wire #3. The console should keep this polarity correct all the way through to the tape machines, the cue amps, the reverb sends, and to the monitor amps as these are the main console outputs. All the console outputs should go positive for the kick drum event. If the polarity is maintained right through to the speakers, then the woofers should move outwards towards the mixer, not backwards into the cabinet, for the polarity to be proper. As far as the tape machine goes, it wouldn't really matter if the tape got recorded out of polarity (ie. which way do the north ends of the tape molecules face - up or down the tape path - for a pressure wave hitting a mike) as long as the same machine played the recorded event back in polarity so that the speakers again moved outwards. This, of course, does not take into account any lacking of polarity considerations by the designers and builders of professional or other audio recording equipment that will come into play when changing machine types after tracking and before mixing. Making a recording on one manufacturer's 24 track and playing your tape on the machine of another's does not at this time guarantee correct polarity. There is now just beginning to be a growing awareness to the proper maintenance of polarity.

To keep polarity correct for your projects, simply record some "clicks" from a polarity test kit on the 24 track machine during the initial session where the project tones are recorded. By testing the output of any subsequent 24 track or 2 track machines that are reproducing your "test clicks", and by testing the click polarity coming from the monitor systems while playing your test clicks in any room that you do mixing or overdubbing in, you will be able to assure correct polarity maintenance.

A concern for polarity is just now being seen in professional recording. Sometimes you will see a notice on a direct to disk or other carefully prepared album saying, "When you play this album, please turn your stereo speaker leads around so that the + amplifier wire goes to the - terminal and vice versa." This is a polarity correcting effort (a shot in the dark really) on the part of the engineering staff who are responsible for the album.

THINK BEFORE SPEAKING

This saying is especially true for a person who is supposed to really know what's going on. Many times you, as the engineer, will be the only one who is sure that the project will turn out O.K. and that everything is going to be all right. For you to open your mouth and put your foot in it is extremely harmful to maintaining your credibility and respectability. The best time to speak authoritatively is when you have seen through a problem confronting the project or when you have realized a difficulty (far enough in advance to prevent it I hope) that no one else has yet realized. By doing this you will cause a flurry of excitement and amazement centered around you. This is good for you and the project.

Don't make rash statements without being sure that you're right. Always have a reason backing your pronouncements. You need the aura about you that you know what you're doing, and that you know what you're talking about. If you're careful, then you can stick by your guns, secure that you have it together.

ENVIRONMENTAL IMPACT AND CONTROL

The physical appearance, atmosphere, and lighting of a studio or control room have a definite affect on the persons working therin. Red light makes people angry. Green lights make people mellow. Flourescent lighting makes people jittery. A grubby mess makes the clients feel that being grubby is O.K. A nicely kept, neat work area is highly recommended. If some grubby guitar player begins to butt out his cigarettes on the polished control room floor, have your assistant follow him around with a neat looking ashtray until he gets the message. Cigarette smoke is really very harmful to electronic equipment especially contacts like card edge connectors, and all consoles and outboard racks should be positively vented so that air conditioned filtered outside air pours out of all the cracks and slots in the equipment preventing smoke from entering. Discourage blowing of smoke towards consoles and tape machines.

Use candles for an elegant lady singer's vocal overdub. Use a small table stand for wine, water, etc. Use nice rugs.

CONTROL ROOM ETIQUETTE

There is a definite etiquette you must observe when working. Never:

1. Put down your studio
2. Interrupt or argue with the producer except if you have a long relationship with him and then only in private
3. Have a musical opinion about the song being worked on unless you ask if you can have one (until you get to know the artist and producer very well)
4. Produce the recording except through the producer (by telling him your ideas)
5. Have somewhere else you have to go when the artist or producer feels like working late in the studio even past the usual stop time
6. Argue with visiting outside engineers about how they want to do some audio operation in your studio unless you know that your equipment won't work their way, and then explain the situation in private

7. Let non-mixer persons ever touch any audio equipment or tape without an immediate instruction to the contrary by you
8. Answer for anyone in the band to the producer or vice versa

CONFIDENCE

You must work very, very hard to perfect your skills such as:
1. Editing tape.
2. Figuring out tricky console troubles or techniques that may have gotten out of hand, without showing any alarm or causing any noticeable delay
3. Working very fast
4. Putting on a "show" so that it looks like you're working hard on the engineering even if you're not working that hard. Many times a producer will feel slighted if the engineer seems to be underworked. The sound is the thing though, and when you develop a clientele, you can relax and make your seeming at ease-ness seem important.

It is only when you really are skilled that you can have true confidence. There is nothing as sad as observing a person who seems confident, but does not have the experience or skill to back it up.

If you have a serious doubt about your being successful at doing some proposed project or operation say, "I don't know how that's done" if you feel that it's true. The most difficult problem some new engineers face is not being able to call for help. A timely call for help is much better than a disaster caused by an engineer who didn't admit that he had gotten lost.

BEING AUTHORITATIVE

Simply stand firm on subjects that you know; say, "I don't know" when you don't, and your authority will never be questioned. You must try to get the hard and fast rules spelled out at the studio where you work so that you can put your foot down at the proper times. The most common of the authority testing adventures will be over whether the client can take his tape with him after the session. In this example, if the work order says C.O.D., then it's cash or no tape, period. Not releasing the finished 2 track mix tape to the client before he pays his total bill is a very important authority consideration.

The studio employees must be properly trained and then the management must stand firmly behind the employee from then on. If the employee makes a blunder, the corrective actions with the employee should happen in private, not in front of a client.

EGOS

Artists of all kinds need their egos desperately. It is necessary for an artist to have a healthy or slightly beyond healthy ego. Build up the ego in your artist. Be subtle. If they feel secure, they will be much more creative. When something isn't going well with a vocal, for example, don't say somthing disheartining. Somehow, always make what you have to say come out sounding positive. Always use the ploy, "I liked that one (last overdub or whatever), but I know you have a better one in you. We'll save that one though." This gives the artist a "saved" performance to "beat," which is a good project for him. In addition, if fatigue or disgust (also a possibility) overcomes the artist, you can always fall back on the overdub you saved.

Use the ploy of "adjusting" or replacing the mike after a take that lacked luster of some sort from the artist. If you seem to be trying hard to get a better" sound, the artist will probably be encouraged to keep trying for a better performance, which will have your now "improved sound."

Never joke around like, "That's getting better, let me know when you want to start recording" when the artist thinks that you have been recording. This kind of false alarm panic generating technique will lead to depressed or angry artists.

As stated before, always try to record the first pass of lead instruments like the lead guitar or vocal because, invariably when you don't record it, the first take is the only "hot" one you will get. Never suppose that a guitarist can play repeated solos with the same exuberance as the first. In cases such as this, it is best to have the run down be "a few (vocal, guitar) licks just for level before we run the tape." If the artist says, "That's O.K., just run the tape", you say, "I really like to record all first takes."

SPEEDINESS

By using the streamlined technique this book presents as a starting point, you should realize that speediness and accuracy should be among your most desired goals. You cannot become too speedy as long as your artistry and accuracy remain intact. The technique aspects are for your non-speedy times too, like after you've been working for 18 hours straight and everybody's still going strong. In this case the technique helps you to run on "automatic pilot" without too much mental effort being committed to "what's next," but instead thinking more on the musical/creative aspects of what you're doing.

THE "TURN IT UP PHENOMENON"

Artists are used to playing loud and they will probably have to be made aware of the fletcher munson problems. Explain that very few consumers will be able to listen to the music at the volumes available in the control room, etc. From time to time, give a loud volume playback for the enjoyment of the troops; you just turn it up and then put in your earplugs or leave the room (this is better).

THE INVERSE CUE SYSTEM VOLUME PHENOMENON

It is very useful when working with professional artists to realize that most of them try to overdub their instruments or vocals at a volume relationship (perspective) to the track that seems proper to them artistically. Therefore some artists are very sensitive to the mix as it is in the phones, and you can make the artist play or sing louder by turning their sound down in the phone mix slightly. To get them to play or sing softer, turn up their sound in the cue mix. This technique worked perfectly with Tina Turner. Once the recording level and EQ has settled down, just change their overdub volume in the cue in order to keep a perfect VU meter indication.

To understand horn players is another matter altogether. Horn players use tremendous interior air pressure to produce the proper lip noises to make their horns work. This interior pressure plus the obvious volume "wrapping around" to the player's ears from the horn itself cause a large amount of volume in the player's ears. Some of the horn energy comes from the outside, and some comes from the inside. In order to have the horn players "hear" their horns at proper volumes, you will have to have a pretty good level from the cue system. Any track or other instrumentation to be heard by the horn players, in addition to their horns, will require additional cue power and capability.

THE TWO MICROPHONE TRICK

When trying to get the artist to use the "right mike" it is sometimes necessary to give him a wrong mike to "use" while you do your recording with the other mike. Vocalists who are used to touching their lips to a mike will not be able to take advantage of the far superior sounds available by using the more expensive Neumann or other mikes. In this technique, use a Neumann U-47 FET to record with and supply a Shure SM-57 with a wind screen for the vocalist to use as a pacifier. With the artist singing directly into the element of the SM-57, record with the U-47 instead.

DRUGS, SLEEP AND FATIGUE

Do not take drugs while you are engineering. You must rely on your hearing ability, and if you "go out" you may enjoy yourself more it's true, but the project will suffer. If you're working for a studio as their engineer and you "get high" while you work, then at the project's end, you may have the experience of hearing "your" client haggling with the traffic office or the president of the company over his studio bill. Using your getting high as an excuse for your having "taken so long" to finish the project, the producer will now want a substantial studio time discount. Nothing will ruin your feeling of accomplishment at having done a good, competent job with optimum speed and good attitude maintenance than hearing something like this. The best of "friends" in the control room will part easily when dollars come between the producer and the engineer. Producers will go to any ends to save money during a project. Many times (hopefully always) the re-dos, start overs and other time wasting adventures will be caused by the inexperience of the producer, not the engineer.

When you are fatigued, let everyone know right away. As the evening wears on and you're getting even more tired, let everyone know again and again. This is better than letting yourself get worn to a frazzle and then suddenly announe that you've been exhausted for hours and have to stop. Warm up the client to your intentions of stopping the session. You can continue however, until you see that you're getting unsafe (like almost erasing a track), and then stop. Better yet, call for another engineer before you have to quit.

Above all remember: recording engineering, although a demanding occupation, should nearly always be enjoyable. So enjoy yourself! Practice being an expert listener. Practice being extra alert. Read up on the new products which are constantly being invented. Always do your best work for ALL your clients and eventually, you will be working with only the best artists and producers. Being in demand to work with excellent artists and producers is worth all the effort you will have to put forward to earn such status.

I intended to use these next four chapters in my upcoming book on computer assisted mixing. Even in its synoptic form, this information will be of great help to every engineer. Keep in mind that these four chapters should be studied along with the manufacturer's operating manuals.

NAMES OF THE CONSOLE SECTIONS AND THEIR USE *

* Note: These are my names, not Neve's *

Channel System

The channel is the input module system. It consists of the "channel block" at the top of the input module, and the multitrack block just below the aux sends.

Mixdown System

The mixdown is the multitrack tape monitoring system. It consists only of the monitor block, located just over the main faders.

The Main Faders

The main faders are used either to control mike input levels (when the console is switched to track mode), or multitrack playbacks (mix mode). Main faders are switchable with the secondary faders by pressing the track or mix buttons. All faders switch from one task to the other (either monitor volumes or console input channel level controls) on a console wide basis, not individually in the unmodified 8108. There is a modification available that uses the QUAD button on the Mixdown Block of each console channel as an individual channel fader reverse control. Whichever fader is performing the channel function receives either mike, tape, or line inputs, according to the selection at the channel block at the very top of the console module. The other fader will always be the monitor fader (for listening to the tape machine), which is known on this console as the "mixdown" fader, since it is associated with the mixdown block and the mixdown patch bay positions called "mixdown insertions."

The Secondary Faders

The secondary faders are used to do the function that the main faders aren't doing. During tracking (mikes to main faders), the secondaries are for tape monitor. During mixing (tape playbacks to main faders), the secondary faders are for mike input channel use, as is the case for overdubbing in the remix mode. If the secondary faders are on the channel (mix mode), their inputs are determined by the mike/line/tape switches at the channel block. During twin coupled 24 track mixing therefore, it is possible (even recommended) to do submixes of multitrack playbacks by patching the extra multitrack playbacks patched to the console's line in and then, using the small faders switched to the line mode, assign the small faders to two busses using the buss pan feature. These stereo busses can then be cross patched to two main faders and thus into the mix. With this method there will be fewer main faders in use and the secondary faders can be considered a "submix" console all to itself. Of course, if there are Necam mixes coming up, it is best to leave as many tape playbacks coming up on the main faders as possible so that their movements will be remembered. This function can be achieved by patching console buss out to channel insertion (or mixdown insertion) at the producer's bay. Use of the channel or mixdown PRE or POST switch associated with the INS (insert) button will determine whether the signal will enter pre or post the EQ on its way to the fader you have selected.

The EQ Block

The EQ block has an EQ and a filter. They may be used in the channel or in the mixdowm systems by pushing CH (for channel use) or by not pushing CH (non-ch), which puts them in the mixdown system.

In the track mode, the CH function EQs the module's output to the 24 track tape machine; non-CH makes a flat 24 track recording with the EQ then being in the mixdown (monitor).

In the mix mode, the non-CH function makes an EQ'd 2 track; CH makes the EQ or filter (separate channel switches) go to the small fader where it will EQ a mike input overdub going to the matrix and thereby to the 24 track tape.

In the direct mode the stereo output of the console is the main concern and the 24 track record system is really a backup system. In direct mode, using the filter or EQ in the CH makes an EQ'd 2 track with an EQ'd 24 track backup; in non-ch, the EQ or filter will only affect the 2 track mix and will make a flat 24 track recording.

The Aux Send Block

The aux block is used to derive sends out of either the channel or mixdown systems. Each send can be fed from the channel system (CH) or from the mixdown system (non-CH).

The aux mixes (monitored at the main monitor block) then generate either cue sends (cue matrix block + cue master) or reverb sends (reverb send block + reverb send master).

During tracking (track mode), the aux sends used for monitor reverb should be set to non-CH and will be related to the small faders, which relate to the tape track numbers (not the mike input fader number scheme). The aux sends used for cue should be set to non-ch as well and will also be related to the tape tracks. If it is desired to have the cue come from the channel system (and thus be related to mike input numbers instead), the CH buttons should be pushed for all those aux pots.

In almost any case the PRE button should be pushed for cue use so that fader motion will not upset the cue mix (monitored at monitor block). In almost every case, an AUX SEND used for reverb should be non-pre so that the reverb sounds will follow the fader motion and be killed should the fader be pulled down.

Aux send CH and PRE switch Settings	Signal Comes From
CH & PRE	pre channel fader after multitrack ON switch
non CH & PRE	pre mixdown fader after mixdown ON switch
CH & non PRE	post channel fader after multitrack ON switch
non-CH & non-PRE	post mixdown fader after mixdown ON switch

The Channel Block

The channel block consists of the TAPE and LINE switches, the PAD, and TRIM. The PAD is really a gain control and will bring to console zero any signal from -60 to fully "0" level mike signals, in 6 dB steps. The trim has a + and - 10 dB range. If the trim function is not needed, it should be centered.

LINE is a separate input that is only accessible by patching something into the input module's line in XLR

TAPE is fed from the 24 track playback.

A non-TAPE and non-LINE is (by default, so to speak) the console MIKE input. Non-LINE + non-TAPE = MIKE.

The Reverb Send/Receive Modules

The rev snd/rcv modules have the rvb send masters, rvb metering selection (send or non-send which is return), "SPIN" and insert (INS) functions, send = return (S=R), the reverb eqs, and the filters.

To send reverb to the chambers turn the send masters up, use (SEND) on rvb modules 1-4 so that these reverb meters will be on the send and RCV on the rest (5-8); otherwise sending via reverb send #1 will derive a reverb send on rev out 1 and 5; sending to 2 will activate reverb outs 2 and 6, etc. These multed reverb outputs are available at the patch bay.

To return reverb, bring the reverb up on the proper fader:

The reverb system EQ and FILTER can control the send if the SEND buttons are pushed; to the return if non-SND is used.

There are two patching areas for the reverb system. One is the patch bay's "reverb insertion," the other is the patch bay's reverb return insertion at the return modules insertion patch positions.

To insert an effect or delay at the producer patch bay, patch insertion out (top) to effect, and effect (or delay) out to insertion back (bottom). To include the effect, press INS and turn the SPIN pot up until the proper amount of effect is heard mixed together with the clean reverb send already in existence. If the effect is a delay, the SPIN switch may be used for regeneration, according to the SPIN pot's setting.

Be aware that when you insert (INS), any reverb send EQ that you may be using on that channel will be removed and switched instead to the inside of the regenerative loop.

*** Do not use the SPIN switch except with delay devices, since horrible chamber feedback can result, possibly damaging the chamber speakers. ***

To cause regenerative delay to the reverb send, press ins and spin, turn spin pot up cautiously while listening for the proper number of repeats.

To use a reverb send effect (perhaps a flanger) without the reverb (to kill the reverb suddenly) press SEND = RETURN. This connects the reverb send (with effects) to the reverb return fader, bypassing the reverb device.

ZEROING THE CONSOLE

Remember that zeroing is the operation where all audio pathways through the console are canceled.

Main and secondary faders - down
Mixdown block on buttons - off
Mixdown block pans - fully to back
*Mixdown block mute a on for modules 1-32 and 33-40
*Mixdown block mute a off for modules 41-48
Mixdown block all mute b's off
Multitrack block all buttons out
Multitrack block pans full left
Auxiliary send block all buttons out
Auxiliary send block all pots down
EQ all right knobs centered
EQ all buttons out
Filter all buttons out
Mike/line/tape block mic pad centered
Mike/line/tape block mic trim centered
Mike/line/tape block all switches out
Reverb return modules spin pots down
Reverb return modules the first four modules set to meter the send the last four set to read return
Buss masters full up
Matrix block clear console
Matrix block speaker mutes on
Matrix block small speakers selected
Matrix block cues and reverb matrix canceled
Matrix block cue masters killed
Oscillator off
Oscillator pot killed
Control room monitor pot killed
Studio monitor pots at noon

*Note: the reason for "arming" all but the reverb return mute A's, is so that a console "destructive" or "muting" type solo function of sorts can be performed, by releasing the mute "A" switch on the module desired to be soloed and activating the master mute "A" button at the matrix block. Doing so will then mute all modules except for the one you released the mute on, giving you a muting solo. Since you didn't pre-mute the return modules (41-48) they won't mute, giving you a "destructive mixing solo with effects" mode in the absence of one having been designed into the console by the makers.

8108 MODES OF OPERATION

MODE 1 - Tracking Sessions (Track Mode)

Microphone Signal Flow:
mikes -> mike/line/tape block (mike pad and trim) -> large fader -> multitrack block -> matrix -> busses -> tape

Tape Playback Signal Flow:
tape playback (or input) -> small fader -> mixdown block -> stereo buss

Cue Auxiliary Use: non-ch, pre, on, - cue matrix block (or monitor 5,6 use)

Cue System Signal Flow:
tape pb (or input) -> small fader -> aux block (via non-ch) -> matrix (aux to cue) -> cue master -> cue system

Reverb Aux Use: non-ch, non-pre, on

Reverb System Signal Flow:
tape pb (or input) -> small fader -> aux block (via non-ch) -> matrix (aux to rvb) -> rvb master -> chamber

MODE 2 - Overdub While Mixing (or just mixing)

Tape Playback Signal Flow:
tape playback - large fader - mixdown - stereo buss

Microphone Signal Flow:
overdub mikes -> channel block (pad and trim) mike -> small fader -> multitrack block -> matrix -> multitrack buss -> tape

Cue Signal Flow:
cue auxiliary use: non ch, pre, on -> cue matrix -> cue master

Reverb Signal Flow:
non ch, non pre, on -> reverb matrix -> reverb master -> chamber

MODE 3 * - Direct (Broadcast Mode)

Microphone Signal Flow:
mikes -> channel block -> large and small faders

large fader -> monitor block -> stereo buss -> two track or stereo FM etc.

small fader -> multitrack block -> matrix -> busses -> multitrack tape

* no multitrack playback possible under normal use

VU AND PEAK METER USE

The 8108 console has both VU and peak metering. The VU metering ballistics are very much like the ballistics of a VU meter. Peak metering is something altogether different.

The peak meter will allow a careful engineer to record as hot as possible without fear of saturating the tape. When using a VU meter it is not possible to "see" the exact level of any signal, except for a steady state sine wave. Any other waveform will always read too low on a VU meter. Any short lived signal (transient in nature) will also ALWAYS read too low. A clave, for example, when indicating a proper VU reading, will move the needle very slightly since the signal duration is so short. The weight of the VU meter pointer (an electronically created "weight" in the case of the totally weightless plasma display used in the 8108 metering system) prevents it from rising instantly to show the full level of the transient type sound. The peak meters in the 8108 have no weight; they move instantly to show the exact levels of transient type program material.

The peak meter will only indicate the same reading as a VU during one special case: during measurement of a steady state sine wave tone. Any other signal (like a steady state triangle wave) will indicate higher on the peak meter. Percussive material will show greatly higher levels on the peak meter. This is as it should be.

Since you now can see peak program material, you no longer have to allow for headroom during tape recording for protection against peaks distorting, since you can "see" the entire signal dynamics INCLUDING the peaks.

If the peak meter is adjusted to read mark 5 when the VU meter is reading at zero (sine wave only), a reading of mark 7 peak is equal to a VU reading of +8. Each increment on the peak scale, known as a "mark," is 4 dB.

When using peak meters, align the machine to at least +6 using the 5 of the peak meter (equal to 0 VU) for setting the oscillator levels. Record all signals at least as hot as mark 6.5 on the peak meter (or record at +12 peak).

MONITOR TO CUE

Use the producer phase scope out and patch to reverb insertion 7 and 8. Push INS and S=R on reverb modules 7 and 8. Turn up the SPIN pots all the way on 7 and 8. Put AUX 1 and 2 from reverb modules 7 and 8 on PRE and turn up the AUX levels to suit. Matrix AUX 1 to cue left and AUX 2 to cue right.

MONITOR OR MIXDOWN "PRE" SWITCHES

These switches do not really put the audio pre the associated ON switches so when pre is selected, turning off the ON button of the channel or mixdown block will turn off your AUX SEND too, causing the associated sound to disappear in the cue system to the dismay of the artists.

To kill some of the mixdown feeds to the control room monitor without dumping the AUXes, use the QUAD PAN button with the quad panner panned to back instead of using the ON button switched to off. This way you are able to kill a sound in the mixdown system (the tape monitor system) and still have it in the aux (cue) system. This ploy will not work if you're using any reverb send from the monitor system though, since the real sound will go away but not the derived reverb.

SAVING CONSOLE ASSIGNMENTS IN THE MEMORY

Four memories are provided which store:

1. channel assignments to buss (48 modules x 32 busses)
2. aux 1-4 assignments to cue
3. aux 1-4 assignments to reverb

Session buss, cue and reverb assignment situations should be stored in one of the four memories as soon as they are settled on:

Press CONSOLE STORE, STORE 1 (or 2-4); this stores the console's assignments into the memory.

If you switch to a new assignment plan for a special tune, save the new assignment in store 2, etc.

To recall an assignment from memory:

1. make sure you save the present one
2. press CLEAR CONSOLE
3. press STORE 1 (or2-4), whichever one has the assignments you want, and then press STORE CONSOLE. This puts the memory assignments into the console.

THE STUDIO MONITOR POT

The studio monitor pot doubles as the talkback volume, so the normalizing routine shows the control room monitor master killed, but the studio monitor pot set to 12 o'clock, so that the talkback will be usable.

RED AND GREEN MUTES

The red mute (adjacent to the a/b preset to mute master mute buttons - above the monitor masters) kills the studio monitors.

The green mute kills all the cue systems. It is activated whenever the "tone to 4 track" or "tone to multitrack" buttons are pushed.

THE MIXING SYSTEM MATRIX AND MASTERS

The mixdown blocks send their signals to the four track mixing system. From there the "quad" signal goes to the four track masters: for channel one and two, the Necam masters, and for channels three and four, the red RR/LR master. This is the leftmost red mix master fader. From these three masters, the four track signals go directly to the regular (not plasma) four track buss meters and to the four track connectors used to connect the tape machines. The four track mix from the four track master faders are also presented to the 4/2 matrix, which is to the right of the two Necam stereo master faders. The four track mix is presented to four pots, each with a L and R button. From the selection and levels set there, a two track mix is derived. The two track mix is sent to the two track master fader (red -- on the right). From this fader the stereo matrixed mix is sent to the two track machines. From the two track mix that you matrixed comes the mono mix, which goes through the rotary mono fader directly to the mono machine. The mono mix is affected by the four track masters (if the particular four track buss is contributing to the stereo and therefore mono mix), the settings of the four pots in the four to two matrix and the setting of the rotary mono master.

THE FOUR TRACK METERING SYSTEM

There's more to the four track metering system than meets the eye!

First of all the plasma four track metering system follows the settings of the machine selected in the white monitor source box just above the master faders. Whichever machine your're monitoring will run the meters. If you monitor a four track, four meters will indicate; if a two track, only two meters will indicate. If you monitor a mono, then two meters will still indicate, but they are indicating the same levels (they're mono-ized). If you monitor tape in (console out) then the meters are buss meters; if you monitor tape out, then the meters are tape machine level indicating meters. The VU meter system (center console) always follows the output of the four track buss post the four track masters (the Necam masters for channel 1 and 2 and the RR/LR master).

SYSTEM TURN ON

Turn on the disk drive unit (power switch on front panel), making sure that there are no disks inserted! Turning on the disk drives with the disks inserted can cause loss of information -- possibly total disk failure!

Do not turn off the main computer as this may cause fader mute malfunctions.

If "Necam computer assisted mixer" or "waiting for tape machine" is not in the controller readout:

Insert the Necam disk in drive zero

Holding the ENABLE button, press the RESET button

Holding the ENABLE button, press the START button

Wait until the drive zero door access light goes out and then remove and store the Necam disk away from magnets and moisture.

Load a properly formatted scratch disk into drive zero.

Load your data disk (or a scratch or formatted disk if you are just starting out) into drive 1.

Press the WRITE PROTECT buttons for each disk drive (so that the computer can record on the disks)

If "waiting for tape machine" is displayed, put on a tape with time code present, patch the code track to the Necam input, press play and then stop (at the tape deck); display should say "Neve computer assisted mixer" and a software identification number like "6."

If Neve message is already displayed, position tape using tape controls, patch time code to Necam, press NEW TAPE button.

THE THREE STATES OF NECAM

1. Clear down: (by pressing CLEAR button) allows you to:

A. Safely run the tape machine manually -- using the tape deck's command buttons without confusing the computer,

B. Find out the SMPTE code position of any assigned label -- press CLEAR, nnn (label number), LABEL button,

C. And find out the from/to information for any mix -- press nnn (mix number), MIX.

2. Manual mode: (by pressing PAUSE or PLAY from clear down mode)

A. Allows you to shuttle or play the tape using the Necam tape function buttons.

3. Assist mode: (by pressing GO from clear down or manual modes)

A. Assist play allows computer to store mixes, if you press KEEP after a mixing event that began with a FROM label number which the computer recognizes.

B. Assist mix allows the computer to perform a mix -- move the faders.

MIXING -- THE NO PAPERWORK METHOD

This is not the best method, but it is handy to know in cases where mental fatigue is present or when operating with an engineer who is new to Necam or doesn't want to know about the complexities, but wishes only to get some help remembering his moves. Such is usually the case when speed is of the essence or when the visiting engineer doesn't want to waste any time "playing" with the mixing system. Here are the steps:

Using clear down mode......

Print SMPTE code throughout the song using tape deck controls to control tape deck movement and function. Do not let the time code pass 00:00:00 while printing SMPTE code, otherwise the computer may be fooled into thinking it has shuttled way too much tape while searching and head mistakenly back up the reel in the wrong direction.

Locate the beginning of the song manually using tape deck controls -- hit GO button -- this sets the search point with the Necam and allows you to begin remembering moves that you make with the mutes and faders.

Press PLAY - tape will be in assisted play (where it can remember fader movements, operate groups, etc.).

When you have finished a section of the mix to your satisfaction, simply press KEEP just before you reach the point where you begin to "lose it" in the mix (the point where you will want to "cut in" to begin the next piece of the mix). You can push KEEP during PLAY or STOP tape modes.

Now press GO and then PLAY and let the computer move the faders just as you did up to the point where you pressed KEEP.

While in assisted mix mode (faders move), any fader you move will become new information leading to a new (improved?) mix, if you just remember to press KEEP when the computer stops the tape at which time the display will say "tape has reached 'to' label (a number)."

To mix beyond the point you have gotten to so far, and to allow the tape to continue past the completed spot, press the TO button to clear the display of its TO number. When you press GO and PLAY, the part of the mix you have finished will still play back with fader motion, but when you reach the spot where you pressed KEEP last, the tape will play on allowing you to move further down the song. When you have successfully completed another piece (or the rest) of the song, press KEEP at the bailout point or end of fade point.

As the end of the song is reached, at the very completion of the fade press KEEP. You now may very well have your finished mix.

Make a two track of the finished mix and you are done.

To work on the next tune, press CLEAR, record SMPTE code on the next song, position the tape at the start point, press MIX, FROM, TO (to clear these from the display), press GO, and begin the same procedure over again.

That's it. No muss, no fuss. No records to keep.

MIXING -- THE FULL DOCUMENTATION METHOD

Positioning the tape at a song's starting point.

Play the tape, hitting the LABEL button just a half beat or so before the down beat of each section of each song. Use pause if you want to give yourself time to write down the needed information on the Necam legend. Be sure to set an "end stop" label so that the tape won't play off a song and onto the tail leader. Make a note on the Necam legend of all the labels you assign. You do not have to note any label that was created by accident or one that you won't be using.

After you finish labeling the song or master reel you should begin to get a rough mix on the loudest part of the song (perhaps one of the final choruses). You can always put in more labels just by pressing LABEL, GO (without a defined FROM point), or KEEP (without a defined TO point). All of these occurrences will create a new label.

Having gotten a rough mix at the hot spot of the tune, mark the faders in one of your grease pencil colors and procede to work on the intro using, if you wish, the levels on the faders as left from the song hot area. Enter the intro label number using the keyboard and press FROM and the label of the section where you want the tape to stop and then press TO. Press GO, PLAY and you will begin assisted play.

When the intro is perfected (or as much of the song as you are mixing), press KEEP after the tape reaches its to label.

Write the mix number the computer assigned, the song section, the from and to, and any improved attribute(s) of this last completed section (or mix) on the Necam legend.

After deciding that the completed section is O.K., and before moving on to each new section, press STORE so that the present fader settings (splice levels) will be remembered. This allows you to launch mixes from the "splice" (actually called butt merge) point after having moved the faders from these positions during "trys" at the next section you are mixing. To reset the faders for each re-try at a section just press SET. Therefore, before each pass at the next section to be mixed, press SET to put the faders where they belong. Not pressing STORE will lose your jumping off fader setting and, not having the faders where they should be, will probably make your merged mixes jump in volume at the "splice." You can, of course, always re-play the previous "kept" mix to the current splice point and press STORE in order to store the fader settings (you will have to do this if you forget to press store at the proper time -- before moving on the the next piece). Another more elegant method of setting the faders where they belong (where they were at the end of a previously completed mix piece when you forgot to press SET before continuing) is to just type in the piece's mix number, press MIX, and then change the FROM number to be the same as the existing TO number, which came into the display when you called up the mix. If you now press GO, PLAY, the tape will not move, but the faders will jolt to the proper settings whereupon you may press SET.

When you finish mixing the pieces (or at any time when you want to hear what you have so far), assemble your pieces together using the notes from the legend and press KEEP, which will make a new mix from the pieces you selected.

When you assemble a complete mix from piece mixes (the little pieces you mixed), write the contributing mix numbers on the "FROM MIXES" block of the Necam legend. This will be helpful in cases where there are many piece mixes of each song section from which you can assemble many types of complete mixes.

Keep a well documented master disk and make copies of only the master mix data for each song you carry forward to a master copy disk. In this way you will come up with several original disks containing piece mixes and one master copy disk containing only the finished mixes. Do not use the method of deleting mixes from the original disk; if you do, you risk damaging the original disk and losing your master mixes or piece mixes due to a computer housekeeping malfunction.

The master copy disk (with only the master mixes on it) does not need all the documentation that the original mix disk has. Just the mix number, the song name, and the from and to information of the mixes you copied forward.

MERGING MIXES

There are two types of merges: one where you merge different pieces from different mixes into one complete mix (butt merging), and merging faders from one mix into the fader activity of another mix (fader by fader merge).

Butt Merges

Butt merges involve a song that has mixes of small pieces of itself and/or several complete mixes of the entire song, which may have been formed by merges of little pieces earlier.

*** Please refer to the Necam legend example ***

MIX#	SONG	AREA	FROM	TO	FROM MIXES /IMPROVED	LABEL	BEGINS
====	====	====	====	==	==========	=====	======
1	OUCH	INTRO	1	2		1	INTRO
2			2	5		2	VERSE 1
3		SOLO	5	6		3	VERSE 2
4			6	9		4	CHORUS
5		CHORUS	- TAG				
			9	10		5	SOLO
6			1	10	1+2+3+4+5	6	CHORUS
7					6/VOCALS	7	RELEASE
8					7/GUIT MIX	8	VERSE 3
9					8/INTRO	9	CHOR TAG
10					9/INTRO	10	END STOP
11					MASTER MIX		
	HELP ME						
						11	BEGINNING
12			11	12	KB GOOD	12	END STOP
13			11	12	GOOD VOCALS		
14			11	12	MASTER MIX		

For the song "Ouch", we made mix 1 (just the intro), then mix 2 (the first verse to the lead solo), mix 3 (just the solo), mix 4 (from the second chorus to the third chorus), and mix 5 (the ending of the song up to the protective end stop, to avoid having the tape run onto the tail leader).

To actually make the merge press 1, MIX, MERGE-MERGE (rapidly to include all faders), 2, MIX, MERGE-MERGE, 3, MIX, MERGE-MERGE, 4, MIX, MERGE-MERGE, 5, MIX, MERGE-MERGE, KEEP. The computer will make your merges into mix 6. Mix 6 = mix 1 + mix 2 + mix 3 + mix 4 + mix 5.

Mix 6 is a merge of the first 5 mixes -- the complete song as it stands now. Then we did mix 7 by playing mix 6 and improving the vocal blends. Mix 8 was from mix 7 with an improved mix on the guitars. Mix 9 is from mix 8 with an improved fade out (master left and right were put on manual near the fade point and re-faded). And mix 10 is from mix 9 with an improved intro and mix 11 is undocumented but is the current master for the song.

When the data disk is nearly filled up, we would transfer mix 11 and all other master mixes so far to a master copy disk using our current scratch disk as a "proved disk" onto which the copy is made. We would then load a new formatted disk as the new scratch disk. This means that we would now have the original disk to which the Necam legend relates, a master copy disk with only one master mix on it so far, and another data disk started for the next song(s), until it gets filled and so forth.

FADER BY FADER MERGES

The next song, "Help Me," was mixed in a straight pass to make mix 12 (good keyboards) and mix 13 (good vocals).

It is important to keep in mind that mutes should be used for mixing where you intend to do "borrowing the faders" type merging. If you use fader motion to mute an unwanted fader, then when the computer pushes it up (where it is supposed to), all the rest of the relatively grouped faders will rush to the very top of the scale in response. It is all right to have a grouped set of faders drive down to infinity together (always use the fader set highest on the scale to do the "muting") but using the mutes is much better. The relative settings are not lost even though some of the faders bottom out at infinity before others; as you push them back up you will see this. Remember that you will want to re-level faders in groups (saving the relative motions already perfected between the faders of the group during the mix from which these fader motions were stolen), and that if a fader had been muted by being pulled to infinity during the mix, the entire group will be pulled down as well during the grouped relative stage of the fader merge. In addition, remember that the computer will want to move the faders and if your finger is touching a fader (or if you have mistakenly left a fader in manual), the proper fader motion will not happen. If you do keep a mix in which you accidentally blocked any proper fader motions that the computer tried to make, then the resulting KEPT mix will be contaminated (have missing moves).

When merging faders from different mixes, press the MERGE button only once to enable fader by fader selection (by touching the proper faders as prompted by Necam).

A typical merge faders command string would be like this:
12, MIX, MERGE, (touch faders 1-12 one by one)

13, MIX, MERGE, (touch faders 13-24 one by one)
KEEP

We now merge mixes using mix 12 and press all the faders, except the vocals and merge mix 13 pressing only the vocal faders. When finished press KEEP and this will make mix 14, which has the fader activities of 12 except for the vocals, which were "borrowed" from the moves made during mix 13. If the vocals are too soft in the mix, we must make a grouped, relative shift of them to bring the vocals up into proper perspective in the mix:

1. Position the tape at the top and press PAUSE
2. Press GROUP, and then touch each vocal fader
3. Press GROUP again to "seal off" the group
4. Set the vocal faders to relative
5. Run the mix, and at the first opportunity, using one of the vocal faders only, push it up until the vocal mix levels are proper in the context of the foreign mix in which they now reside. Remember: you must not have pulled any of the vocal faders to infinity during the source mix; rather you must have used mutes, otherwise the computer now pulling down one of the faders will pull all the vocal faders down due to their being grouped.
6. Observe the amount of movement that was required
7. Return to the top and just as the tape begins to play, move any one of the vocal faders the required amount
8. KEEP the mix
9. Put the vocal faders back on N -- Necam position
10. Keeping the group, play back the mix (mix 14 in our example) looking for spots where the relative vocal levels can be improved. If any are needed make them and KEEP the mix
11. Play the last mix looking for individual fader motion improvements and make them (KEEP these mixes too)
12. Kill the group: (in assist pause) GROUP, touch one of the grouped faders, GROUP.

MERGING MIXES WHILE CHANGING THE FROM AND TO INFORMATION

If you fool around with the TO or FROM info that comes from the computer when merging mixes, then you are asking for trouble: sections of the merged, kept mix may have dead zones with no computer information on what to do with the faders.

The only correct way to merge mixes is by doing an assisted play between a "customized" FROM and TO and then KEEPing the passage for use in MERGEing:

Let's say that you put SMPTE on a shop stock tape (while erasing all tracks at the same time), then go back and label it from 1 to 10. Then you do mix 1 from 1 to 10 with the faders at -10. You do mix 2 from 1 to 10 with the faders set to -40. Now you attempt a merge like this:

1 MIX (faders at -10)
1 FROM 5 TO
MERGE, MERGE (save all faders)
2 MIX (faders at -40)
5 FROM 6 TO
MERGE, MERGE

1 MIX
6 FROM 10 TO
MERGE, MERGE
KEEP

When you play this mix you would expect that the faders would start out at -10, go to -40 at label 5, and return to -10 at label 6 out to the end, but this will not happen! The faders will remain motionless.

What you would have to do is this:

2 MIX
5 FROM 6 TO
GO, PLAY, KEEP = mix 3
1 MIX
6 FROM 10 TO
GO, PLAY, KEEP = mix 4
1 MIX, MERGE, MERGE
3 MIX, MERGE, MERGE
4 MIX, MERGE, MERGE
KEEP = mix 5

Mix 5 will have the desired moves.

Remember always: the Necam updates the mutes every 1/30 of a second; Necam only updates faders when you touch and move them, or when they are set to manual as this is the same thing to the computer as touching them. If you merge a part of a mix (by altering the from and to numbers) with a different mix that has no fader movement, the Necam will not recognize the faders as having been moved or needing movement.

COPYING DISKS USING NECAM TESTS 39 -- COPIER

With master disk (with mixes you wish to copy) in drive 1, load the Necam tests disk (drive 0) and press ENABLE/RESET and ENABLE START. Now enter 39 at the controller. Load a suitable formatted or scratch disk in drive 0 and safely store the Necam tests disk.

When you initiate the copy function you are given the choice: type 1 to copy all mixes or 2 to select mixes. Type your preference.

If you choose to copy all:

1
GO - copy function automatic

If you choose to select mixes *:

2
KEEP - makes a copy of the mix shown in the display onto disk 0 from disk 1
DELETE - skips the current mix shown in the display
CLEAR - finish the current copy which enables you to load another data disk (to be copied) in drive 1 and another formatted or scratch disk in drive 0 (use go to begin new copy)

* Note *

Copy will only copy the mixes you request, but ALL labels are copied to the new disk whether or not they are a part of the master mixes you're interested in.

SPECIAL WIDE BAND REPRO CARD

When using 1200 recorders, substitute the repro card with the Ampex wide band card on the tape track that has the SMPTE code on it. When switching to another recording having the SMPTE code on a different track, replace the repro card in its original position and substitute the wide band card for the proper new SMPTE track. Adjust the wide band card's volume pot so that the SMPTE level on the machine's VU meter reads about -7 VU.

Never attempt to pass the time code to the Necam system through a de-processing Dolby unit as this modified code is not recognizable by Necam. Use the bypass switch on the M24 Dolby unit for the track that the time code is on.

FADERS ON MANUAL WHILE GROUPING

Be sure to set any faders to be grouped to "N" (Necam) before setting up a group. If you do not, then the first time you try to move the group, the presence of any manual set fader in the group will cause the fader you move to move all by itself, thereby ruining the group's original relative settings. If any fader in a group is moved by the computer, the same upsetting of the group will occur.

USING NEW TAPE COMMAND

Before using the new tape button, be sure that the machine is set to the right speed, and that the time code is patched to the Necam time code input. Failure to do this may result in the computer "hanging up," causing you to have to "re-boot" the Necam program.

USING PAUSE BEFORE STORE

The computer will not "store" during assist play; you must hit PAUSE before hitting STORE. You may hit SET at any time however.

PROPER EDP TECHNICS

If you only intend to use the Necam system in a transient fashion (just to get the mix to the two track), without being concerned about storing any mixes for use later, then this section will not be of importance. Instead, you could just switch the disks from drive 0 to drive 1 and the reverse, and just begin again on the next mix with an entirely fresh disk.

EDP (electronic data processing) methods must be considered when intending to store Necam mixes for re-use later on. A little knowledge about how the computer handles data is quite important.

Frequent and Redundant Saves

The computer may error and destroy the data disk at any time. This is unlikely but still possible. Very simply: if you have been making backups, your mixes will not get "ripped off"; if you have not made backups, you will get ripped off someday. It's as simple as that. No person can "repair" a destroyed data disk.

There are two ways in which backup copies are made:

1. Once each time period (like each 30 minutes) in a revolving fashion (rotating backups) or

2. Before any alteration of the master disk is performed (like deleting mixes from it for example).

The best technique for Necam use is the latter since "backing up" is quite time consuming - requiring re-booting of the computer first to "Necam tests", and then back to "Necam program." For this reason, the best EDP backup method (rotating backups) is really not the best Necam backup method; rather use the technique of making a backup each time you are nearly out of disk space.

BACKING UP WHEN DISK SPACE REQUIRES IT

Only copy the mixes you want causing the copy disk to be a second generation copy of the master disk with all unused disk space available again.

If you make x deletions on a data disk using DELETE, you will change the data disk from an original into an (x times 2) + 1 generation copy (you lose 2 generations for each deleted mix). The reason for this is that when you command DELETE, the computer copies the contents of the data disk onto the scratch disk (one generation loss), and then copies everything, except the mix you asked to be deleted back to the data disk again, erasing the original data disk (two generations lost). Bear in mind that these "generations" are digital data and don't get degraded copy after copy like audio does. The big risk here is suffering a dropout or a computer malfunction, which might be only one tiny character, but in a critical place on the disk pack. This one wrong or missing character could "crash" the entire disk. From then on Necam would not know that there were any mixes on the disk even though most of the mixes were actually there and probably O.K.

Since there is really only one reel of multitrack tape that can be associated with any data disk (because of the tape's time code relationship to the disk's labels), and because you cannot copy a mix from one data disk onto an existing data disk already having mixes on it (the copy program only works with formatted or scratch disks), the backup procedure to use to fit all of a master reel's songs (perhaps 5 songs) onto one disk is the following:

1. Mix the first song using as many mixes as you wish (total mixes allowed = 999)

2. If you run out of disk space, trade your "proved" scratch disk for the data disk, and set the data disk aside. Use a clean disk, unwanted data disk (see item 7 below), or scratch disk in drive 0 and continue mixing until you have the mix of song one completed. Using a disk in drive 0 as a scratch disk successfully "proves" that disk.

3. Using the copy program, copy just the master mixes you want to a new "proved" scratch disk

4. Using the new disk with the master mix copy(s), mix the next song(s)

5. When the next song is finished, make a copy of the current disk

onto a proved disk copying only the master mixes you want to "carry along" with the Necam mix project (perhaps the master mix and one or two alternate mixes for each song)

6. When all the songs on the tape are mixed, you should make at least one backup of the entire finished master disk.

7. Along the way, you can re-use those data disks that lost importance when their mixes were copied (carried forward) to the current data disk generation in use. Putting any data disk back into the scratch disk slot (drive 0) during Necam program operation will cause the Necam system to offer you this warning, "PRESS GO TO ZAP DRIVE 0." This means that if you press go, the computer will "erase" the data disk and make it ready for use as a scratch disk. You could, of course, remove the disk in drive 0 and save it from destruction before you press GO.

COPYING WHEN DISK BECOMES FULL

Leaving the disk you wish to copy in drive 1, remove the current scratch disk and replace it with the Necam Tests disk.

Hold down ENABLE and momentarily press RESET and then START.

Go to the controller box and type in 39

This will load the copier program into the computer

Remove the Necam tests disk, and insert a formatted or scratch disk into drive 0

The controller will give you the copy options of:

Type 0 for all mixes

Type 1 to select mixes

The controller will scroll backwards up the listing of mixes on the data disk in drive 1 starting with the highest numbered mix. If you press DELETE, the computer will not copy this mix. If you press KEEP, the computer will copy the mix now indicating to the disk in drive 0. If you press CLEAR, the copy program will end and the disk in drive 0 will be completed as a new data disk with only the mixes you selected on it.

You cannot exchange either of the disks during the course of a copy. Do not remove either of the disks until you finish the copy and get the "copy done" message from the controller.

HOW TO USE NECAM TESTS (19)

The Necam Test Disk has three programs on it: 19, 29 and 39. These programs are accessed by "re-booting" the computer by inserting the test disk, and using enable/reset and enable/start. This action (re-booting) gives you the message, "Necam Tests." From here you select the program you want. You get only two button pushes at this stage of the operation: either 19, 29 or 39. 19 is Necam tests (of which there are 3 functions); 39 is the copying program; 29 is the formatting program.

The three Necam Tests

re-boot Necam tests and type 19

from here you can select these functions:

1 for tape machine/computer interface tests

2 for fader output tests

3 for fader input tests

commands for function 1 and 2 are the same:

KEEP = go to next step in the test

MERGE = go through the test steps automatically

GROUP = stop automatic stepping of tests (return to manual operation using keep button)

CLEAR = end current test and return to "Necam tests" message -- from which you can select function 1, 2 or 3 again

Test 3 has no commands; it only operates all the faders in a triangle wave motion while turning on and off the fader mutes.

Use test 2 to test fader sensitivity prior to important Necam mix sessions: step through all the faders.

Use test 3 to check the computer's ability to move the faders smoothly and turn on and off the fader mutes.

To test the system's fader setting acccuracy, using Test 2 set the fader indicated in the controller window to a digital output reading (like 180). Now press SET. Check that all faders are within +/- 2 digital increments of the proper amount. The easiest and quickest way to infinity all faders and turn off all mutes after observing test 3:

press CLEAR

Move the number one fader to infinity and mute off, press set, press clear.

HOW TO USE FORMATTER (29)

Load a disk that you wish to check (and erase completely) into drive 1

Re-boot Necam tests, press 29 on the controller and press go

The computer will either report, "DISK FORMATTED CORRECTLY", or "DISKETTE OR HARDWARE FAILURE." On a failure, reject the disk as being no good (only give the disk one try). Return the bad disk to the manufacturer. A disk getting a good report is suitable for use as a scratch or data disk (only use "proved" disks for data disks where possible).

To format another disk load a disk into drive 1, and press GO, GO.

TESTING THE 224

To make sure the 224 is running properly there are two tests: the input output test (not testing the computer) and the computer test.

When testing the input and output circuitry, the entire audio chain including the entire delay memory system is connected together so that the audio is digitized (analog to digital converted), delayed, recovered, analoged (digital to analog converted) and presented to the 224's outputs as quad -- or stereo delay. If the delayed signals A/B with the input properly, then this part of the 224 test is O.K.

*** Note ***

Before launching and ending the 224 tests, make sure that you turn the monitor and cue masters down to protect the speakers and phones from a possibly loud thump, which may come from the 224 at the beginning and end of both tests.

To test the input/output system, hold SHIFT while you press PRE DELAY. The display should clear and a lower case "d" should appear. Now check the audio quality at the 224's outputs -- the sound quality should be somewhat duller than its inputs. To end the input/output test, press any program button (top row).

The other test checks the computer and the program modules. It also tests all the LED's (by lighting them all) and shows you which program version is installed (in the readout window). This test takes about two seconds to run, and if all's well, the test ends with the 224 re-setting itself to normal mode, ready to be used. When it resets itself at the end of a successful check, the #1 program LED, the #8 program LED, the IMMED, and the two reverb time LED's light. If the unit is not all well, but you want to try to use it anyway, press any program button as for the other test.

If the unit will not power up properly, it may be possible to "force" it to power up with a reverb program by turning it on while holding down the #1 program button. This will cause the unit to power up without self-testing. If you do this, carefully check the outputs to make sure they're clean and clear since the device technically is malfunctioning. It may sound great even with the malfunction so this attempt is worth it.

STORING A SETUP'S SOUND

There are two levels of activity with the 224: the level of settings presently set by the sliders and program select buttons and the settings actually in use by the computer (the running program). The only time the settings of the program switches and the sliders have any thing to do with the sound processing coming from the unit is when the IMMED (immediate) button is pushed. If the SET or CALL buttons are in use (lit), then the switch and slider values in use by the computer are not affected by the controls. You can pull the sliders completely off without it having any effect on the current sound.

To "get a sound", set the function to IMMED. When you have a sound you like, press SET and then the memory of your choice -- register A, B, C or D. Once you press SET, make no further adjustments, if you do, the changed values will be stored while the changes you are making will not be affecting the sound (since IMMED is not pressed). Do not let your stored sounds come back alien to you by thus "working in the dark"; only adjust the sound with the IMMED switch activated.

Once you have some (up to four) reverb "setups" stored in the registers, you can call these sounds out of memory by pressing CALL, and then A, B, C or D. The previously setup sound, consisting of your choice of program buttons and slider settings, will be heard. As long as you select only SET or CALL, the sound you hear will be just like the one you stored, but if you press IMMED, then the program buttons and slider's current settings will "ruin" the called setup you had from memory. That is, the current (not the same most likely) button and slider settings will be put into effect in the "running program" and this sound will be heard, not your stored sound. You can, of course, just press CALL A-D and get back to your stored sound.

During a mix you may be able to change from one stored sound to another by first calling one then another mix, or by jumping into IMMED and changing the settings. Understand that you cannot expect to jump between vastly different setups without some odd sounds from the 224; this is especially true if the two setups have different amounts of pre delay. In fact, however, the amount of pre delay is really the only setting that will make the jumps sound horrible (since the change of delay causes Doppler pitch change).

GOING FROM A STORED SOUND BACK TO THE CONTROLS

Two fancy functions not mentioned in the 224 manual:

1. Checking a registered sound against an "improvement" of the same sound now set on the buttons and sliders:

Suppose a setup is sounding pretty good and you want to save it before you try to conjure up something better. Save what you have for now by pressing SET, A. Now that your setup is "saved", press IMMED and procede to try to "beat" the sound. When you have something that you think is better, check it against the sound stored in register A. To do this, (compare the current setup against the sound in register A) press CALL A and then alternately press IMMED. If you like the new sound better, save it in another register, or re-record (write over) register A again with the better setup by pressing SET A.

2. Updating a sound that is in a register (once the setup has been changed drastically):

Suppose you have given up trying to get a better type of sound for your acoustic requirements, and you wish to re-set the buttons and sliders to the exact positions they were in when you saved an interesting setup a while back which is in register C, for example. Once you can get the sliders and buttons set where they're needed to be to duplicate the sound in register C, you can then transfer control over to IMMED and make any slight improvements necessary. Here's how to do this:

First press CALL C. Note which program lights are lit (top row of switches/LED's), and whether program 8 (decay time enhancement -- good for all programs) is on or off.

Now press IMMED and press the proper program switches to fit the setup shown when you CALLed C.

Now CALL C again. Press the switches for and note the readouts from the bass and mid as they are stored in the computer so that you can set the sliders to these same readings.

Now press IMMED and by pressing the bass button, set the bass slider to the proper number; do the same for mid.

Continue across the row in groups of two setting crossover/treble decay, and depth/pre-delay in the same fashion.

Now the sliders are set exactly where they should be in order to "take over" IMMED control and have the same sound as stored in register C (check, if you want to, that they're the same by pressing CALL C and then IMMED alternately).

SETTING THE CROSSOVER SLIDER

The crossover control is very important. There seems to be two ways to "think" about this control:

1. Set the cross over frequency to a low frequency (perhaps around 300 Hz) so that you can use the bass slider to carefully control the decay time of low frequency content in the sound (to reduce the low frequency "rumblings" to taste).

2. Set the crossover to a high frequency (perhaps around 2-3 KHz) so that the mid slider can carefully control the decay time of the brightness of the audio.

Either way the combination of the two gives rise to the amount of overall decay time. You cannot use just one or the other, but rather a relationship between the two should be slid up and down together while adjusting overall decay time.

SETTING THE TREBLE DECAY

The treble decay control has an effect similar to that of treating the acoustic surfaces (walls, floors, ceilings) with cement (treble decay at max), or with fuzzy carpets, drapes and ceiling hangings (treble decay at minimum). A reverb chamber properly treated should sound something like treble decay set to nearly maximum.

SETTING THE DEPTH CONTROL

Depth has the effect of a person walking away through the acoustic environment from the source (the stage). As you push up the depth slider, your distance from the stage increases along with your ability to hear the beauty (or horridness) of the acoustic environment that you have created.

STARTING A NOTEBOOK

If you started now and began a small black book for the 224, then one year from now you would have a considerable "bag of tricks" worked out. I have found it very handy to name all the unusual (or usual, for that matter) sounds that I come across in my electronic wanderings. You should start to develop these settings right away:

1. A short, dull slightly tubby reverb sound, which makes the bass drum sound louder in the mix, enabling you to actually turn the bass drum down in the mix and still hear it perfectly. There is a magic to the amount of pre-delay you use, which in part depends on the speed of the beat and the

nature of the kick part the drummer is playing. If you find this sound, you might consider recording the sound of the 224 on a track of the 24 track while tracking, which would free the 224 during mixing for other tasks.

2. A bright medium reverb sound, which makes vocals stand out in the mix without calling very much attention to the fact that there's reverb on the voice.

3. A really nice "concert hall sound," which will put any reasonably recorded music (not recorded in "Baffle City") into a live recorded sound situation.

UNDERSTANDING THE MONITOR MEMORY

The monitor memory consists of:

1. The display area, which has the track by track LED readouts of what the monitor state of each track is at any given time.

2. The monitor command switches READY, SAFE, INPUT, REPRO, SYNC, MUTE) and the track assign switches (along the bottom of the display area -- tracks 1-24)

3. The GROUP buttons, which are used to assign tracks to groups

The "membrane" switches and readout leds do two things:

1. They allow you to make monitor assignments and

2. They show you what assignments are presently in use

This makes the panel a memory display/edit system: it shows you what is already in memory, and allows you to change it. Any change you make is instantly updated in the memory and actually becomes the situation in the memory -- if you examine another tape function (play, stop, fast, or record) and return to the previous one, you will see your just inputted alterations have already been stored.

The entire display/edit system is stored in each of the four monitor memories by use of the A B C D buttons and the R P F S (record, play, fast, and stop) buttons. After finishing one complete set of track by track functions in one memory (like memory A), you can make up a completely different set of display/edit assignments in B, C or D. Each monitor memory contains the necessary information to provide a completely different set of deck monitor modes for each: Record, Play, Fast (shuttle) and Stop (RPFS). That means 24 (tracks) x 4 (memories) x 4 (machine modes - RPFS) x 5 functions (mute, ready, safe, repro, sync) = 1920 separate monitor situations can be saved in the machine's memory. This is not counting the groups and the AUX play settings.

To see what is in any of the four RPFS memories, press the RPFS button of your choice: the 24 channel display/edit system is now displaying and ready to edit the deck function that you just selected.

Setup memory A is on when the machine wakes up. The machine wakes up with the RPFS off so the machine is really in manual mode when it wakes up. To get any automatic functions, engage the RPFS by pressing one of the membrane switches. All setup memories will be in whatever condition they were in when the last engineer finished using the machine (the memories don't reset when the ATR is turned off). To set all the machine's monitor memory to square one, press MASTER RESET and the deck STOP button at the same time.

To be certain that you don't "inherit" pesky monitor setups from the previous operator, it is a good idea to hold master reset down while you press stop. This blanks the entire ABCD of all RPFS's so that the entirety of the setup is reset to safe and repro.

PROGRAMMING THE MACHINE'S MONITOR MEMORY

Hold down MASTER RESET and press STOP to clear all present monitor memory setups.

Now set up monitor memory A for mixdown use playbacks in REPRO and SAFE:

Press F MUTE ALL this puts all tracks in mute for silent rewinds and fast forwards (F = fast)

This programing will give you an all safe, quiet rewind setup, which will be in setup memory A and be suitable for playbacks and mixing.

Now to set up a "blank slate" overdub memory out of memory B:
press B
S SYNC ALL
P SYNC ALL
R SYNC ALL

This makes B an almost overdub setup with audible rewinds and sync mode except that we don't have the overdub track selected yet. To do an overdub on track 19, for example, we transfer our blank overdub setup to C so that we don't imprint any specialized, unwanted information on our overdub blank slate, which is in B :

B TRANSFER C

Now C is just like B -- a blank slate overdub setup. Since C will still be lit, we are now displaying/updating setup memory C. To ready track 19 for overdubbing:

READY 19
INPUT 19 F 19 STOP STOP

Now track 19 has a ready to go completely automatic overdub program waiting to be used. All you have to do to activate the automatic functioning is to hit STOP again, which will wake up the RPFS. For now, we don't want to wake up the RPFS, which is why we pushed STOP STOP at the end of the programming sequence above. What we want now is for the ATR to "freeze" in the current monitor situation and NOT switch to sync on track 19 while we "run it down." For rolling punch ins and outs, simply press S to reactivate the RPFS. Input switching on record commands is automatic as long as you have the deck in sync on the track punched in on; otherwise the deck will remain in repro.

Let's say you finish the overdub on 19, and wish to overdub next on 21, 22, 23, and 24. This operation will illustrate the use of the groups.

Press B (we want to set any group information in B so that the group information will "ride" into memory C with the rest of the "blank slate" for overdubbing).

Press G1 21 22 23 24 G1 (this seals off the group)
Now transfer B to memory C: B TRANSFER C
READY G1
INPUT G1 F G1 S S

Some Helpful Hints:

Remember that you can transfer monitor memories of one deck function to other deck functions -- EG.

P TRANSFER R transfers the current display/edit assignments in play to the record deck function.

When the RPFS is indicating the current deck function, the RPFS LED will glow steadily. When a RPFS is selected for a deck function other than the one currently in use, the RPFS will operate but the LED will flash. The RPFS may be reselected and re-defined at any time during any deck function.

All you get in the way of any kind of automatic switching is the automatic switching to input function when recording with the monitor in sel sync. Perhaps we can get Ampex to get the computer to be more creative later. Any kind of automatic operation is possible since the entire machine is driven by software, which can be infinitely changed given time and a willing programmer.

THE AUX OUTPUTS

The aux vertical row programs the aux male XLR connectors at the rear of the machine. There is only one function provided for all tracks for each RPFS, no track by track differences are possible. The aux would be quite handy during heavy live to 24 work since you could press R REPRO AUX and have a separate tape out -- repro mode -- during tracking

For mixing, the aux could be commanded to: P SYNC AUX providing an additional output of a track for sending to a delay machine to provide easy flanging -- with the same track on repro for the mix.

SHUTTLE

This is a combination "on" and "which way" control. Putting your finger firmly down anywhere on the bar will cause the machine to respond in shuttle function. The shuttle speed is set to whatever energy is present when you remove your finger. Shuttling to the center area will stop the tape. You can shuttle with the lift defeat engaged to find your way to a spot on the tape just as easily as if you got up and went to the machine and spun the capstan by hand.

THE MULTI POINT SEARCH TO CUE (MPSTC)

The Keyboard

The keyboard entry system allows entry of numbers from the keyboard to whichever number window has the KE lamp illuminated

To re-enter a missed tape time or event time press its KE and begin again

To do album side tune timing addition with minutes and seconds arithmetic (use locate time window):

1. Enter first song time, press + and then the next song time, etc.
2. Add the number of leaders and their times
3. Press = to get the running time

To change the cue number, press KE at the cue number window and then type the number

The Tape Time Window

The tape time window follows the tape location at all times. The master reel should be put on the machine, taken to the head, the tape time set to zero KE TAPE TIME 0. Now spool down the master tape, making a note on the master box legend of the tape time window indication for each tune leader. Make a note of the tape time of the last tail leader.

When putting on the tape next time, position the tape at the correct spot on the tail leader for which you have the tape time and enter the correct tape time from the master box legend:

KE TAPE TIME and then enter the proper time as shown on the tape box.

If during the session the tape time window drifts off slightly, position the tape at any leader, look up the proper tape time for that leader and re-enter it.

Event Time

The event time windows show:

1. The cue point locator time. There are 99 cue points so you must check the cue number window to see which of them is indicating in the cue point time window.
2. The record start time
3. The record end time
4. The search - play time
5. The search - stop time
6. The stop time

These are the tape times when the deck will automatically perform the function showing in the window, according to some rules and safety precautions.

To find out which of these times is in the window, observe which function LED lamp is on (these are the lamps in the upper right part of the function windows). If the search play led is on, then the event time window is showing at which tape time the deck will enter search play. To view cue points it is necessary for the cue point window to have its led on -- press cue point.

There are 99 (#01 to #99) storable cue points -- 99 event time readings of hours, minutes, seconds, and frames.

There is only one tape time storable in each of these however:

Record start
Record end
Search play
Search stop
Stop

An Example of Use

1. Load and position the master reel at the tail leader
2. KE the tape time into the window
3. For each tune time starting at the head leader of the first tune:

a. KE cue no 01 do not use cue no 00 for a cue point
b. KE cue point the cue time of the first song
c. KE cue point #02
d. KE cue point the next song etc.
e. Set the cue point of the last song

Now to play any song press KE at the cue no. window, and keyboard enter the cue number of the song 1,2,3, etc. and press SEARCH PLAY

While the song plays, select record start. When the punch in point occurs press the down arrow in the tape time window.

Select record end. When the punch out time occurs, press the down arrow again.

Select the search play, search stop, or stop (whichever function you wish the deck to do). When the tape reaches the proper spot press the down arrow again.

To accomplish an automatic search, press either SEARCH PLAY or SEARCH STOP.

Press EVENT ENABLE, RECORD START, EVENT ENABLE, RECORD END, EVENT ENABLE then whichever ending function you wish -- SEARCH PLAY, SEARCH STOP, or just STOP.

When you EVENT ENABLE a MPSTC function its lower left LED will glow. Make sure these LEDs are not glowing if you do not want the deck jumping to search mode or stopping suddenly. Use all usual safety precautions (safe during playbacks etc.) -- use the all safe situation stored in setup memory (A).

To work on another song, press (KE) on the cue no. window and enter the proper number for the song, press SEARCH, PLAY and the deck will locate and play the song.

For Persons Who Like Dealing With Numbers

You may place tape times into the record start and other registers by using the keyboard directly, if you wish. The down arrow method, however, is greatly recommended for safety and ease.

If you wanted to go to a song, play 2 minutes of it, then go to the top and stop automatically:

(KE) the proper cue number into the event time window, assign the keyboard to the locator window, press +, keyboard enter 00:02:00:00, press =, press EVENT TRANSFER, press SEARCH STOP.

Press EVENT DISABLE, RECORD START, EVENT DISABLE RECORD END, EVENT DISABLE SEARCH PLAY, EVENT DISABLE STOP and EVENT ENABLE SEARCH STOP

Now checking, press CUE POINT and check the cue point to see that this event time is the song you had in mind, and press SEARCH PLAY to check that this time is two minutes past the cue point.

Now press SEARCH PLAY and the deed will be done.

Decrement DEC and Increment INC

These simply make larger or smaller the numbers in the window for which the KE LED is lighted. The increments are units of one. Holding down the DEC or INC button for longer than about 1 second, causes the numbers to advance or retreat automatically.

X -> Y -> X

This is for use in case you error during keyboard entry to a window especially the tape or event time windows. This function allows you to recall the last number in the display so that the incorrect number you just entered will be corrected.

Auto Preroll

This is a special function that uses cue no. 00. This is why you should not use cue no. 00 for search to cue. The cue number is immediately set to 00 by the machine when you press AUTO PREROLL.

AUTO PREROLL takes the setting you dialed in on the auto preroll card -- at the bottom of the ATR electronics bay (white colored vertical thumb wheels) -- and automatically loads cue point 0 with the tape time, which is the number of seconds (your selected pre roll time) before the current RECORD START point. Thus, if you press the down arrow at the actual spot where you want to punch in with the record start window selected, then press AUTO PREROLL, the machine will search to the punch in point minus the pre roll time so that there will be a preroll to the punch in point. The punch in point is never changed by the machine.

AUTO PREROLL also sets the SEARCH PLAY, the SEARCH STOP, and the STOP windows to the same number of seconds past the RECORD END time (which you set) -- this is the post roll. Whether the SEARCH PLAY, SEARCH STOP, or STOP function is performed depends on which function the user has EVENT ENABLED.

DIFFERENT DECK COMMAND STYLE DURING MPSTC USE

When punching in during manual use, PLAY and RECORD are pressed together. During MPSTC use, press SEARCH, PLAY, and then the RECORD button. You can also push the REHEARSE button if you want a monitor switching to occur without actually recording and erasing.

Most deck function buttons are dual purpose: the upper lamp indicates the function selected, the lower lamp indicates the actual present function in effect. During an MPSTC automatic SEARCH, PLAY, PUNCH IN, PUNCH OUT, etc., when you punch in at the start of the command sequence, only the upper part of the record button will light. When the machine actually punches in, the lower lamp will light too.

To do a "fake" punch in, press the rehearse button instead of the record button.

Punching out of record during non-automatic deck use requires holding the record button while you momentarily press the stop button (old style); to punch out during automatic punch in by the MPSTC or when rehearsing, just the play button is pressed.

Please note: manual punch outs cannot be done when in automatically initiated punch in recording -- you can only punch out by hitting stop.

BUTTOM PUNCHING RAPIDITY

If you punch the deck or other function buttons too rapidly, you may cause malfunctions or non-functions of the machine. Take your time.

If you press SEARCH, PLAY, RECORD too quickly, you may cause a non record pass when doing automatic punch in work.

You cannot change search number locations once you have engaged a search operation. To do this you must change to a manual search (press STOP, then quickly REWIND), change the cue number and then press SEARCH again.

You cannot change monitor memory situations while in search as the processor is too busy to recognize your button pushing.

ERR 1

Error 1 means that you have selected a speed (perhaps by hitting MASTER RESET, which reverts the machine to 15 ips where there is no EQ selected). There must be an EQ selected for the machine to move tape.

OPERATING GROUPS

The commands for setting up GROUP 1 are:

G1, then touch the track switch membranes for the tracks you wish to be in the group.

There is no way to remove a track from an established group except for clearing the group eg:

GC G1

And then reassigning the correct members to the group.

Absorbers

Absorbers are meant to be poor reflectors and thus absorb into their surfaces or sub surfaces some or most of the sound energy striking them.

ACA

Active combining amplifier. A console combining system that enables an engineer to sum or add audio signals without affecting their quality (very much).

ACN

Active combining network.

Acoustic energy

Pressure / rarefaction phenomenon in air perceivable by humans.

Acoustic signal

Musical sounds the engineer wishes to record.

Acoustic leakage

Musical sounds the engineer does not wish to record (at least with the mike in question). Leakage: unexpected stray sound pickup due to improper mike aiming, placement, or pickup pattern use.

Ambient noise

Ambient noise is that noise that exists without anyone present making any sounds. It is usually comes from street rumble (cars, trucks -- tire and exhaust noise), air conditioning rumble, motors, power hum from large building power circuits, etc.

Assign switches

A switch with one input and many outputs. Eg. assigning input module #1 to any one of 16 possible busses.

Attack

The part of a musical sound that occurs at the initiation of each new note.

Audible noise

An audible noise is a musical noise that is obvious and expected.

Baffles

Baffles are (usually) medium and large sized, sound absorbing, moveable, visually attractive studio room dividing "furniture." Sometimes large 4 ft. x 6 ft. x 2 ft foam pieces are used. Engineers, determined to isolate all sounds and musicians from one another and their instruments while they are attempting to play together in the studio, place these sound dividers/absorbers in strategic locations between various instrument areas so that stray sounds cannot easily get into mikes placed near other perhaps quieter acoustic sources.

Balanced inputs

A balanced input means that there are two signal wires plus a shield, and that common mode rejection should be expected with balanced sources feeding the balanced input.

Bel

The unit of sound measurement. It is a huge quantity of volume change, however, so the dB or decibel (1/10 of a bel) is used instead. 10 dB is perceived as being twice as loud (depends on acoustic and other conditions).

Boom

The boom is the horizontal part of the mike stand that makes the "reach" to the spot where the mike is desired to be located. A mike stand is really only the vertical part; the boom mounts on the stand and provides horizontal mobility.

Buss

My definition of a buss is a console signal combining system including a VU meter and a position in the multitrack monitor system. If there is no combining possible, or if there is no VU meter, or if there is no way to conveniently monitor the signal in the monitor system, then the audio hookup is not a buss. The term buss comes from the old days when a buss bar (heavy unshielded solid wire) was soldered just behind the buss assign switches themselves to "buss the signals together." Behind the buss one switches, for example, was soldered a piece of buss bar, which would contain all buss one signals (by pushing other buss one switches).

Channels (console)

A console channel is a vertical strip of components which accepts the console's input signals and processes them to the console's outputs. A 32 input console has 32 vertical input modules each of which can accept a mike or line signal. This console has 32 channels.

Close miking

Close miking is the use of mikes within several inches of the acoustic source. Close miking results is less leakage (due to the higher level from the source by being so close), higher signal levels (same reason), and peculiarities of the sound's EQ due to proximity effect.

Common mode rejection

The process performed by balanced inputs whereby the signals from the mikes (or other transducers) are passed through to the console where interferences (radio interference and hum mostly) are not. This is because the signals are of a push-pull nature, where the interferences are of a push push, pull pull nature (interference is common to both wires). The common signals (never coming from a mike) are blocked from proceeding because they are pushing in the same direction on both sides of a revolving door (the input transformer) so to speak, and cannot make any headway. Mike signals, on the other hand, co-operate at the input to the mike transformer, and push and pull electrons easily through the transformer windings (like two people pushing on opposite sides and halves of the revolving door).

Cue distribution box

A cue distribution box is a (usually) small metal box that the headphones are plugged into. They are generally used for two reasons. (1) They enable studios to use "odd ball" or uncommon connectors for their headphones thereby making the prospect of stealing a pair of phones less attractive to musicians and other recording studio visitors. (2) They enable the use of distribution resistors to isolate each headphone from all other headphones. This allows a set of phones to be shorted out (possibly by mistreatment) without all the other phones in use going dead.

Decay

The decay is the nature of the instrument's note to fade away after a short time. Decay can be increased by recording at a slower tape speed, and by expanders.

Decibel

1/10 of a Bel.

Diffusers

Diffusers are meant to reflect whatever sounds they reflect in many directions, rather than sending off a more perfect reflection. Two reflectors (like hard, un-absorbtive, parallel walls) will support standing waves. Two parallel diffuser type surfaces will not support standing waves.

Electrical signal

An electrical signal is a desirable electrical waveform. These are to be found within audio equipment that is properly connected, operated within normal operating limits, and not malfunctioning.

Envelope

An envelope is a picture of volume across time. It is the graphic representation of the instrument's volume behavior.

Far field

Distances from an acoustic source where the sound waves cease to act like spheres and the inverse square law ceases to be in effect.

Frequency

Frequency is a scientific term meaning that some event is re-occurring at a regular rate, and it (the frequency) consists of so many occurrences per given time. In audio, the given time is always one second, and the occurrence is almost always a regularly repeating waveform caused by some instrument.

Frequency distribution

A picture of volume over frequency. The chart of sound energy pictured over the range of audible frequencies so as to show where the instrument's sound enengy is located.

Frequency spectrum

This is the picture of an instrument's sound energy with the volumes of the frequency areas vertically over the frequencies themselves shown horizontally with the lowest lows to the left and the highest highs (that humans can hear) to the right.

Ganged

Ganged usually refers to volume controls that are connected electrically or physically so that one can change several console buss levels by moving only one volume control. In modern consoles, buss masters (ganged together in the old days in groups of four or eight) are electronically coupled. This means that any number of volumes can be controlled by one volume control (the group master).

Generation loss

Any time a tape recording is recorded and then played, this constitutes a generation loss.

Harmonics

Special case overtones that happen to have an even (2,4,6,8, etc.) or odd (3,5,7,9, etc.) numerical relationship to the tonic (main instrument note). When solid state distortion occurs, odd "harmonics" are generated; when transformers or tube electronics distort, even "harmonics" are generated. Even harmonics sound "better" than odd harmonics as a rule.

Headroom

Headroom is the distance in dB between the nominal level and the maximum output of an audio device.

Hertz

This is a shortening of the expression of frequency most commonly used in audio engineering: the number of cycles per second. CY/SEC = hertz. 440 hertz = 440 cycles per second = concert A.

Hz

Some clever person invented this abbreviation for the expression cycles per second. Thus Cy/sec = Hz. 10,000 CY/SEC = 10 KHz.

Impedance

The electrical equivalent of gear ratio. Consoles operate at low impedance. This is like low gear: there is little actual motion going on, but the electrical activity is hi in strength. Hi impedance is like hi gear: there is a lot of electrical activity, but the signals are very low strength. A hi impedance mike expects a hi Z (impedance) console input. If a hi Z mike is inputted to a low Z console, the load of the lo Z will reduce the mike signal tremendously, and change the eq significantly. A lo Z mike expects a lo Z console input. If a lo Z mike is inputted to a hi Z console (like a PA console hi Z mike input) the signal will have the proper EQ but be so weak that console noise will result.

Input module

The input module is the part of a single input system of a console where the fader, EQ, reverb send and buss assign switches are located for that numbered input. The input module is also known as a console channel.

Intervals

A musicians term for the musical distance between the pitches of two notes.

K (as in Kohm or KHz)

Metric prefix for thousand. A kilo ohm (KOhm) is one thousand ohms. 1 KHz = 1,000 hertz or 1,000 cycles per second.

Line level

This is the level that the signal remains at after the mike input stage of modern consoles. In the "old days," the signal would be reduced to about -30 or less by the old fashioned mix networks in use, but todays modern active ACN's or ACA's (active combining networks or amplifiers) hold the signal up much better. If the signal is below line level just before it is "delivered" to the patch bay, then the common practice is to boost the signal up to line level. This means that all patch points post the mike pre amp are at line level and enables the audio engineer to patch any outboard equipment into any patch position.

Masking

Masking is the loss of audibility of one instrument's sound by there being another instrumet's sound in the mix that has the same frequency spectrum.

Master reel

The master reel is the collection of all the master takes having been removed from their original reels. It usually requires two master reels when using 30 ips recording speed. It is both very handy and very dangerous to construct (edit together) a master reel. It takes very much less time to work on all the album tunes, but any magnetic or other damage (dropping the reel) will effect all or part of your album. A safety copy should be made when the master reel is first made up, and again when all overdubs are completed.

Maximum output

The highest level that an audio device can put out without distortion.

Mults

Mults are the opposite of y cords: they have one input and many outputs. The main function of a mult is to connect the outputs of a two track to the inputs of several cassette machines or quarter tracks for making copies. This is an acceptable audio procedure since the professional tape playback machine is lo Z out and the cassettes and quarter tracks are hi Z in.

Musical noise

Non tone, non transient scrapings, grindings, and screechings that are a necessary part of several instrument's sounds (eg. the violin).

Near field

This is the acoustical term for the distances from an acoustical source where the sound waves still act like expanding spheres. It is in the near field that the inverse square law applies. When the sound spheres are very large (at a great distance), they behave more like walls of sound than spheres; at this distance and beyond is the far field where the inverse square law does not apply.

Noise floor

This is the level at which the electronic noise of the device exists at all times. It is perfectly normal for a device to have electronic noise, but this noise must be "so many dB down." This means that the noise floor is so many dB down from the nominal level. At the mike pre out stage of a console, the noise floor can be as low as -130 dB from nominal level. After the equalizers, the combining, and program amplification stages within the console, the noise floor is usually around -90 dB. Professional tape machines usually have a non - noise reduction noise floor of about -62 dB. Cassette machines have a noise floor of about 50 dB. Digital recording has a noise floor of almost 100 dB.

Noise prevention

This is the practice of operating audio equipment properly so that unwanted noises are not added to the audio signal. This means using properly sensitive mikes, recording with properly aligned tape machines, using the console with the proper fader, sub master, and master settings, recording with the proper sounds (rather than going for desirable brightness at mix down time and "bringing out" lots of tape hiss), and recording with the proper levels.

Nominal level

The nominal level is the level (operating voltage) at which the equipment designer wants you to operate the equipment. To operate at higher than nominal level risks distortion; to operate at lower than nominal level risks equipment noise (electronic hiss).

Overtones

Frequencies above the main note of an instrument that give it (in part) its characteristic sound.

Pad

A pad is a special volume reducing device that is carefully designed to change only the volume and not the sound quality. There are adjustable pads and fixed pads within almost all recording devices.

Parametric equalizer (EQ)

This is the most complex EQ used. With a parametric EQ, the frequency of boost (or cut), the width of the affected area (the "Q" of the

EQ), the amount of the boost or cut, and whether a peak or shelf is adjustable continuously over a usually wide range. The higher the Q, the narrower is the area of affected audio. Parametric EQ is the most powerful and dangerous of all the EQ varieties. The simplest and least dangerous to audio quality is the simple tone control found on electric guitars and basses. This type is known as a simple R/C tone control (resistance/capacitance).

PCN

Passive combining network. This was the first type of combining system that consisted of a resistor network followed by an amplifer of approximately 30 dB which was necessary to make up the loss from the combining resistors.

Pitch

This is a musicians' term for what an audio engineer would refer to as frequency. It is the number of occurrences of a musical waveform per second. A note that was out of tune (sharp) would be too high pitched; it would be at too high a frequency.

Proximity effect

Proximity effect results from a mike be used very close to the acoustic source so that the bass frequencies are accentuated out of proportion causing the engineer to use "corrective" EQ, which causes further peculiarities in the sound.

Reflectors

A reflector is meant to reflect sound into a new direction similar to a billiard ball striking the rail of the billiard table.

Release

A synthesizer term for the speed at which a note will fade away after it has attacked, decayed, and sustained.

Run down

A run down is a musical event for the benefit of the engineer in which the musicians try to provide the sounds (at the volumes and tonal qualities) that they will be using for the "real take." Always try to record run downs (if you're prepared) because sometimes these "trail runs" are better musically than the "actual performances" for the tape.

Select switches

A switch with many inputs and one output. (Eg. a console with two line inputs and two mike inputs would have a mike 1, mike 2, line 1, line 2 select switch on each console channel -- input module.)

Shield

The third wire of a balanced input system that is "braided" into a jacket used to contain the hi and low signal wires.

Sine wave

The only audio signal that constantly and smoothly changes

voltage from its plus excursion to its minus. The sine wave never pauses or changes abruptly in voltage.

Slate

A slate comes from the movie industry where a cameraman's assistant holds out a small slate with chalk writing on it indicating the take number, the date, etc. When the camera is rolled, he slaps the little wooden clacker at the top of the "slate board" for future use in syncing up the picture to the sound track. In audio engineering, a slate consists of a low frequency tone with the engineer's voice mixed in. Its purpose is to enable the recordist to locate takes quickly (before a master reel is made up -- or even before any leaders are used at all) by playing any one of the slates to find out where he is on the tape, he then can find any other take by counting the beeps (sound of a slate at rewind or fast forward speed) until the desired take is positioned.

S/N ratio (signal-to-noise ratio)

The is the ratio expressed in dB of the signal (while at nominal level) to the noise floor. A plus four signal (+4 dB) to a console noise floor of -90 would be a signal-to-noise ratio of 94 dB.

Spooling

Spooling is the act of playing a tape at normal or perhaps at as much as 60 ips without it touching the heads for the purpose of getting a "flat wind" on the takeup reel prior to storing the tape away for a period of time. Since all tapes are to be stored tails out, all tapes should be found in a spooled condition except when they are in use. Not having properly spooled tapes risks experiencing edge damage due to a ragged pack.

Standing waves

Standing waves result when a sound is initiated between two reflecting parallel surfaces -- the sound is reflected again and again back and forth and is audible as a series of very quick repeats of the original sound.

Subliminal noise

A musical noise that would never be heard unless the engineer were to play the instrument himself or be very close to the musician during the playing of the music. It is a delicate sound that if missing is not really noticed. When a subliminal noise is captured, monitored, and then removed, it is clearly noticed, however, and missed.

Sustain

An instrument that does not decay is said to sustain. Sustain is the absence of decay. With sustain, notes last longer. Sustain can be created by overdubbing at a higher than normal tape speed, and by compressers and limiters.

Tails out

Tails out means that the reel of tape was not rewound after its last playback. To play the tape you must first rewind it. All tape should be stored tails out.

Tape legend

A tape legend is the "take sheet" to which I refer. This should be taped to the tape reel box and should contain the following information: client, producer, engineer, date, studio, tape type, tape speed, tape recorder EQ (CCIR, NAB, or AES), whether there are tones recorded on the reel, the names of the songs recorded, take numbers, tune times, and the recordist's code information.

Track allocation

Track allocation refers to the process of deciding how many tracks to use up of the number available on the multitrack machine. For producers or engineers who dislike doing any mixing until the very last minute, as a form of putting off committing themselves, many tracks are allocated for recording multi mike instruments: drums, pianos, electronic organs, etc. For bolder engineers and producers or under circumstances where many overdubs will have to be recorded (like many vocals), fewer tracks may be recorded by doing some mixing of mikes at the tracking and overdubbing sessions, which leaves more tracks available. The Abbey Road album was reportedly recorded on a four track where every recording was a many mike mix to only two of the tracks. 24 track machines were not used for "having options" about the mix or quality of the sound in the future because none were needed. The mixer did his job right the very first time, and that was that.

Tracks (tape)

The individual recording channels (pardon the use of channel here) of a tape machine are called tracks. Consoles have channels and busses, and tape machines have tracks.

Transducer

A transducer is a device that changes one kind of energy into another kind. A microphone changes acoustical energy into electrical energy. A tape machine changes magnetic energy into electrical energy. A speaker changes electrical energy into acoustical energy. All are transducers.

Transients

A transient is a very short sound. Transients are usually associated with percussion instruments: pianos, drums, bells, etc. The striking noise usually associated with a transient is more difficult to record than a smooth waveform.

VCA

A VCA is a voltage controlled amplifier. This means that the gain of the amplifier is controllable by a simple DC voltage and that the gain of a console channel (for example) can be adjusted by: automation, another fader, a group master, or a synthesizer DC control voltage.

Waveform

A waveform is a picture of instantaneous voltage across time. The picture is one presented by a common oscilloscope. On a graph, the vertical axis is the audio waveform's volts, and the horizontal is the time axis.

XLR

This is a type of connector used almost universally now, for mike and line level signals in recording, sound reinforcement, and television studios. For cables carrying many signals, larger connectors such as 90 or 120 pin Elco connectors are more practical, but for universality, XLR is the connector. There are pictures of the XLR connector in Yamaha's cable wiring instructions in the Basic graphics section.

Y cords

These are dubious devices that are intended to "combine" audio signals. Audio signals should only be combined with console combining networks, never with Y cords or with patch bay mults. A Y cord accepts two signals in, and provides one signal out. Y'ing mike signals together because the console cannot accomodate the required number of mikes due to too few input channels is an audio engineering compromise situation to be avoided if at all possible.